	I B	II B	III A	IV A	V A	VI A	VII A	Gases 0
								4.0026 $1s^2$ He 2 Helium
			10.81 2.0 [He]$2s^22p$ B 5 Boron	12.011 2.5 [He]$2s^22p^2$ C 6 Carbon	14.0067 3.1 [He]$2s^22p^3$ N 7 Nitrogen	15.9994 3.5 [He]$2s^22p^4$ O 8 Oxygen	18.9984 4.1 [He]$2s^22p^5$ F 9 Fluorine	20.179 [He]$2s^22p^6$ Ne 10 Neon
			26.9815 1.5 [Ne]$3s^23p$ Al 13 Aluminum	28.0855 1.7 [Ne]$3s^23p^2$ Si 14 Silicon	30.97376 2.1 [Ne]$3s^23p^3$ P 15 Phosphorus	32.06 2.4 [Ne]$3s^23p^4$ S 16 Sulfur	35.453 2.8 [Ne]$3s^23p^5$ Cl 17 Chlorine	39.948 [Ne]$3s^23p^6$ Ar 18 Argon
58.70 1.8 [Ar]$3d^84s^2$ Ni 28 Nickel	63.546 1.8 [Ar]$3d^{10}4s$ Cu 29 Copper	65.38 1.7 [Ar]$3d^{10}4s^2$ Zn 30 Zinc	69.72 1.8 [Ar]$3d^{10}4s^24p$ Ga 31 Gallium	72.59 2.0 [Ar]$3d^{10}4s^24p^2$ Ge 32 Germanium	74.9216 2.2 [Ar]$3d^{10}4s^24p^3$ As 33 Arsenic	78.96 2.5 [Ar]$3d^{10}4s^24p^4$ Se 34 Selenium	79.904 2.7 [Ar]$3d^{10}4s^24p^5$ Br 35 Bromine	83.80 [Ar]$3d^{10}4s^24p^6$ Kr 36 Krypton
106.4 1.4 [Kr]$4d^{10}$ Pd 46 Palladium	107.868 1.4 [Kr]$4d^{10}5s$ Ag 47 Silver	112.41 1.5 [Kr]$4d^{10}5s^2$ Cd 48 Cadmium	114.82 1.5 [Kr]$4d^{10}5s^25p$ In 49 Indium	118.69 1.7 [Kr]$4d^{10}5s^25p^2$ Sn 50 Tin	121.75 1.8 [Kr]$4d^{10}5s^25p^3$ Sb 51 Antimony	127.60 2.0 [Kr]$4d^{10}5s^25p^4$ Te 52 Tellurium	126.9045 2.2 [Kr]$4d^{10}5s^25p^5$ I 53 Iodine	131.30 [Kr]$4d^{10}5s^25p^6$ Xe 54 Xenon
195.09 1.4 [Xe]$4f^{14}5d^96s$ Pt 78 Platinum	196.9665 1.4 [Xe]$4f^{14}5d^{10}6s$ Au 79 Gold	200.59 1.5 [Xe]$4f^{14}5d^{10}6s^2$ Hg 80 Mercury	204.37 1.4 [Xe]$4f^{14}5d^{10}6s^26p$ Tl 81 Thallium	207.2 1.6 [Xe]$4f^{14}5d^{10}6s^26p^2$ Pb 82 Lead	208.9804 1.7 [Xe]$4f^{14}5d^{10}6s^26p^3$ Bi 83 Bismuth	(209) 1.8 [Xe]$4f^{14}5d^{10}6s^26p^4$ Po 84 Polonium	(210) 2.0 [Xe]$4f^{14}5d^{10}6s^26p^5$ At 85 Astatine	(222) [Xe]$4f^{14}5d^{10}6s^26p^6$ Rn 86 Radon

158.9254 1.1 [Xe]$4f^96s^2$ Tb 65 Terbium	162.50 1.1 [Xe]$4f^{10}6s^2$ Dy 66 Dysprosium	164.9304 1.1 [Xe]$4f^{11}6s^2$ Ho 67 Holmium	167.26 1.1 [Xe]$4f^{12}6s^2$ Er 68 Erbium	168.9342 1.1 [Xe]$4f^{13}6s^2$ Tm 69 Thulium	173.04 1.1 [Xe]$4f^{14}6s^2$ Yb 70 Ytterbium	174.967 1.1 [Xe]$4f^{14}5d6s^2$ Lu 71 Lutetium
(247) ≈1.2 [Rn]$5f^97s^2$ Bk 97 Berkelium	(251) ≈1.2 [Rn]$5f^{10}7s^2$ Cf 98 Californium	(254) ≈1.2 [Rn]$5f^{11}7s^2$ Es 99 Einsteinium	(257) ≈1.2 [Rn]$5f^{12}7s^2$ Fm 100 Fermium	(258) ≈1.2 [Rn]$5f^{13}7s^2$ Md 101 Mendelevium	259 [Rn]$5f^{14}7s^2$ No 102 Nobelium	260 [Rn]$5f^{14}6d7s^2$ Lr 103 Lawrencium

Electronic Properties of Engineering Materials

MIT Series in Materials Science & Engineering

In response to the growing economic and technological importance of polymers, ceramics, advanced metals, composites, and electronic materials, many departments concerned with materials are changing and expanding their curricula. The advent of new courses calls for the development of new textbooks that teach the principles of materials science and engineering as they apply to all classes of materials.

The MIT Series in Materials Science and Engineering is designed to fill the needs of this changing curriculum.

Based on the curriculum of the Department of Materials Science and Engineering at the Massachusetts Institute of Technology, the series will include textbooks for the undergraduate core sequence of courses on Thermodynamics, Physical Chemistry, Chemical Physics, Structures, Mechanics, and Transport Phenomena as they apply to the study of materials. More advanced texts based on this core will cover the principles and technologies of different materials classes, such as ceramics, metals, polymers, and electronic materials.

The series will define the modern curriculum in materials science and engineering as the discipline changes with the demands of the future.

The MIT Series Committee

Electronic Properties of Engineering Materials

James D. Livingston
Massachusetts Institute of Technology
Cambridge, Massachusetts

John Wiley & Sons, Inc.
New York • Chichester • Weinheim • Brisbane • Singapore • Toronto

To my teachers and my students.

ACQUISITIONS EDITOR Wayne Anderson
MARKETING MANAGER Katherine Hepburn
PRODUCTION EDITOR Ken Santor
DESIGNER Kevin Murphy
ILLUSTRATION COORDINATOR Jaime Perea
PHOTO EDITOR Tia Jones
BACKGROUND PHOTO © Dr. Jeremy Burgess / Science Photo Library / Photo Researchers, Inc.
FRONT COVER PHOTO INSET © Alfred Pasieka / Science Photo Library / Photo Researchers, Inc.

This book was set in 10/12 Times Roman by UG and printed and bound by Courier Companies, Inc.
The jacket was printed by Phoenix Color Corporation.

This book is printed on acid-free paper. ♾

The paper in this book was manufactured by a mill whose forest management programs include sustained yield harvesting of its timberlands. Sustained yield harvesting principles ensure that the numbers of trees cut each year does not exceed the amount of new growth.

Library of Congress Cataloging-in-Publication Data

Livingston, James D., 1930–
Electronic properties of engineering materials / James D. Livingston.
p. cm.—(MIT series in materials science and engineering)
Includes bibliographical references and index.
ISBN 0-471-31627-X (cloth: alk. paper)
1. Electronics—Materials. 2. Conductors. 3. Resistors. 4. Capacitors. 5. Semiconductors. I. Title. II. Series.
TK7871.L58 1999
620.1′1297—dc21 98-24461
CIP

Printed in the United States of America

10 9 8 7 6 5 4 3 2 1

Preface

In a display in one of the halls of MIT, the field of Materials Science and Engineering is represented by a tetrahedron, with the four corners labeled PRinciples, PRoperties, PRoducts, and PRocesses. (It was clearly the work of an expert in PR.) This text is PRimarily concerned with one edge of that tetrahedron, the connection between the basic PRinciples of chemistry and physics and the PRoperties of engineering materials–metals, semiconductors, polymers, ceramics, glasses, and composites.

This text was prepared for a core course of the MIT undergraduate program in Materials Science and Engineering that introduces students to the "electronic," i.e., electrical, optical, magnetic, and elastic properties of materials. (Other basic materials-science topics, including crystallography, thermodynamics, kinetics, strength, fracture, and processing fundamentals are covered in other core courses.) Most of our students take this course in the spring semester of their sophomore year, although some delay it until their junior year. The prerequisites include two courses in freshman physics (classical mechanics plus electricity and magnetism), one course in freshman chemistry, and three courses in calculus (through differential equations). However, our students have had little or no practice with applying mathematical concepts such as vector calculus and complex numbers to physical problems, have no familiarity with the mathematics of waves, and have had little exposure to quantum mechanics. As a result, most existing texts dealing with the electronic properties of materials, largely developed for upperclass or graduate courses, were found to be inappropriate for our students.

We have found it effective to delay the introduction of quantum mechanics, and our first seven chapters approach the electronic properties of solids from the viewpoint of classical physics. This approach shows that many phenomena of solids can be adequately explained classically, but that many cannot. Furthermore, it allows the student to become familiar with some of the mathematics necessary for quantum mechanics before being exposed to its bewildering fundamental concepts. The various chapters are largely independent, however, allowing instructors to present the topics in whatever sequence they feel appropriate for their students.

Many engineering properties of solids depend sensitively on microstructural features such as grain size, dislocation concentrations, second-phase particles, etc. Such effects could have been discussed in many places in the text, but, based on my personal research experience, I have chosen to emphasize this aspect primarily in our discussion of the coercivity of ferromagnets (Chapter 5) and the critical currents of superconductors (Chapter 6). Although understanding the fundamental origins of both ferromagnetism and superconductivity requires quantum mechanics, much of the phenomenology important to engineering applications can be treated without it.

After introducing the fundamentals of quantum mechanics in Chapter 8, we approach the properties of electrons in solids from two complementary approaches: the atoms → molecules → solids approach popular with chemists and the nearly-free-electron approach popular with physicists. After justifying the existence of energy bands and energy gaps with both approaches, we consider electronic properties of solids from the viewpoint of elementary band theory, and end with a brief treatment of semiconductors and some semiconducting devices.

In most college courses, the material covered is about half of what the professor would like to cover, and about twice what the students believe they can absorb. Our course is no exception. Although this text includes fewer topics than most introductory textbooks on solid-state physics or solid-state chemistry, it is still a challenge to teach this entire text adequately to materials-science sophomores in a one-semester course. With 16 chapters but only 36 lectures, our schedule allows only two lectures for most chapters. (For some students this is a very rapid pace, reinforcing the old adage that learning at MIT is "like trying to drink from a waterhose.") We have usually chosen to devote four lectures to Chapter 8, and only one each to Chapters 6 and 7, but students in some colleges may have already had an introductory course in quantum mechanics, which would allow spending less time on Chapter 8 and more time on other chapters.

Although this text was developed for an undergraduate course, many graduate students have found it helpful as a refresher text when taking our graduate course in electronic properties, which uses a much more advanced text. Any other readers who wish an introductory electronic-properties text to help bridge the gap between freshman physics and chemistry and graduate texts may also find it useful.

I acknowledge with special thanks the assistance of Professor Eugene A. Fitzgerald, who has taught with me from a draft of this text for three years and contributed significantly to its organization. Professors Robert M. Rose and Lionel C. Kimerling have also taught the course (both without me and with me), and helped to develop the syllabus. Thanks are also due the many students who contributed suggestions about which parts of the text needed further clarification, and especially to graduate students Kevin Chen and Deborah Lightly for carefully checking through the manuscript to improve my wording and correct numerous careless errors. Finally, I thank my wife, Sherry, and my daughters Joan, Susan, and Barbara for their continued love and support.

INTRODUCTION FOR STUDENTS

The major player in this story is the *electron*. Gravitational forces hold the solar system together, and nuclear forces hold the protons and neutrons together in atomic nuclei. But it is the Coulomb force of electrical attraction between positive and negative charges that holds atoms together in solids, and this job of chemical bonding is done primarily by the outer or valence electrons. Most of the properties of solids—electrical, optical, magnetic, and mechanical—depend on these same outer electrons. The better we understand those electrons—their distribution of position, energy, and momentum—the better we'll understand the properties of solids.

Although some important properties of solids depend on structural features of larger dimensions (e.g., grain size), most of the action is on the scale of atoms—10^{-10} meters. Through most of this course, we must THINK SMALL! You may recall that the size of the nucleus is MUCH smaller yet—about 10^{-15} meters. If we were to scale up an atom so that its diameter stretched from California to Massachusetts, the nucleus would be about the size of a farmhouse in Kansas—like Dorothy's. So although most of the mass of solids is in their nuclei, nearly all of their volume is taken up by the electrons. (If you want to lose some weight, shed some nuclei. But if what you really want to lose is volume, which is usually the case, shed some electrons!)

With the simple model of a solid made up of atoms with positively-charged nuclei surrounded by negatively-charged electrons, and the use of some Newtonian mechanics and Maxwellian electromagnetics, we can start tackling some of the properties of engineering solids that are important to their use—in particular, their response to electric and magnetic fields and their response to light and other forms of electromagnetic radiation. We'll do that in the first few chapters, and we'll make some useful progress. But along the way we'll find out that there are lots of things we *can't* explain.

The problem is that, on the scale of atoms, Newtonian mechanics is not enough. And in considering the interaction of light with matter, Maxwellian electromagnetics is not enough. In the 1890s, scientists treated electrons as *particles*, light as *waves*. Early in the 20th century, however, they learned that light also had a particle nature (this particle, called a *photon*, is the major supporting character in the plot). And soon thereafter, they learned that electrons (and protons and other matter) also had a wave nature. Both light and matter exhibited *wave-particle duality*. To explain many of the properties of atoms, molecules, and engineering solids, we'll have to take this into account, and to do so, we'll have to understand some of the basics of *quantum mechanics*. (In fact, to be honest, quantum mechanics is necessary to understand even the *existence* of atoms!)

So HERE'S THE PLAN OF THE TEXT: in Part I, using only "classical" (pre-quantum) ideas, we'll tackle some of the electrical, optical, magnetic, and mechanical properties of solids—basic properties like electrical conductivity, light transmission, index of refraction, dielectric constant, magnetic permeability, modulus of elasticity—that are pertinent to the use of materials as components in various products. For these first seven chapters, we'll treat electrons only as particles and light only as waves. We'll be able to at least *partly* explain a few properties that way, and we'll see just what properties of solids we *can't* explain without quantum mechanics. This will also give us the opportunity to firm up some of the basic math and physics that we'll need later, especially the mathematics of waves and the use of complex notation to treat quantities that vary periodically in space and/or in time (like light and other electromagnetic waves).

In Part II, we'll finally move into the 20th century and introduce some of the basic concepts of quantum mechanics, including *Schrödinger's wave equation* and *electron wave functions*. (Now electrons are not only particles—they're also waves. And now light is not only waves—it's also particles.) After considering a couple of model problems, we'll apply quantum mechanics to atoms and molecules, leading gradually to solids as we go to larger and larger molecules. Among many other things, we'll learn why neon lights are red and why grass is green (something you've wanted to know since childhood). We'll then revisit one case we treated in Chapter 1—that of free electrons in metals—and see what changes quantum mechanics introduces.

Our focus will continue to be on the outer or valence electrons—the electrons that hold solids together and control most of their properties. Both from the above atoms-to-molecules-to-solids approach and from an alternative "nearly-free-electron" approach, we'll come to the concepts of electron *energy bands* and *energy gaps*. Based on this *band theory of solids*, we'll be able to explain aspects of the properties of solids that we couldn't explain using only classical physics, and we'll develop some understanding of the electrical and optical properties of *semiconductors*, the materials that serve as the basis for today's microelectronics. We'll end with semiconductor *p-n junctions*, which are the basis for optoelectronic devices like light-emitting diodes (LEDs) and photodetectors, as well as transistors and semiconductor lasers. Students who wish to learn more about any of the topics covered in this text should consult some of the suggestions for further reading that appear after Chapter 16.

That's the overall plan. We'll start by applying classical physics to the electrical and optical properties of conductors (Chapters 1 and 2), the electrical and optical properties of insulators (Chapters 3 and 4), magnetic properties (Chapters 5 and 6), and mechanical properties (Chapter 7). We'll see what we can and can't explain without resorting to quantum mechanics.

One more point: the chapter titles often include reference to materials applications pertinent to the chapter's contents—conductors, insulators, capacitors, inductors, optical fibers, magnets, light-emitting diodes, solar cells, etc. This is to remind you that materials science is an applied science, and that the ultimate use of engineering materials is in engineering devices. Where appropriate, the text indicates the relevance of various materials properties to device applications.

Contents

PART ONE Semi-Classical Approach 1

Chapter 1 Conductors and Resistors 3

1.1 Ohm's Law and Conductivity 3
1.2 Colliding and Drifting Electrons—Mobility and Relaxation Time 8
1.3 Resistance as Viscosity 11
1.4 The Hall Effect—Counting Free Electrons 11
1.5 Mobility and Carrier Density in Metals and Nonmetals 13
1.6 Using the Hall Effect 15
1.7 AC/DC 16
1.8 Heat Conductivity and Heat Capacity 17
Summary 18
Problems 19

Chapter 2 Windows, Doors, and Transparent Electrodes (Optical Properties of Conductors) 23

2.1 Seeing the Light 23
2.2 Making Waves with Maxwell 26

2.3 Only Skin Deep 29
2.4 Plasma Frequency and Transparent Electrodes 31
2.5 Cyclotron Resonance and Effective Mass 34
2.6 Complex Conductivity 34
Summary 36
Problems 37

Chapter 3 Insulators and Capacitors 39

3.1 Polarization of Dielectrics—Storing Electric Energy 39
3.2 DC Insulator, AC Conductor 43
3.3 Relaxation and Heating Tea 45
3.4 Electronic and Ionic Resonance 47
3.5 The Dielectric ''Constant'' 50
Summary 51
Problems 52

Chapter 4 Lenses and Optical Fibers (Optical Properties of Insulators) 55

4.1 Refraction: Slowing the Waves 55
4.2 Dispersing the Waves 59
4.3 Attenuating the Waves 62
4.4 Diffracting the Waves 64
Summary 65
Problems 66

Chapter 5 Inductors, Electromagnets, and Permanent Magnets 69

5.1 Inductors—Storing Magnetic Energy 69
5.2 Magnetization—In a Spin 73
5.3 Magnetic Domains and M(H) 77
5.4 Rotating Against Anisotropy 80
5.5 The Domain Wall 83
5.6 Harder and Softer 84
Summary 87
Problems 88

Chapter 6 Superconductors and Superconducting Magnets 91

6.1 The Big Chill and Superconductivity ($R = 0$) 91
6.2 Meissner Effect—Superdiamagnetism (B = 0) 95
6.3 High-Field Superconductors—The Mixed State 97
6.4 Superconductivity at "High" Temperatures 101
Summary 103
Problems 104

Chapter 7 Elasticity, Springs, and Sonic Waves 107

7.1 Stress and Strain 107
7.2 Hooke'd on Sonics—Elastic Waves 109
7.3 Vibrating Atoms and Dispersion 112
7.4 Group Velocity and Brillouin Zones 114
7.5 Lattice Heat Capacity and Phonons 116
Summary 117
Problems 117
Halftime Review—The Story So Far 119

PART TWO Quantum Mechanical Approach 123

Chapter 8 Light Particles, Electron Waves, Quantum Wells, and Springs 125

8.1 Thank Planck 125
8.2 Photoelectrons and Photons 128
8.3 Electron Waves 129
8.4 A New Wave Equation 131
8.5 Free Electron Waves and Tunneling 133
8.6 Wave Packets and Uncertainty 137
8.7 Boxed-In Electron Waves: Energy Levels 138
8.8 The Harmonic Oscillator—A Quantized Spring 142
Summary 146
Problems 148

Chapter 9 The Periodic Table, Atomic Spectra, and Neon Lights 151

9.1 Degeneracy and Symmetry 151
9.2 Schrödinger Does Hydrogen 153
9.3 Atomic Orbitals 156
9.4 Beyond Hydrogen—Screening and the Periodic Table 160
9.5 The SCF Approximation 162
Summary 164
Problems 165

Chapter 10 The Game Is Bonds, Interatomic Bonds 167

10.1 The Simplest Molecule 167
10.2 MOs from LCAO 169
10.3 The Neutral Hydrogen Molecule 173
10.4 Beyond Hydrogen—Covalent Diatomic Molecules 175
10.5 Unequal Sharing—Polar Bonds 180
Summary 183
Problems 185

Chapter 11 From Bonds to Bands (and Why Grass Is Green) 187

11.1 Adding More Atoms 187
11.2 The Versatile Bonding of Carbon 190
11.3 Delocalization 193
11.4 Living Color (Why Grass Is Green) 199
11.5 From Two to Many—Energy Bands 202
11.6 Secondary Bonds 204
Summary 206
Problems 207

Chapter 12 Free Electron Waves in Metals 209

12.1 Electron Waves in a Big Box—The Fermi Energy 209
12.2 The Density of States 216
12.3 Heating Things Up 218
12.4 Electronic Heat Capacity 220
12.5 Measuring Fermi Energy with X-Ray Spectroscopy 223
12.6 Photoelectron Spectroscopy 224

12.7 Emitting Hot Electrons 227
12.8 Contact Potential 228
Summary 229
Problems 230

Chapter 13 Nearly-Free Electrons—Bands, Gaps, Holes, and Zones 233

13.1 Less Than Free 233
13.2 Atoms in the Box—Diffraction and the Gap 234
13.3 Effective Mass 239
13.4 The Hole Truth 243
13.5 Divalent Zinc—Overlapping Bands and Hole Conduction 244
13.6 Fermi's Surface Meets Brillouin's Zone Boundaries 246
13.7 Phase Stability and Properties 251
Summary 252
Problems 253

Chapter 14 Metals and Insulators 255

14.1 Metals 255
14.2 Insulators 260
14.3 Inorganic Color 262
14.4 Beyond Bands 265
Summary 266
Problems 267

Chapter 15 Semiconductors 269

15.1 Crossing the Gap 269
15.2 Intrinsic Semiconductors 270
15.3 Extrinsic Semiconductors 275
15.4 Recombination and Lifetime 279
15.5 III-V and II-VI 280
Summary 282
Problems 283

Chapter 16 LEDs, Photodetectors, Solar Cells, and Transistors 285

16.1 Where *p* Meets *n* 285
16.2 Junction Width 288

16.3 Diffusion Meets Drift (Fick vs. Ohm) 290
16.4 Bias and Rectification 292
16.5 Creating Light 294
16.6 Detecting Light 296
16.7 Energy from the Sun 297
16.8 Transistors—Amplification and Switching 298
16.9 Tunnel Diodes 300
Summary 302
Problems 303
Final Review: The Story So Far 304

Suggestions for Further Reading 307

Index 311

PART ONE

Semiclassical Approach

Chapter 1

Conductors and Resistors

1.1 OHM'S LAW AND CONDUCTIVITY

In many applications, one of the most important properties of solids is their response to the application of electric fields, as measured by the material's *electrical conductivity* (or its inverse, electrical resistivity). For the wires carrying current between the wall socket and your desk lamp, you want a material with high conductivity. Surrounding those wires and separating them from each other you want a material with extremely low conductivity—you want an *insulator*. For the filament of your incandescent bulb, you want a conductor, but one that offers enough resistance that the current will heat it to a high temperature—and radiate enough light for you to read this text. In the integrated circuit inside your calculator, you have materials with high conductivity to carry current from place to place, materials with very low conductivity to block current flow in other directions, and materials with intermediate conductivity—called *semiconductors*—to perform the calculations (with the help of the others).

As any electrician can tell you, $R = V/I$, where V is the voltage applied across a material (in volts), I is the resulting current carried through it (in amps), and R is the material's electrical resistance (in ohms). This is sometimes *called* Ohm's law, but it's not quite. As it stands, it's simply a definition of electrical resistance.

$$R = \frac{V}{I} = \text{constant} \quad \textbf{(1.1)}$$

Now, *that's* Ohm's law! Ohm's law states that the ratio of voltage to current remains *constant*, independent of the magnitude or direction of the current. Plotting V as a function of I will give you a straight line through the origin. But like many laws, it is not always obeyed. Take that lamp filament we referred to earlier. If you decrease the current through it by a factor of ten, the voltage across it will decrease by *more* than a factor of ten. The resistance, V/I, will decrease because the temperature of the filament will decrease.

Okay, you say, of course Ohm's law applies only at constant temperature. (We just wanted to warn you that varying current might vary temperature, and Ohm's law will break down at high currents by that fact alone.) But even if we limit current to levels that cause insignificant heating, some materials violate Ohm's law. Some ceramic materials are specifically designed to have a very nonlinear voltage-current (V-I) curve, with a greatly reduced resistance at high voltages. They're called *varistors*, short for variable resistors, and are useful for lightning protection, among other things. Some materials are designed to have a very different resistance if the direction of current is reversed and can thereby convert AC to DC. They're called *rectifiers* and, in each direction, usually have a strong variation of resistance with current, that is, nonlinear V-I curves. Any material with a nonlinear V-I curve is called *nonohmic*.

That's enough nitpicking. Most materials are ohmic over at least a limited range of current and, as materials scientists, we'd like to understand Ohm's law in terms of what's happening at the atomic scale. First, we have to convert the electrician's Ohm's law, $R = V/I$ (= constant), to the scientist's Ohm's law. The resistance R is proportional to the length L of the sample, and inversely proportional to the cross-sectional area A of the sample. So we write $R = \rho L/A$, where ρ is the material's *resistivity* (in ohm-m).

Next, we note that for a uniform sample, the electric field in the sample is just the voltage divided by the length, that is, $E = V/L$ (in V/m). Finally, we define the *current density* as $J = I/A$ (in A/m^2). Plugging in those relationships, Ohm's law becomes

$$\rho = \frac{\mathbf{E}}{\mathbf{J}} = \text{constant} \tag{1.2}$$

This form of Ohm's law, the materials scientist's version, has several advantages. It's expressed in terms of a material parameter ρ, the resistivity, that, unlike the resistance R, is not dependent (in most cases) on sample dimensions. It's also an expression that is *local*, that is, it is valid at every point in the material (we haven't proven that, but it is). Finally, it's a *vector* relation, showing that the two vector quantities—the electric field **E** and the current density **J**—are parallel and proportional everywhere in the material. (I've foreseen their vector nature by using bold letters for these two quantities, a practice I'll continue for all vectors except when I refer only to the magnitude of the vector.) In this local vector form, Ohm's law can be used even in complicated situations in which the direction and magnitude of **E** vary with position.

There's one more minor modification to make in our equation. Rather than focus on resistivity, those who think positively would rather deal with its inverse, *conductivity*, for which the symbol σ ($=1/\rho$) is used. The units for conductivity are $(\text{ohm-m})^{-1}$ or siemens per meter (S/m). After much ado, we have arrived at our preferred form of Ohm's law. We write

$$\mathbf{J} = \sigma \mathbf{E} \tag{1.3}$$

remembering that Ohm's law is obeyed if σ is independent of $\mathbf{E}$.

Sample Problem 1.1

A metal has a conductivity at room temperature of 5.92×10^7 S/m. What is the resistance of a wire 1.8 mm in diameter and 50 cm long? If the wire is carrying a current of 1.3 amps, what is the voltage applied to the wire and the electric field within the wire?

Solution

$\rho = 1/\sigma = 1.69 \times 10^{-8}$ ohm-m
$L = 0.5$ m, $A = \pi d^2/4 = 2.54 \times 10^{-6}$ m^2, $R = \rho L/A = 3.33 \times 10^{-3}$ ohms
$V = IR = 4.32 \times 10^{-3}$ V, $E = V/L = 8.64 \times 10^{-3}$ V/m

Check

$J = I/A = 5.12 \times 10^5$ A/m^2, $\rho = E/J = 1.69 \times 10^{-8}$ ohm-m

---

For simplicity, we have assumed here that the material is *isotropic*, that is, it has the same properties in every direction. In many materials, this is not strictly true. In noncubic crystal structures, conductivity is usually different in different crystal directions. Although a randomly oriented polycrystal of such a material could generally be assumed isotropic, a single crystal or a polycrystal with a preferred orientation—a "texture"—would have an anisotropic conductivity. Many polymers have essentially linear molecules, can be easily processed to align the molecules, and have a different conductivity parallel and perpendicular to the molecules. Aligned-fiber composites and laminated composites also commonly have anisotropic conductivity. In all such cases, conductivity is not a simple scalar, as we have assumed thus far, but is a nine-component *tensor*, and Ohm's law becomes

$$\begin{aligned} J_1 &= \sigma_{11}E_1 + \sigma_{12}E_2 + \sigma_{13}E_3 \\ J_2 &= \sigma_{21}E_1 + \sigma_{22}E_2 + \sigma_{23}E_3 \\ J_3 &= \sigma_{31}E_1 + \sigma_{32}E_2 + \sigma_{33}E_3 \end{aligned} \tag{1.4}$$

where the J_i and E_i represent the components of current density and electric field along the 1, 2, and 3 axes. In an anisotropic material, the various σ_{ij} may differ, and it can no longer be assumed that **E** and **J** remain parallel. A material with anisotropic resistivity might make a good homework problem (see Problem 1-7 and next sample problem), but we'll keep things isotropic for now.

Figure 1.1 shows the electrical conductivity measured at room temperature for a wide variety of materials. Values range from nearly 10^8 S/m for silver and copper down to less than 10^{-16} S/m for diamond, polyethylene, and teflon—a difference of 24 orders of magnitude! (The difference between maximum and minimum conduc-

Electrical Resistivity (ohm-m)	(log)		Electrical Conductivity (S/m)
	−8	8	
intercalated graphite	−7	7	copper, silver iron
	−6	6	stainless steel, metallic glass
graphite (in-plane) graphite (out of plane)	−5	5	$YBa_2Cu_3O_7$ (ab plane)
polyacetylene (doped)	−4	4	$YBa_2Cu_3O_7$ (c-axis)
	−3	3	silicon (doped) ZnO (doped)
TTF-TCNQ	−2	2	seawater
	−1	1	Fe_3O_4
	0	0	germanium
	1	−1	
	2	−2	silicon
	3	−3	InSb
	4	−4	
	5	−5	water AgCl
polyacetylene (undoped)	6	−6	
	7	−7	
	8	−8	ZnO (undoped)
	9	−9	
	10	−10	
Bakelite	11	−11	NaCl
polypyrrole	12	−12	
	13	−13	
Lucite (PMMA)	14	−14	Al_2O_3
polyvinyl chloride	15	−15	mica silica
polyethylene, teflon	16	−16	diamond

Figure 1.1. Resistivity and conductivity of a variety of materials at room temperature. (Note: values are approximate, and for insulators and semiconductors, *very* sensitive to impurities.)

tivities is even greater than this, since superconductors have infinite conductivity and vacuum has zero conductivity, but getting either of those points on the scale would have required much more paper!)

In addition to the huge range of values, several other points are shown in Fig. 1.1. Note that graphite and $YBa_2Cu_3O_7$, which each have layered crystal structures, have *anisotropic* conductivity; in each, the conductivity is higher if measured parallel to the layer planes than if measured perpendicular to them. As noted earlier, in single crystals or textured samples of such materials, conductivity is a tensor and **E** and **J** are not always parallel.

Sample Problem 1.2

You have constructed a lamellar composite consisting of alternating layers of phases A and B, with volume fractions f_A and f_B and conductivities σ_A and σ_B. What is the conductivity of the composite measured parallel to and perpendicular to the layers?

Solution

If the electric field and current are parallel to the layers,
$E_A = E_B = E$, and therefore $J = f_A J_A + f_B J_B = \{f_A\sigma_A + f_B\sigma_B\}E$
So $\sigma = f_A\sigma_A + f_B\sigma_B$ in the parallel case.

If the electric field and current are perpendicular to the layers,
$J_A = J_B = J$, and therefore $E = f_A E_A + f_B E_B = \{f_A\rho_A + f_B\rho_B\}J$
So $\rho = f_A\rho_A + f_B\rho_B$ and thus $\sigma = (f_A\rho_A + f_B\rho_B)^{-1} = (f_A\sigma_A{}^{-1} + f_B\sigma_B{}^{-1})^{-1}$
(In the parallel case, the conductivity of the composite is the weighted average of the two conductivities. In the perpendicular case, the resistivity of the composite is the weighted average of the two resistivities. The two cases correspond to parallel and series electrical circuits, respectively.)

Note also that the conductivity of stainless steel, an iron-based alloy, is less than that of pure iron, while the conductivities of "doped" (alloyed) silicon, zinc oxide, and polyacetylene are much *higher* than those of the undoped materials. Adding alloying elements to pure metals usually decreases conductivity, but adding other elements to many semiconductors and insulators often *increases* their conductivity. Although not shown in Fig. 1.1, metals, semiconductors, and insulators also differ in the *temperature dependence* of conductivity. Increasing temperature usually decreases the conductivity of metals but *increases* the conductivity of many nonmetals. There can be similar differences in the effects of microstructural variables, such as grain size or dislocation content, on conductivity of various materials. It would be helpful to develop some understanding of these various effects, and of the huge range of conductivities shown in Fig. 1.1, if we, as materials designers, want to be able to control conductivity for various applications.

We therefore want to see if we can explain Ohm's law and what controls electrical conductivity by thinking small—picturing what's happening inside a material on the atomic scale. We picture electric current as the motion of many infinitesimal charged particles—electrons—in response to electric fields. Electrons were discovered by Thomson in 1897, and soon thereafter Drude developed the *classical free-electron theory of metals* to explain the electrical conductivity of metals. It is a highly simplified model, partly inspired by the prior success of the molecular theory of gases in explaining the ideal gas law. In that theory, the relation between macroscopic pressure, volume, and temperature of a gas is explained in terms of the dynamics of an assembly of fast-moving microscopic molecules, with a Maxwell-Boltzmann thermal distribution of randomly directed velocities, colliding with each other and with the walls of their container.

1.2 COLLIDING AND DRIFTING ELECTRONS—MOBILITY AND RELAXATION TIME

Drude's classical free-electron theory of metals has often been called the first theory of materials science. However, it is not only of historical interest. His concepts appear, in modified form, in the quantum-mechanical free-electron theory of metals (Chapter 12) and, almost unchanged, in the modern treatment of conduction electrons in semiconductors (Chapter 15).

Drude assumed that in a metal one or more valence electrons per atom were completely delocalized and were free to bounce around inside the metal, much like the air molecules bounce around in your room. Pursuing the analogy, he assumed that the average kinetic energy ($mv^2/2$) of the electrons was $k_BT/2$ per degree of freedom, or $3k_BT/2$ in a three-dimensional free-electron "gas." In this model, the randomly directed, fast-moving free electrons in the metal can collide with each other, the surfaces of the metal, and with the positive ions they left behind, and their average thermal velocity is

$$v_{th} = \left(\frac{3k_BT}{m}\right)^{1/2} \quad \textbf{(1.5)}$$

Here k_B is Boltzmann's constant (which was so important to him that it actually appears on his gravestone) and m is the electron mass. For a temperature T of 300 K, v_{th} is about 10^5 m/s, a very high speed! (We shall find later from quantum mechanics that the velocity of conduction electrons in metals is actually significantly greater than this, although it remains a fair approximation for conduction electrons in semiconductors.)

Each of the rapidly moving free electrons carries an electric charge of $-e = -1.6 \times 10^{-19}$ coulombs. (In nonscientific usage, one of the meanings of "free" is "no charge," but this does not apply in the case of electrons.) With no applied electric field, there is no preferred direction for the electron motion, and therefore no current.

When an electric field **E** is applied, however, the free-electron gas achieves an average *drift velocity* $\mathbf{v}_D$ and a net current density **J**. As with a breeze in a normal gas that results from a pressure gradient, the drift velocity will be much less than the randomly directed thermal velocity $\mathbf{v}_{th}$. If the number of free electrons per unit volume, called the *charge carrier density*, is represented by N_e, the current density will be given by

$$\mathbf{J} = -eN_e\mathbf{v}_D \tag{1.6}$$

If N_e is independent of **E**, which seems reasonable, Ohm's law will be satisfied if the drift velocity is proportional to field. If we define that proportionality constant as the *mobility* μ_e of free electrons, $\mathbf{v}_D = -\mu_e\mathbf{E}$, and the conductivity is

$$\sigma = eN_e\mu_e \tag{1.7}$$

The units for mobility can be seen to be $m^2V^{-1}s^{-1}$ (m/s per V/m). Note also that the drift velocity is opposite to the field direction because the charge carriers, the electrons, have negative charge. So far all we've done is say that the conductivity depends on the product of the charge carrier density and the mobility. Now we have to derive some relation for the mobility in terms of some microscopic properties. We go back to Newton's laws of motion, remind ourselves that the force on each of the electrons produced by the field is $-e\mathbf{E}$, and find that the resulting acceleration is

$$\mathbf{a} = \frac{-e\mathbf{E}}{m} \tag{1.8}$$

If the electron were totally free, as in a vacuum, this constant acceleration would lead to a steadily increasing velocity, and the electron would gain more and more kinetic energy from the field. That won't give us Ohm's law. But in the metal, the free electrons have frequent collisions with the lattice ions, and Drude assumes that in each collision the electron loses its field-induced drift velocity and transfers its field-induced energy to the lattice (in the form of heat). The average drift velocity of the free-electron gas will therefore equal the acceleration times the average time τ between collisions, called the *collision time*, or sometimes, the *relaxation time*. From this it follows that

$$\mu_e = \frac{e\tau}{m} \tag{1.9}$$

(Without a more precise definition of "average time between collisions," you might think that the proper expression should be $e\tau/2m$, but proper consideration of the statistics yields $e\tau/m$ as the average drift velocity of the electron assembly at any instant of time. See Problem 1-14.) With that expression for mobility, we've reached another expression for conductivity:

$$\sigma = \frac{N_e e^2 \tau}{m} \tag{1.10}$$

Ohm's law is satisfied only if all the terms in that equation are independent of electric field. Is that reasonable for τ? It's probably reasonable to assume that the average *distance* between collisions, the *mean free path*, is independent of field, and that should equal the product of τ, the average *time* between collisions, times the *total* electron velocity. Although drift velocity of course depends on field, the *total* velocity will be little affected if it's true, as we claimed earlier, that the drift velocity is much less than the average thermal velocity. It's time to put in some order-of-magnitude numbers to check that out.

Consider a metal carrying a current density of a thousand amperes per square centimeter (10^7 A/m^2). That's a lot of current. Assuming a typical density of atoms and about one free electron per atom would give a charge carrier density of about 10^{29} m^{-3}. Approximating the electron charge as 10^{-19} C, these values correspond to an average drift velocity of only 10^{-3} m/s, a factor of 10^8 lower than the random thermal velocity! It looks as if τ will usually be independent of field, and Ohm's law will be obeyed. As impressive as a thousand amperes per square centimeter may seem, it represents only a small perturbation on the velocity distribution in our classical free-electron gas.

Some other order-of-magnitude calculations might be instructive. Figure 1.1 shows that the conductivity of a typical metal is about 10^7 S/m. With the approximations used earlier, that corresponds to an electron mobility of about 10^{-3} m^{2}V^{-1}s^{-1}, which in turn corresponds to a relaxation time of about 10^{-14} seconds and a mean free path of about 10^{-9} m. So although those free electrons are moving pretty fast, they have so many collisions per second at room temperature that they travel only a few atom distances between collisions.

Sample Problem 1.3

A conductor has a conductivity of 5.92×10^7 S/m, is monovalent (one free electron per atom), and has an atomic volume of 7.14 cc/mol. Calculate the electron mobility, collision time, mean free path, and electron drift velocity when a 1.8-mm diameter wire is carrying 1.3 amps of current.

Solution

$N_e = (6.02 \times 10^{23}$ atoms/mol$)/(7.14 \times 10^{-6}$ m^3/mol$) = 8.43 \times 10^{28}$ m^{-3}
$\mu_e = \sigma/(eN_e) = 4.39 \times 10^{-3}$ m^2/V-s, $\tau = \mu_e$m/e $= 2.50 \times 10^{-14}$ s
$(mfp) = \tau v_{th} = 2.92 \times 10^{-9}$ m, $J = 1.3/(2.54 \times 10^{-6}) = 5.13 \times 10^5$ A/m^2
$-v_D = J/(eN_e) = 3.79 \times 10^{-5}$ m/s

1.3 RESISTANCE AS VISCOSITY

By bringing in the collision or relaxation time τ, we moved from Newton's law (acceleration is proportional to electric field) to Ohm's law, which corresponds to drift *velocity* (current) proportional to electric field. Alernatively, we could have written Newton's law incorporating the collision processes in the form of a viscous drag force proportional to velocity:

$$\mathbf{F} = m\mathbf{a} = m\frac{d\mathbf{v}_D}{dt} = -e\mathbf{E} - \frac{m\mathbf{v}_D}{\tau} \tag{1.11}$$

The second term on the right represents the frictional or viscous drag force associated with the collisions. (Electrical resistance can be looked at as a frictional resistance to electron motion.) After application of a DC electric field to an assembly of electrons with an initial drift velocity of zero, solution of this differential equation shows that electrons will accelerate, in a time on the order of the collision time τ, up to a *terminal drift velocity* of $\mathbf{v}_D = -(e\tau/m)\mathbf{E} = -\mu_e\mathbf{E}$. In this "steady state" ($d/dt = 0$) condition, the two terms on the right cancel, no further acceleration occurs, and we have Ohm's law.

This phenomenon of a friction-set terminal velocity is what saves you from severe damage when, as in the old song, "raindrops keep falling on your head." Without frictional resistance from the air, raindrops would continually accelerate in their long fall from the clouds. However, the frictional drag on the falling drop increases with velocity until it just balances the gravitational force, no further acceleration occurs, and the raindrop hits you with its modest terminal velocity.

For a static electric field and a constant unidirectional current, equation (1.11) is simply a reminder that electrical resistance can be viewed as a frictional or viscous resistance to electron drift, and is fully consistent with Section 1.2. However, we will find this equation useful in later chapters when we specifically consider the dynamics of electron motion under time-varying AC fields.

1.4 THE HALL EFFECT—COUNTING FREE ELECTRONS

We saw in Fig. 1.1 that materials have a wide range of conductivities, and noted that, in any given material, conductivity can have various dependences on temperature, alloying additions, and microstructural defects. But conductivity depends on the product of charge carrier density and mobility, and it would be helpful if we could somehow get a separate measure of one of these variables. We could then get a quantitative test of Drude's model, and this would also enable us to determine which effects depended on changes in carrier density and which depended on changes in mobility, information that could be useful to us in devising ways to control conductivity. We can do this by using the *Hall effect*, discovered by graduate student E. H. Hall in 1879.

You'll recall that a particle of charge q moving with a velocity $\mathbf{v}$ in a magnetic field $\mathbf{B}$ experiences a force $\mathbf{F} = q\mathbf{v} \times \mathbf{B}$, which is perpendicular to both $\mathbf{v}$ and $\mathbf{B}$.

Hall applied a perpendicular magnetic field to a sample carrying current and observed a voltage perpendicular to both magnetic field and current, a voltage he found to be proportional to both magnetic field and current. As shown in Fig. 1.2a, the transverse force on electrons moving with an average drift velocity $\mathbf{v}_D$ leads to a build-up of net charge on the longitudinal surfaces of the sample—an excess of electrons on one side, a deficiency of electrons on the other.

This charge build-up creates a transverse electric field E_H, known as the *Hall field*, which in turn creates a transverse electric force $-eE_H$ that opposes the transverse magnetic force $ev_D B$. In equilibrium the two transverse forces are equal and opposite:

$$-eE_H = ev_D B \tag{1.12}$$

that is, when $E_H = -v_D B$. If we define the *Hall coefficient* R_H by $E_H = -R_H JB$, and substitute the expression for current density $\mathbf{J}$ in terms of drift velocity, we find

$$R_H = \frac{v_D}{J} = \frac{-1}{eN_e} \quad \text{and} \quad \sigma R_H = -\mu_e \tag{1.13}$$

We have chosen the signs in the preceding equations with the assumption that the charge carriers were negatively charged electrons, yielding a Hall coefficient that is negative (by convention). As we shall soon see, however, the sign of the Hall coefficient is not the same for all materials.

We now have a method to separate the two variables in the expression for conductivity. From the Hall coefficient (which has units of m^3/C) we can calculate the carrier density, and its product with conductivity tells us the mobility.

Let's see how Drude's classical free-electron theory compares quantitatively with direct measurements. The measured Hall coefficient for lithium at room temperature is -1.7×10^{-10} m^3/C. (By convention, the negative sign corresponds to the direction of Hall field expected for electron flow, as shown in Fig. 1.2a.) Lithium, an alkali metal, is monovalent. From the measured density of lithium, and assuming one free

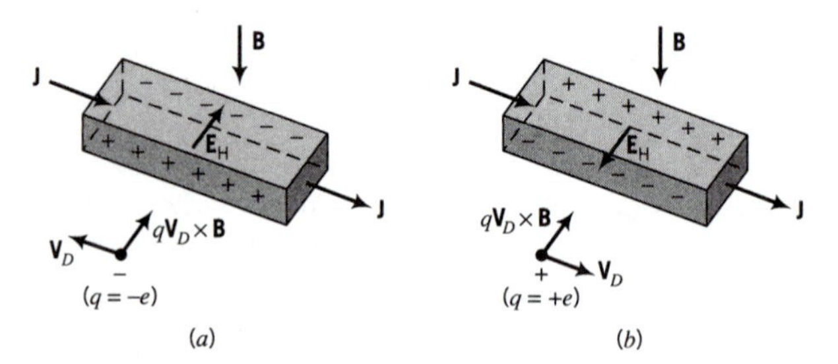

Figure 1.2. Hall effect for (a) negative charge carriers (b) positive charge carriers. Because the signs of both the charge and the drift velocity are changed, the magnetic force on both types of charge is in the same direction.

electron per atom, the calculated value of Hall coefficient is -1.4×10^{-10} m^3/C. That's pretty good, considering how simple the model was. In the same units, the measured Hall coefficient for aluminum is -0.3×10^{-10}, and the calculated value, assuming three free electrons per atom for this trivalent metal, is in good agreement with experiment (even better than the data for lithium). Several other metals give similar agreement between measured and calculated Hall coefficients. So far, so good.

Now for the bad news—or perhaps it is the interesting news. For divalent zinc, the calculated value is -0.5×10^{-10} and the measured value is $+0.3 \times 10^{-10}$. That's not a typo. Although the magnitude of the Hall field isn't that bad, it is *in the wrong direction*! Several other divalent metals have positive Hall coefficients, as do many semiconductors. As we show in Fig. 1.2b, the Hall field will be in the opposite direction if the current is somehow being carried not by electrons but by *positively charged particles*. Protons? Positrons? Positive ions? This remained a mystery for many years, and it will remain a mystery in this text for many chapters. We'll learn later that one of the many interesting features of quantum mechanics and the band theory of solids is that, although current is really being carried by electrons, it can sometimes behave as if it were being carried by positively charged "holes." We'll dig into these holes later.

Drude's theory gives a reasonable agreement with measurement for most monovalent and trivalent metals, but it gives the wrong sign of the Hall coefficient for some divalent metals and for many semiconductors. Semiconductors and other nonmetals have Hall coefficients many orders of magnitude higher than those of metals (making them much easier to measure), indicating that the charge carrier density may be as low as one free electron (or hole) per million (or billion or more) atoms. However, you can't blame Drude for not being able to explain that, since he was trying to explain only the conduction of *metals*.

1.5 MOBILITY AND CARRIER DENSITY IN METALS AND NONMETALS

Being able to use the Hall effect to separate the two major factors determining conductivity clarifies the main source of the huge range of conductivities seen in Fig. 1.1. Semiconductors and insulators have much lower conductivities than metals not because of lower mobilities (which actually are often higher than mobilities in metals), but because of much lower charge carrier densities, as revealed by their much higher Hall coefficients. An insulator may have a *total* electron density that is little different from that of a metal, but it has a much lower density of *mobile* (free) electrons. *Thus a materials designer interested in controlling electrical conductivity should usually focus on controlling charge carrier density if working with nonmetals and on controlling mobility if working with metals.*

What about the effects of temperature and alloying on conductivity (which, as we noted earlier, are often opposite in metals and nonmetals)? The Hall coefficient (and therefore the charge carrier density) of most metals doesn't change much with temperature, with moderate alloying, or with changes in grain size, dislocation density, and so on. But the conductivity sure does! Cooling pure copper from room temperature to 4.2 K, for example, can increase the conductivity by a factor of many thousands. The increase in conductivity with cooling therefore must be a result of an increase in mobility (and collision time and mean free path). Apparently lattice vibrations, which increase rapidly with increasing temperature, decrease electron mobility. Figure 1.3 shows the increase of resistivity of copper with increasing temperature, and also the increase with solid-solution alloying (with tin, creating bronze) and plastic deformation (increased dislocation density). *To generalize, any departures from a perfect lattice decrease electron mobility in metals (and thereby decrease conductivity).*

Suppose that you, as a materials engineer, must select an alloy for the windings of an electromagnet designed to create very high magnetic fields. The windings must sustain stresses much higher than pure copper can handle, but alloying copper to increase its strength usually also decreases its conductivity, which leads to undesirable increased heating of the windings. To get an increase in strength with minimum decrease in conductivity, should you use soluble or insoluble alloying elements?

Even one percent of an alloying element soluble in Cu, such as Zn, Sn, or Be, substantially decreases the mean free path and mobility (see Fig. 1.3). However, Cu can also be strengthened by alloying with an element like Nb, which is essentially insoluble in Cu, and will be dispersed as separate particles of Nb within the Cu matrix. The finer the dispersion of the Nb, the greater the strength, but the conductivity will be only moderately decreased (decreased largely because of the decreased volume fraction of copper) as long as the spacing between Nb particles is much larger than the electron mean free path in Cu determined by lattice vibrations. Finely dispersed Cu-Nb composites are therefore a good choice for the windings, because they have greatly increased strength but only a moderate decrease in conductivity.

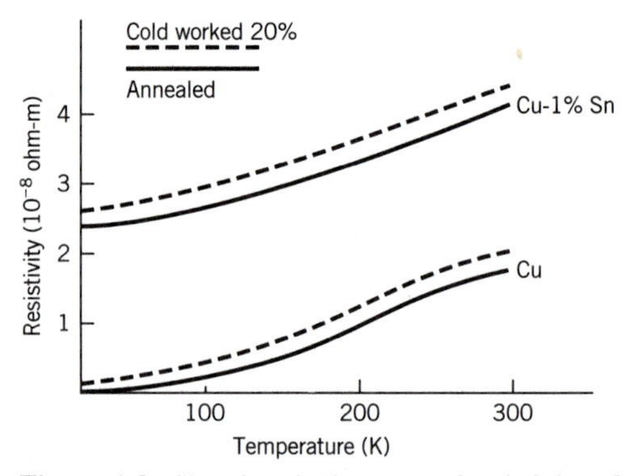

Figure 1.3. Showing the increase of resistivity of copper with increasing temperature (lattice vibrations), alloying with Sn, and cold work (plastic deformation). Various contributions to resistivity are additive. (Adapted from L. H. Van Vlack, *Materials Science for Engineers*, Addison-Wesley, Reading, Mass., 1970, p. 287.)

Semiconductors behave very differently from metals. Hall coefficients decrease very rapidly with increasing temperature, and usually with increasing doping, indicating a strong increase in charge carrier density. There are small changes in mobility as well, but the changes in conductivity in nonmetals are usually dominated by changes in carrier density rather than by changes in mobility. We'll need quantum mechanics to explain these effects, something we'll get to in Chapter 15.

Of course if mobility were *extremely* high, a material could have a respectable conductivity even with an extremely low charge carrier density. An example is interstellar space, which is believed to have conductivities well above 10^2 S/m, higher than many semiconductors despite a much lower charge carrier density. Here the rarity of collisions leads to very long mean free paths and relaxation times, and charged particles can accelerate to high velocities between collisions.

What about superconductors? They have infinite conductivity, but they certainly don't have infinite charge carrier density. Apparently they have *infinite mobility*. We know now that some of the conduction electrons in superconductors form electron *pairs* via a subtle electron-electron attraction that is capable, at sufficiently low temperatures, of overcoming the Coulomb repulsion between them. These so-called "Cooper pairs" do not transfer energy to the lattice via collisions as single electrons do, and can therefore carry current without resistance. We'll learn more about superconductors in Chapter 6.

1.6 USING THE HALL EFFECT

The Hall effect is useful for much more than enabling separate determination of charge carrier density and mobility. One important use is in "*Hall probes*," which use semiconductors to measure magnetic field by measuring the Hall voltage. It is easy to show from earlier equations that the ratio of the Hall field to the longitudinal electric field E_L driving the current is equal to the magnetic field times the mobility.

$$\frac{E_H}{E_L} = \mu_e B \tag{1.14}$$

In the presence of a magnetic field, the *total* electric field (which has transverse and longitudinal components E_H and E_L) makes an angle with the current known as the *Hall angle*. The optimum materials for Hall probes are clearly materials with high mobilities, such as InSb.

Another important use is in Hall-effect sensors. Suppose you want to measure the speed of rotation of a shaft. If you have a small permanent magnet mounted on the shaft and a stationary Hall-effect sensor mounted nearby, the sensor will deliver a voltage pulse with each rotation as the magnet sweeps by, and the pulse rate can therefore be used to measure rotation rate. Many automotive, aircraft, and industrial control systems use Hall-effect sensors to measure various parameters.

Before leaving the Hall effect, a word about units may be appropriate. Many students feel more comfortable with the units for electric field, volts per meter, than with the units for magnetic field, teslas. (Tesla is the preferred unit in the SI system, although gauss is still frequently used. One gauss is 10^{-4} tesla, and equals about twice the earth's field.) Mathematical relations between various physical quantities can often be helpful in deriving the relations among various units, thereby expressing unfamiliar units in terms of more familiar ones. For example, from equation (1.14), the units for **B** must be the inverse of the units for mobility (since the ratio of two electric fields must be unitless). Thus one tesla is equivalent to one volt-second per square meter (V-s/m^2). You can reach the same relation between these units by examining Faraday's law of induction (which we encounter in the next chapter).

Sample Problem 1.4

(a) In the conductor of the previous sample problems, what is the Hall coefficient? (b) What would the Hall electric field be when a magnetic field of 2.3 tesla is applied normal to the wire while it is carrying 1.3 amps of current? (c) What would be the Hall angle?

(As determined in the earlier problems, charge carrier density is 8.43×10^{28} m^{-3}, mobility is 4.39×10^{-3} m^{2}V^{-1}s^{-1}, and the longitudinal electric field when carrying 1.3 amps in a wire 1.8 mm in diameter is 8.64×10^{-3} V/m.)

Solution

$R_H = (eN_e)^{-1} = 7.41 \times 10^{-11}$ m^3/C,
$E_H = E_L\,\mu_e B = (8.64 \times 10^{-3}) \times (4.39 \times 10^{-3}) \times (2.3) = 8.72 \times 10^{-5}$ V/m,
Hall angle $\theta = \tan^{-1}(E_H/E_L) = \tan^{-1}(\mu_e B) = 0.01$ radians $= 0.58°$
(Hall coefficients, Hall fields, and Hall angles are very small in metals because the charge carrier density is very high.)

1.7 AC/DC

Until now, we've implicitly assumed we were dealing with static electric fields and direct current. From the preceding analysis, would we expect Ohm's law, considered in the form of a local vector relation between electric field and current density, to be modified if we considered time-varying fields and alternating current? There certainly will be changes in the spatial distribution of fields, since Maxwell's equations (which we'll remind you about in Chapter 2) show that time-varying magnetic fields yield space-varying electric fields, and vice versa. But from the microscopic viewpoint, will $\mathbf{J} = \sigma\mathbf{E}$ still hold locally?

Suppose the local electric field is varying as $\mathbf{E} = \mathbf{E}_o \sin(\omega t)$, where ω is the angular frequency (in radians per second, or s^{-1}) and equal to $2\pi\nu$, where ν is the frequency (in cycles per second, or Hertz). Our microscopic explanation of Ohm's law was based on the field-induced energy of the electrons being transferred to the lattice via a collision about 10^{14} times each second. That's pretty often. Compared to that pace, a field that cycles 60 times each second ($\nu = 60$ Hz), or even a million times each second ($\nu = 1$ MHz) is pretty sluggish. It seems likely that, barring heating effects, the local AC conductivity will be nearly the same as the DC conductivity as long as the period of the oscillating field is long compared to the collision time, that is, as long as the product $\omega\tau$ is much less than one.

$$\text{AC:} \qquad \mathbf{J} = \sigma\mathbf{E} = \sigma\mathbf{E}_o \sin(\omega t) \qquad \text{provided } \omega \ll 1/\tau \tag{1.15}$$

Thus the *local* current density will vary as $\sin(\omega t)$, in phase with the electric field, and the *local* AC conductivity will be equivalent to the DC conductivity, provided $\omega \ll 1/\tau$. For τ of the order of 10^{-14} seconds, that will be satisfied at frequencies up to as high as 10^{12} Hz.

However, visible light consists of electric and magnetic fields that oscillate at frequencies greater than 10^{14} Hz, frequencies at which the subject of conductivity becomes more complex—in both the common and the mathematical sense of the word. That's the topic of the next chapter, the interaction of light with conducting materials.

1.8 HEAT CONDUCTIVITY AND HEAT CAPACITY

Now let's put some heat on Drude by seeing how well his theory did in calculating some thermal properties of metals. Materials that conduct electricity well usually also conduct heat well, as you may have discovered by grabbing the handle of a silver teapot. Drude's theory tackled heat conductivity and did pretty well, with those same free electrons that conduct electricity carrying heat from the hot end to the cold end of a piece of metal. The equivalent to Ohm's law in one-dimensional heat flow is

$$J_Q = -K\frac{dT}{dx} \tag{1.16}$$

Here J_Q is the rate of heat flow per unit area (W/m^2), dT/dx is the temperature gradient (K/m), and K is the material's thermal conductivity (Wm^{-1}K^{-1}). (Sorry about that. The standard symbol for thermal conductivity is K, but K is also the standard symbol for degrees in the Kelvin temperature scale.) In 1853, Wiedemann and Franz had reported an empirical "law" relating electrical and thermal conductivity, namely that $K/\sigma T$ *was constant* and approximately the same for most metals. Drude's classical free-electron theory in fact yielded this *Wiedemann-Franz law*, although the predicted constant was a bit off.

So despite a bit of a problem explaining why some materials have a Hall coefficient of the wrong sign, the classical free-electron theory of metals doesn't look too bad—until you consider the *specific heat capacity* expected from those free electrons. From the assumption that the average kinetic energy of each free electron is $3k_BT/2$, it's easy to see that N_e free electrons per unit volume should yield a contribution to heat capacity per unit volume (the temperature derivative of energy per unit volume) of

$$C_e = \frac{3k_B N_e}{2} \tag{1.17}$$

For a *monovalent* metal (one free electron per atom), the number of free electrons would be N_A, Avogadro's number, per mole, and since $k_B N_A = R$, the gas constant, the predicted C_e *per mole* would be $3R/2$. Unfortunately for the Drude theory, experimental measurements of the electronic contribution to the heat capacity of metals yield values about two orders of magnitude smaller than predicted by equation (1.17). In fact, the heat capacities of metals and nonmetals near room temperature are nearly the same, mostly deriving from the thermal vibrations of the lattice. This is a serious flaw of the Drude theory, and explaining it (which we do in Chapter 12) was one of the early triumphs of the quantum mechanics of solids.

A similar discrepancy was found when comparing the Drude theory with experimental measurements of the *thermoelectric effect*. A temperature gradient dT/dx in a metal is found to produce an electric field proportional to the temperature gradient

$$E = Q\frac{dT}{dx} \tag{1.18}$$

where the proportionality constant Q is termed the *thermopower*. As with the electronic heat capacity, the Drude theory predicts a thermopower about two orders of magnitude higher than experiment. Not too surprising, since thermopower can be shown to be proportional to electronic heat capacity.

How could Drude's theory do so well with thermal conductivity of metals and so badly with such closely related properties as thermopower and electronic heat capacity? Detailed analysis of thermal conductivity shows that it is proportional to the product of electronic heat capacity (which is related to how many electrons per unit volume increase their energy with increasing temperature) and the average kinetic energy of those electrons. As we'll learn in Chapter 12, quantum mechanics shows that although the number of thermally excited electrons is actually about two orders of magnitude lower than Drude predicted, their average kinetic energy is about two orders of magnitude *higher*. The two discrepancies approximately compensate each other, yielding a predicted thermal conductivity close to the experimental value and making the Drude theory look better than it actually is. In estimating the contribution of free electrons to thermal conductivity, Drude was just lucky.

SUMMARY

From the viewpoint of materials science, Ohm's law is a local vector relation of proportionality between electric field and current density, that is, $\mathbf{J} = \sigma\mathbf{E}$. The classical free-electron theory of metals assumes a "gas" of fast-moving, randomly directed electrons, with application of an electric field producing an average drift velocity $\mathbf{v}_D$ much smaller than the random thermal velocity. Individual electrons are accelerated by the field, but they frequently collide with the lattice ions and transfer their field-induced energy to the lattice, so that the average drift velocity (and therefore the electron mobility) is proportional to the time between collisions. Because the drift velocity is much less than the thermal velocity, the collision time is nearly independent of field, and Ohm's law holds. Alternatively, the drift velocity can be viewed as the terminal electron velocity achieved against a viscous drag force representing electrical resistance.

The conductivity σ depends on the product of charge carrier density N_e and mobility μ_e, but the Hall effect can be used to measure N_e, and thereby separate the two material properties. Measured Hall coefficients are in rough agreement with classical free-electron theory for monovalent and trivalent metals, but Hall coefficients of some divalent metals (and many semiconductors) are of the wrong sign. Although many nonmetals have charge mobilities significantly higher than those of metals, their conductivities are much lower because they have much lower densities of mobile charge carriers. Decreases of conductivity in *metals* with increasing temperature, alloying, and structural defects result mostly from decreases in *mobility* produced by departures from lattice perfection that reduce electron mean free path. Changes in conductivity in *nonmetals* result primarily from changes in *carrier density*, and are often in the opposite direction, for example, increasing conductivity with increasing temperature.

Under AC conditions, the local Ohm's law and conductivity of materials will be little changed from that for DC as long as the product $\omega\tau$ ($\omega = 2\pi\nu$ = angular frequency, τ = collision time) is much less than one, that is, as long as the electrons undergo a great many collisions within each field cycle. (Chapter 2 addresses the case of higher frequencies.)

The classical free-electron theory does a fair job of explaining the Wiedemann-Franz law, an empirical relation between the thermal and electrical conductivities of metals. However, it predicts a much larger electronic contribution to the heat capacity of metals and a much larger thermopower than are experimentally observed. To explain these discrepancies in thermal properties, positive Hall coefficients of many materials, and the effects of temperature and other variables on charge carrier density in nonmetals, we'll need quantum mechanics.

PROBLEMS

1-1. Because of a shortage of copper during World War II, the electromagnets used in Oak Ridge to separate U-235 isotopes for the Hiroshima atomic bomb were wound with pure silver, which has an atomic volume of 10.3 cc/mol. If 100 m of silver wire 2 mm in diameter required 8.1 watts of electrical power when carrying 4 amps of current, calculate the (a) resistivity, (b) electron mobility, (c) collision time, (d) mean free path, (e) Hall coefficient, and (f) the Hall field in a transverse magnetic field of 9 tesla. Assume $T = 300$ K, and assume silver is monovalent (one free electron per atom). (We remind you that electrical power equals the product of current and voltage.)

1-2. An aluminum wire with a square cross-section 2 mm on a side is carrying a current of 1.5 amps. Aluminum has a resistivity of 2.69×10^{-6} ohm-cm and an atomic volume of 10 cc/mol. Assuming aluminum has three free electrons per atom, calculate the (a) Hall coefficient, (b) electron mobility, (c) drift velocity, (d) collision time, (e) electric field, (f) power dissipated (heat generated) per cm length, and (g) the Hall voltage across the wire in a transverse magnetic field of 1 tesla.

1-3. Copper has 8.5×10^{22} free electrons/cm^3 and a resistivity at room temperature of 1.69×10^{-6} ohm-cm. Calculate (a) electron mobility, (b) average number of collisions per second, (c) average distance between collisions, and (d) average drift velocity at a current density of 1 amp/mm^2. (e) If cooled to $T =$ 2 K, the resistivity decreases by a factor of a thousand. Calculate (a), (b), and (d) for this temperature.

1-4. (a) Derive equation (1.14) from equations given earlier. (b) The electron mobilities of indium antimonide, silicon, and a typical metal are about 7, 0.15, and .005 m^2/V-s, respectively. For each, calculate the Hall angle in a magnetic field of 1 tesla. (c) In each case, what is the net force on the electrons in the direction of the Hall field?

1-5. In a Hall experiment, a semiconductor carrying current in the positive x-direction is exposed to a magnetic field in the positive y-direction. (Assume a right-handed coordinate system.) If the charge carriers are positively charged "holes," in which direction is: (a) the drift velocity, (b) the cross product of velocity and magnetic field, (c) the Lorentz magnetic force, (d) the Hall electric field? (e) Which of these vectors would change direction if the charge carriers were not holes but electrons? (f) If the longitudinal electric field and the magnetic field are held constant, and the charge carrier concentration of the material were doubled with no change in mobility, by how much would the Hall voltage change?

1-6. (a) In normal household wiring, appliances and lightbulbs are connected *in parallel*, that is, each is exposed to the same voltage (120 V). Calculate the resistance of a 60-watt bulb and of a 100-watt bulb, and how much current flows through each in normal household operation. (b) Now suppose these two bulbs are instead connected *in series*, that is, so that the same current flows through each. Which bulb will consume more watts of power in this situation? (c) Neglecting any changes of the bulbs' electrical resistances with filament temperature, what will be the net resistance of the two bulbs connected in series? In parallel?

1-7. A composite material is made of alternating thin lamellae of copper (resistivity 1.69×10^{-6} ohm-cm) and a niobium-titanium alloy (resistivity 7×10^{-5} ohm-cm) of equal thickness. (a) What is the resistivity of this composite measured parallel to the lamellae? If current is passed through the composite in this direction, what fraction of the current will be carried by the copper? How will the electric fields in the two phases compare? (b) What is the resistivity of this composite measured perpendicular to the lamellae? If current is passed through in this direction, what will be the ratio of the electric field in the alloy to that in the copper? (c) At a temperature of 4.2 K, the resistivity of the copper has decreased to 1×10^{-8} ohm-cm, and the niobium-titanium alloy has become superconducting. Answer questions (a) and (b) for this situation.

1-8. A wire of length L and resistivity ρ has a circular cross section with a diameter that tapers linearly from d_1 at one end to d_2 at the other end. Derive a formula for the electrical resistance along this length of wire. (Hint: Since the cross-sectional area changes along the length, the current density and electric field will change, so some integration may be required.)

1-9. Equation (1.11) is a differential equation for drift velocity. (a) At $t = 0$, we suddenly increase the electric field from zero to a finite value. For this situation, solve (1.11) and get a general expression for the time dependence of drift velocity for $t > 0$. (b) How does velocity vary with time for times short compared to the collision time? (c) How does velocity vary with time for times long compared to the collision time? (d) Which of these limits corresponds to Ohm's law?

1-10. A metal wire with 10^{23} free electrons per cm^3 is carrying 60 Hz alternating current with a maximum current density of 100 amps/cm^2. (a) Write an expression for the average electron drift velocity as a function of time. (b) What is the average drift distance of free electrons in one direction in a half-cycle of current?

1-11. The Wiedemann-Franz constant $K/\sigma T$ is 2.4×10^{-8} $J\Omega K^{-2}s^{-1}$. For the composite material described in Problem 1-7, calculate the room-temperature thermal conductivities of the two separate phases, and of the composite parallel to and perpendicular to the lamellae. Also calculate the thermal conductivity of copper at 4.2 K.

1-12. The Wiedemann-Franz law applied to a superconductor implies that the thermal conductivity would become infinite once the electrical conductivity becomes infinite, but the electronic contribution to thermal conductivity actually decreases in the superconducting state. Why?

1-13. In the SI system of units, every unit can be expressed as some combination of meters, kilograms, seconds, and amperes. For example, the unit for force is the newton, which is equivalent to kg-m-s^{-2}. (That can be derived easily by remembering that $\mathbf{F} = m\mathbf{a}$.) Similarly, express in terms of m, kg, s, and A each of the following units: joule, watt, volt, ohm, siemens, tesla.

1-14. Assuming that the probability that a free electron will travel a time t before a collision is proportional to $\exp(-t/\tau)$, calculate the average time between collisions and the average drift velocity.

1-15. For lamellar composites like the one discussed in the sample problem or in Problem 1-7, under what conditions will the electric field and the electric current be nonparallel?

Chapter 2

Windows, Doors, and Transparent Electrodes (Optical Properties of Conductors)

2.1 SEEING THE LIGHT

Most transparent solids, like window glass and Lucite (polymethylmethacrylate, or PMMA), are insulators. In contrast, conducting metals are opaque to visible light, and "make a better door than a window." But is it possible to design a material that can both transmit light *and* conduct electricity?

Liquid-crystal displays, such as those in laptop computer screens, are based on elongated organic molecules that can be aligned by an electric field, thereby going from opaque to transparent when a voltage is applied. But if you had to apply the voltage with metallic electrodes on each side of the layer of liquid crystal, you wouldn't be able to get light through the electrodes. This is one of several important applications where it would be helpful to have a *transparent electrode*, a transparent material that also conducts electricity.

Classically, light is an electromagnetic wave, containing electric and magnetic fields that oscillate at very high frequencies. Thus to examine whether light transmission and electrical conductivity are totally incompatible, we'll have to consider in some detail what happens when an electromagnetic wave impinges on a conductor. This would be a good time for you to have your earlier physics and math textbooks within reach for review, because we'll be using Maxwell's equations, vector calculus, complex numbers, and partial differential equations. The formalism will be useful in this chapter, but will also provide a basis for much of what follows in later chapters, including our treatment of quantum mechanics.

One very useful mathematical shorthand is the "vector operator" *del*, which can represent spatial derivatives of scalar and vector quantities. Nobel-Prize physicist Richard Feynman wrote that del is "extremely amusing and ingenious—and characteristic of the things that make mathematics beautiful." It is written as $\boldsymbol{\nabla}$, and defined in terms of partial derivatives and the unit vectors **i**, **j**, and **k** in the x, y, and z directions, respectively:

$$\boldsymbol{\nabla} = \mathbf{i}\frac{\partial}{\partial x} + \mathbf{j}\frac{\partial}{\partial y} + \mathbf{k}\frac{\partial}{\partial z} \tag{2.1}$$

Del operating on a scalar quantity gives a vector quantity called the *gradient*. For example, if the scalar is electric potential (voltage) V, the negative gradient of V is the electric field

$$\mathbf{E} = -\boldsymbol{\nabla} V \tag{2.2}$$

and Ohm's law becomes

$$\mathbf{J} = -\sigma\, \boldsymbol{\nabla} V \tag{2.3}$$

Similarly, the vector equivalent of equation (1.16), the heat-flow equation, becomes $\mathbf{J}_Q = -K\boldsymbol{\nabla}T$. The analogous equation in diffusion theory is Fick's first law of diffusion, which relates the diffusive flow of an element to the negative gradient of its composition.

Del can also operate on vector quantities. The dot product of del and a vector yields a scalar called the *divergence*. If the vector involved is thought of as representing a flow, the divergence represents the local flow outward from a point. For example, if the vector is the current density **J**, then the divergence of **J**, written as $\boldsymbol{\nabla} \cdot \mathbf{J}$, represents the outward flow of electric charge. The equation of charge conservation is therefore written as

$$\frac{\partial \rho}{\partial t} = -\boldsymbol{\nabla} \cdot \mathbf{J} \tag{2.4}$$

where ρ is here the local density of electric charge in coulombs per cubic meter. (The same symbol is usually used for charge density and resistivity. That's another reason we usually express Ohm's law in terms of conductivity rather than resistivity.) This charge-conservation equation states in mathematical form that a net outward flow of charge from a point, that is, the divergence of electric current density, produces a loss of local charge density. It makes sense.

The first two of Maxwell's equations, which are the Gauss's laws that you may have learned in the form of surface integrals, can be transformed with the divergence theorem (see your calculus text if you've forgotten it) into equations for the divergence of electric fields and magnetic fields at any point. Gauss's law of electrostatics becomes

$$\varepsilon_o \nabla \cdot \mathbf{E} = \rho \tag{2.5}$$

(The constant ε_o is the *permittivity* of vacuum, 8.85×10^{-12} farads/m, and ρ is still charge density.) Gauss's law for magnetism becomes

$$\nabla \cdot \mathbf{B} = 0 \tag{2.6}$$

The first of Gauss's equations relates the divergence of electric field from any point to the local *net* electric charge density, and the second states that there can be no net magnetic equivalent of charge, that is, no separate north or south poles (no "magnetic monopoles"). Magnetic fields are divergenceless, with no sources or sinks of **B**.

The cross product of del with a vector is called the *curl*, which is a vector quantity related to rotational flow in a vector field. You probably learned Faraday's law of electromagnetic induction in a form relating the line integral of electric field around a closed path to the time rate of change of magnetic flux through the surface surrounded by that path. Stokes' theorem (see your calculus text again) relates such a line integral to the surface integral of the curl, which, if we let the surface shrink to an infinitesimal area, allows Faraday's law to be transformed into

$$\nabla \times \mathbf{E} = -\frac{\partial \mathbf{B}}{\partial t} \tag{2.7}$$

Pretty neat! Similarly, you probably learned Ampere's law in a form relating the line integral of magnetic field around a closed path to the current through the surface surrounded by that path. Applying Stokes' theorem to that transforms Ampere's law into

$$\nabla \times \mathbf{B} = \mu_o \mathbf{J} \tag{2.8}$$

Also very neat! (The constant μ_o is the *permeability* of vacuum, 1.26×10^{-6} henries/m. Although permittivity and permeability sound very much alike, the former is related to electric fields, and the latter to magnetic fields.) Maxwell added an important second term to that equation involving the time rate of change of electric field. In differential form, the Ampere-Maxwell equation is

$$\nabla \times \mathbf{B} = \mu_o \mathbf{J} + \mu_o \varepsilon_o \frac{\partial \mathbf{E}}{\partial t} \tag{2.9}$$

Quoting Feynman again: "From a long view of the history of mankind—seen from, say, ten thousand years from now—there can be little doubt that the most significant event of the 19th century will be judged as Maxwell's discovery of the laws of electrodynamics." In differential form, Maxwell's famous four equations, which appear on many T-shirts, are

$$\varepsilon_o \nabla \cdot \mathbf{E} = \rho \qquad \nabla \cdot \mathbf{B} = 0$$
$$\nabla \times \mathbf{E} = -\frac{\partial \mathbf{B}}{\partial t} \qquad \nabla \times \mathbf{B} = \mu_o \mathbf{J} + \mu_o \varepsilon_o \frac{\partial \mathbf{E}}{\partial t} \tag{2.10}$$

In vacuum, both charge density and current are zero, so Maxwell's equations simplify to

$$\nabla \cdot \mathbf{E} = \nabla \cdot \mathbf{B} = 0 \qquad \nabla \times \mathbf{E} = -\frac{\partial \mathbf{B}}{\partial t} \qquad \nabla \times \mathbf{B} = \mu_o \varepsilon_o \frac{\partial \mathbf{E}}{\partial t} \tag{2.11}$$

We see that in vacuum, both electric and magnetic fields are divergenceless, and the curl of each is related to the time variation of the other.

2.2 MAKING WAVES WITH MAXWELL

Something interesting happens if you take the curl of a curl. It's straightforward but a bit tedious to show that $\nabla \times (\nabla \times \mathbf{E}) = \nabla(\nabla \cdot \mathbf{E}) - \nabla^2 \mathbf{E}$, where ∇^2 is $\nabla \cdot \nabla$, the divergence of a gradient. ∇^2 is a second-derivative operator, called the *Laplacian*, that appears often in physics. We'll see a lot of it when we get to quantum mechanics. In x, y, z coordinates, it is

$$\nabla^2 = \frac{\partial^2}{\partial x^2} + \frac{\partial^2}{\partial y^2} + \frac{\partial^2}{\partial z^2} \tag{2.12}$$

Taking the curl of Faraday's equation yields a Laplacian of the electric field on the left (the divergence term is zero), and the curl of $\partial \mathbf{B}/\partial t$ on the right. By inverting the order of the spatial and time derivatives, and using the Ampere-Maxwell equation to substitute for $\nabla \times \mathbf{B}$, the result is

$$\nabla^2 \mathbf{E} = \mu_o \varepsilon_o \frac{\partial^2 \mathbf{E}}{\partial t^2} \tag{2.13}$$

If you take the curl of the Ampere-Maxwell equation and do the same manipulations, you will get the same equation for the magnetic field vector:

$$\nabla^2 \mathbf{B} = \mu_o \varepsilon_o \frac{\partial^2 \mathbf{B}}{\partial t^2} \tag{2.14}$$

These equations were familiar to Maxwell as forms of the *wave equation*, and they correspond to waves of electric and magnetic field traveling at a speed of $(\mu_o \varepsilon_o)^{-1/2}$. Substituting the values of the permittivity and permeability of vacuum indicates that these "electromagnetic waves" would travel at a velocity of $c = 3 \times 10^8$ m/s, the velocity of light!

$$(\mu_o \varepsilon_o)^{-1/2} = c = 3 \times 10^8 \text{ m/s} \tag{2.15}$$

Until then, it had not been realized that light was an electromagnetic wave; Maxwell's discovery merged the fields of electricity, magnetism, and optics. And electromagnetic waves are not restricted to the narrow range of frequencies and wavelengths of visible light. "Maxwell's rainbow" includes waves of much lower frequencies (infrared, microwaves, radio) and much higher frequencies (ultraviolet, x-rays, gamma rays).

There are many forms of waves. For simplicity, we consider a simple cosine wave traveling in the positive z-direction. We assume in addition that it is a *plane wave*, so that **E** and **B** are independent of x and y (i.e., are constant in the *x-y plane*). Since the z-components of the curls of each field (which involve derivatives with respect to x and y, you'll recall) are therefore zero, Maxwell's equations (2.11) indicate that there are no time-varying fields in the z-direction. Thus the oscillating **E** and **B** fields are both in the x-y plane, *directed perpendicular—transverse—to the direction of motion.* Electromagnetic waves are *transverse waves.* (Waves in which the varying quantities are directed parallel to the direction of motion are called instead *longitudinal* waves. We'll see some in Chapter 7.) One possible expression for the electric field in this wave is

$$\mathbf{E} = \mathbf{E}_o \cos(\omega t - kz) \tag{2.16}$$

As before, $\omega = 2\pi\nu$ is the angular frequency. The quantity k is called the *wave number* and equals $2\pi/\lambda$, where λ is the wavelength. Inserting (2.16) into (2.13) shows that (2.16) is a solution of the wave equation if

$$k^2 = \omega^2 \mu_o \varepsilon_o \qquad \text{or} \qquad \omega/k = (\mu_o \varepsilon_o)^{-1/2} = \lambda\nu = c \tag{2.17}$$

A fixed phase of the wave corresponds to a fixed value of the argument $(\omega t - kz)$, which remains unchanged after a time interval Δt if $k\Delta z = \omega \Delta t$, that is, $\Delta z/\Delta t = \omega/k = c$. Thus the wave described by (2.16) is traveling in the *positive* z-direction. If the argument were changed to $(\omega t + kz)$, we would instead have a wave moving at velocity c in the *negative* z-direction. If we had written a similar expression with a sine instead of a cosine, it would have been a wave 90° out of phase with the cosine wave.

Suppose we say the electric field is in the x-direction. Then curl **E** will have only a y-component ($\partial E/\partial z$ is the only nonzero term), proving that the magnetic field is in the y-direction. In an electromagnetic wave, electric and magnetic fields oscillate at right angles to each other (Fig. 2.1). Faraday's equation is now simply $\partial E/\partial z = -\partial B/\partial t$, so

$$kE_o = \omega B_o \qquad \text{or} \qquad B_o = \frac{E_o}{c} \tag{2.18}$$

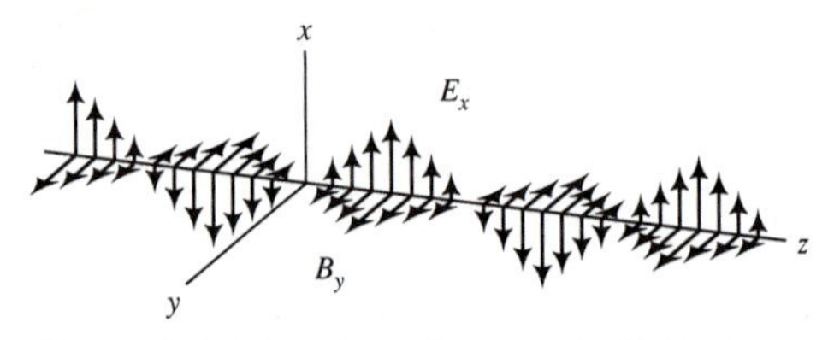

Figure 2.1. Electric and magnetic fields in an electromagnetic wave.

We can see that if an electromagnetic wave encounters an electron that is moving with velocity v, the maximum magnetic force on the electron ($evB = evE/c$) will be much smaller than the electric force ($e\mathbf{E}$) unless the electron is moving with a velocity near the velocity of light. Thus in our consideration of the interaction of electromagnetic waves with solids, we can usually concentrate only on the electric force.

Sample Problem 2.1

A plane wave of violet light ($\nu = 7.5 \times 10^{14}$ Hz) has a maximum electric field E_o of 10^{-4} V/m. Calculate the wavelength, wave number, angular frequency, and maximum magnetic field of this electromagnetic wave.

Solution

$\lambda = c/\nu = (3 \times 10^8)/(7.5 \times 10^{14}) = 4 \times 10^{-7}$ m $= 400$ nm (in vacuum)
$k = 2\pi/\lambda = 1.57 \times 10^7$ m^{-1}, $\omega = 2\pi\nu = 4.71 \times 10^{15}$ s^{-1}
$B_o = E_o/c = 3.33 \times 10^{-13}$ T

Before we let an electromagnetic wave enter our intended target, a conductor, a few more definitions and mathematical points are in order. If we add a plane wave traveling in the positive z-direction to another of the same amplitude traveling in the negative z-direction, a bit of algebra will show that the result is

$$\mathbf{E} = 2\mathbf{E}_o \cos(\omega t) \cos(kz) \tag{2.19}$$

This is no longer a traveling wave. It is a *standing wave*. Note in particular that the *nodes*, positions in space where the field goes to zero, remain at constant z, whereas they moved to the right with velocity c in the expression (2.16). Although we will be interested in traveling waves in the remainder of this chapter, we will encounter both standing and traveling waves in future chapters. It will be helpful to remember that *addition of two equal waves traveling in opposite directions yields a standing wave.*

In dealing with the mathematics of waves, it is often simpler to express things in

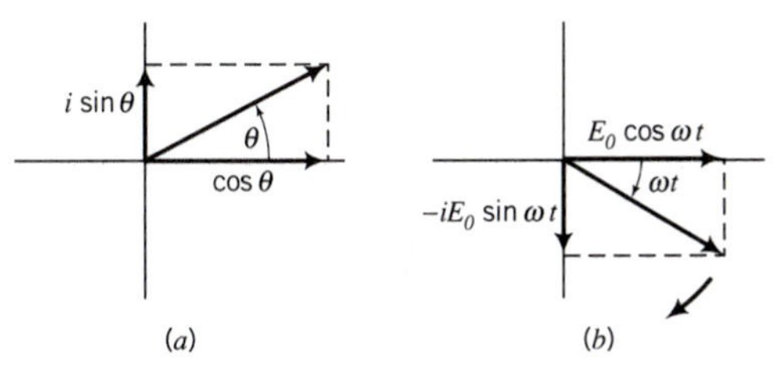

Figure 2.2. (a) Representation of a complex number $e^{i\theta} = \cos\theta + i\sin\theta$ as a vector in the complex plane. (b) Representation of $E_0e^{-i\omega t}$ as a vector rotating clockwise in the complex plane.

complex notation, because exponentials are easier to manipulate through various operations than sines and cosines are. As paradoxical as it may sound, complex numbers usually simplify the mathematics. So as an alternative to (2.16), we prefer

$$\mathbf{E} = \mathbf{E}_o e^{-i(\omega t - kz)} \tag{2.20}$$

If we add to that traveling wave, $\mathbf{E}_o e^{-i(\omega t + kz)}$, a traveling wave moving in the opposite direction, we get $2\mathbf{E}_o \cos(kz)e^{-i\omega t}$, a standing wave. (Even that simple operation was a bit easier than adding the two cosine waves.) In dealing with complex numbers, it will occasionally be useful to remember Euler's formula:

$$e^{i\theta} = \cos\theta + i\sin\theta \tag{2.21}$$

So

$$e^{-i\omega t} = \cos(\omega t) - i\sin(\omega t) \tag{2.21a}$$

You'll recall that a complex number can be represented by a vector in the complex plane, with the horizontal component representing the "real" part and the vertical component the "imaginary" part (Fig. 2.2a). Any quantity that varies as $e^{-i\omega t}$ can therefore be thought of as a vector rotating clockwise in the complex plane (Fig. 2.2b). (If we had instead chosen $e^{+i\omega t}$ to represent the time-dependence, as is common in electrical engineering texts, the vectors would instead rotate counterclockwise.) We will find this concept useful later when considering the relationship between two quantities (say, field and current) that are both varying periodically with time. Then the phase difference between the two varying quantities will be represented by the angle between two rotating vectors.

2.3 ONLY SKIN DEEP

We are now ready to examine, using Maxwell's equations and Ohm's law, what happens when visible light, or any other electromagnetic wave, enters a conductor. To move Maxwell's equations from vacuum into the material, we will replace the permittivity and permeability of vacuum, ε_o and μ_o, by ε and μ, the permittivity and permeability of the material, and *insert Ohm's law,* $\mathbf{J} = \sigma\mathbf{E}$, *into the Ampere-Maxwell equation.* The Ohm's law term will complicate the wave equation for **E**, which now becomes

$$\nabla^2\mathbf{E} = \mu\varepsilon \frac{\partial^2\mathbf{E}}{\partial t^2} + \mu\sigma \frac{\partial \mathbf{E}}{\partial t} \tag{2.22}$$

which is known, for reasons that soon will become obvious, as the *damped* wave equation.

We will assume that a plane electromagnetic wave of the form given in (2.20) is traveling in the z-direction inside this conducting material. A wave of this form will be a solution of (2.22) if

$$k^2 = \omega^2\mu\varepsilon + i\omega\mu\sigma \tag{2.23}$$

For any good conductor, the second term in (2.23) will be dominant. In this limit, and since Euler's formula can be used to show $\sqrt{i} = \pm(1 + i)/\sqrt{2}$,

$$k \approx (i\omega\mu\sigma)^{1/2} = \pm\frac{(1 + i)}{\sqrt{2}}(\omega\mu\sigma)^{1/2} = \pm\frac{(1 + i)}{\delta} \tag{2.24}$$

where we define the *skin depth* δ by

$$\delta = \left(\frac{2}{\omega\mu\sigma}\right)^{1/2} \tag{2.25}$$

The wave number in (2.24) is *complex*! Since we are interested in a wave traveling in the positive z-direction, we choose the plus option of $\pm$. Inserting (2.24) into (2.20), we get

$$\mathbf{E} = \mathbf{E}_o e^{-i(\omega t - z/\delta)} e^{-z/\delta} \tag{2.26}$$

This is still an oscillatory wave, with the $e^{-i(\omega t - z/\delta)}$ term representing the oscillations. However, the amplitude $E_o e^{-z/\delta}$ *declines exponentially*, with the skin depth δ defining the scale of the decline. (The use of complex numbers was very convenient in reaching this conclusion. Although it is possible to derive the skin depth without complex numbers, it is much more difficult.) The factor that indicates an exponential decay arises from the "imaginary" part of the wave number k, but it is very real. Light and other forms of electromagnetic radiation are attenuated as they penetrate an Ohm's-law conductor, as a result of the oscillating currents produced in response to, and in phase with, the oscillating electric field.

Note from equation (2.26) that it is the electric *field* that decays by the factor e in a distance of one skin depth. Since the *intensity* of the electromagnetic wave is proportional to the *square* of the field, it will decay by the factor e^2 in that distance. The relations between field and intensity of an electromagnetic wave appear later as equation (4.7).

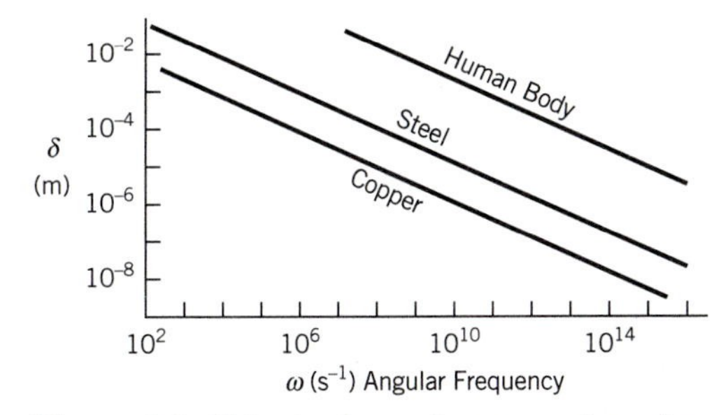

Figure 2.3. Skin depth vs. frequency for a few materials.

Sample Problem 2.2
(a) Calculate the skin depth of an electromagnetic wave with a wavelength in vacuum of 3 m when traveling in a nonmagnetic material with a conductivity of 500 S/m. (b) By how much will the intensity of this wave be decreased in traveling through 1 mm of this metal?

Solution
(a) $\omega = 2\pi\nu = 2\pi c/\lambda = 6.28 \times 10^8\ \text{s}^{-1}$, $\mu = \mu_o = 1.26 \times 10^{-6}$ h/m, $\sigma = 500$ S/m, $\delta = (2/\omega\mu\sigma)^{1/2} = 2.25 \times 10^{-3}$ m
(b) $I/I_o = \exp(-2z/\delta) = 0.411$ (intensity decreased by 58.9% at $z = 1 \times 10^{-3}$ m)

From (2.25) we can estimate δ for a nonmagnetic ($\mu = \mu_o$) metal, like copper or aluminum, as a function of frequency. Skin depths are of the order of a centimeter at power frequencies (60 Hz), about 10 μm at 100 MHz (FM range), and only a few nanometers for visible light. Now we can see why metals make better doors than windows. Visible light penetrates only a few atom distances, and with so little energy entering the metal, most of the incident energy must be reflected.

(The limited skin depth of metals for microwave frequencies explains why you should remove the wrap of aluminum foil before reheating your dinner in the microwave oven—and why the window of the oven has a metal screen to avoid cooking you.)

Figure 2.3 shows skin depth vs. frequency for a few metals and for the human body, which has a conductivity near that of sea water (between six and seven orders of magnitude less than that of pure metals) and a classical skin depth for visible light of a few tens of microns.

2.4 PLASMA FREQUENCY AND TRANSPARENT ELECTRODES

From our skin-depth calculations, conductivity and transparency do indeed look pretty incompatible as long as $\mathbf{J} = \sigma\mathbf{E}$, with σ essentially unchanged from the DC conductivity. In Chapter 1 we concluded that would be valid as long as $\omega\tau$ remained very much less than one. If it isn't, what then? Rewriting and rearranging equation (1.11), we obtain

$$m\frac{d\mathbf{v}_D}{dt} + m\frac{\mathbf{v}_D}{\tau} = -e\mathbf{E} \tag{2.27}$$

This equation represents the dynamics of free electrons in metals, with the drift velocity related to current density through $\mathbf{J} = -eN_e\mathbf{v}_D$. For a DC field and steady state ($d/dt = 0$), we dropped the first term and got Ohm's law. The viscous or resistive term dominated. As we discussed in Chapter 1, that's also legitimate for AC as long as $\omega\tau \ll 1$. In this limit of low frequencies, an electric field varying as $\sin(\omega t)$ will yield a current varying as $\sin(\omega t)$, that is, it is *in phase*.

Consider instead the opposite extreme, $\omega\tau \gg 1$. In this limit, we can drop instead the *second* term on the left side of (2.27), that is, the term related to scattering and the normal electrical resistivity. Rather than experiencing many collisions in each cycle of the field, electrons will experience many cycles of the field between collisions. In this limit, electrons will behave like electrons in free space, and *it will be the acceleration* ($d\mathbf{v}_D/dt$), *not the velocity, that will be in phase with the field.* With the field and electron acceleration varying as $\sin(\omega t)$, the current (proportional to velocity, the integral of the acceleration) will vary as $-\cos(\omega t)$, *90° out of phase with the field.*

Let's carry out the calculation for this case using exponential notation. Between collisions, Newtonian mechanics will apply, and an oscillating electric field $\mathbf{E}$ given by $\mathbf{E}_o e^{-i\omega t}$ will give the electron an acceleration $(-e\mathbf{E}_o/m)e^{-i\omega t}$. Integrating this acceleration to get the drift velocity yields an oscillatory drift velocity of $(e\mathbf{E}_o/i\omega m)e^{-i\omega t}$, which is equivalent to a current density $\mathbf{J}$ of $(-N_e e^2\mathbf{E}_o/i\omega m)e^{-i\omega t}$.

The ratio $\mathbf{J}/\mathbf{E}$ is now $(-N_e e^2/i\omega m)$, the effective "conductivity" to be inserted into (2.22) and (2.23). In terms of the DC conductivity $\sigma_{dc} = N_e e^2\tau/m$, the AC "conductivity" $\sigma_{ac} = \mathbf{J}/\mathbf{E}$ *in the limit* $\omega\tau \gg 1$ is equal to

$$\frac{\mathbf{J}}{\mathbf{E}} = -\frac{\sigma_{dc}}{i\omega\tau} = \frac{i\sigma_{dc}}{\omega\tau} = \sigma_{ac} \qquad (\omega\tau \gg 1) \tag{2.28}$$

What does it mean to have $\mathbf{J}/\mathbf{E}$ imaginary? Since $i = e^{i\pi/2}$, it simply means that $\mathbf{J}$ *and* $\mathbf{E}$ *are* $\pi/2$ *out of phase*. Substituting this expression for conductivity into equation (2.23) yields

$$k^2 = \omega^2\mu\varepsilon - \frac{N_e e^2\mu}{m} = \omega^2\mu\varepsilon\left(1 - \frac{\omega_p^2}{\omega^2}\right) \tag{2.29}$$

where we have introduced the *plasma frequency*

$$\omega_p = \left(\frac{N_e e^2}{m\varepsilon}\right)^{1/2} \tag{2.30}$$

We can see from (2.29) that the wave vector will no longer have an imaginary component, that is, k^2 will be positive, as long as $\omega > \omega_p$. Recalling that it was the

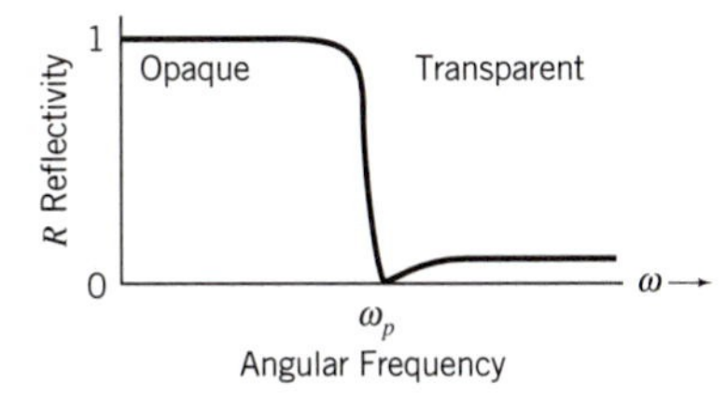

Figure 2.4. Increase in light transmission (decrease in reflectivity) of metals for frequencies beyond the plasma frequency (schematic).

imaginary component of k in (2.24) that led to attenuation, (2.29) shows that electromagnetic waves with frequencies greater than the plasma frequency will not be attenuated! As long as our assumption that $\omega\tau$ is much greater than one is valid, our conductor *will become transparent at sufficiently high frequencies.*

Calculating ω_p for most metals gives plasma frequencies above the frequencies of visible light, so metals still aren't good windows for visible light. But alkali metals show the predicted abrupt increase in transmission in the near-ultraviolet, slightly above visible frequencies, and at frequencies close to those calculated from (2.30). Transmission of electromagnetic waves through conductors above their plasma frequency (Fig. 2.4) is real, and the plasma frequencies of some metals aren't that far from the frequencies of visible light.

To get those transparent electrodes that we wanted for liquid crystal displays, a material with ω_p in the near infrared (just below visible frequencies) would do just fine. The experimental plasma frequencies for metals and equation (2.30) suggest that a conductor with a charge carrier density about 1% of that of metals would have such a plasma frequency. And with 1% of the carrier density of metals, it would still have a pretty good conductivity. The material most commonly used today for transparent electrodes is indium tin oxide, which indeed has a plasma frequency in the near infrared, allowing it to transmit visible light but still have usable conductivity.

The earth's ionosphere has a much lower charge carrier density, yielding a plasma frequency of the order of 10^8 Hz. Thus the ionosphere is opaque to AM radio waves, which are reflected back to earth, but is transparent to TV waves, allowing Martians and other distant aliens to enjoy *The X-Files*.

Sample Problem 2.3

A metal is found to be transparent to electromagnetic waves with a frequency ν greater than 6.82×10^{14} Hz. Assuming the effective permittivity is ε_o, calculate the charge carrier concentration.

Solution

$\omega_p = 2\pi\nu_p = 4.29 \times 10^{15}\ \mathrm{s}^{-1}$, $m = 9.11 \times 10^{-31}$ kg,
$\varepsilon = \varepsilon_o = 8.85 \times 10^{-12}$ f/m, $N_e = \omega_p^2 m\varepsilon_o/e^2 = 5.80 \times 10^{27}$ electrons/m^3

2.5 CYCLOTRON RESONANCE AND EFFECTIVE MASS

When $\omega\tau \gg 1$, free electrons in metals experience many field cycles between collisions and are essentially "ohmless," acting much as if they were in vacuum. That suggests that something very interesting might occur if we applied a constant magnetic field. You may recall that an electron traveling perpendicular to a magnetic field **B** experiences circular motion with the cyclotron frequency:

$$\omega_c = \frac{eB}{m} \tag{2.31}$$

Most of the free electrons in a metal, moving in random directions with an average thermal velocity v_{th}, will have a component of velocity perpendicular to **B**, and the Lorentz force from **B** will induce a component of circular motion at a frequency given by (2.31). If $\omega_c\tau \gg 1$, the electrons will be able to make at least a few laps between collisions. While the electrons are performing this dance, if we then also expose them to an electromagnetic wave and varied its frequency, we would expect an enhanced absorption of energy when the frequency of the electromagnetic wave matched the cyclotron frequency. We would expect what is called *cyclotron resonance*.

Such experiments have been done on a number of materials, usually on pure materials at low temperatures to reach a collision time τ long enough to ensure that $\omega_c\tau \gg 1$. Resonances (absorption peaks) are indeed observed at frequencies proportional to the magnetic field **B**, but the proportionality constant is usually different from the predicted e/m. Apparently even between collisions, the electron knows it is not really traveling in vacuum. It behaves like a particle of charge $-$e, but with an *effective mass* m^* different from that of an electron traveling in vacuum. In fact, experiments on some materials give several cyclotron resonances, suggesting that charge carriers of more than one effective mass are present. Although the presence of cyclotron resonance of electrons in solids is predicted classically, we'll need quantum mechanics to explain why the electrons act as if they have effective masses different from the mass of a totally free electron. We'll learn later about "light" and "heavy" electrons and holes.

2.6 COMPLEX CONDUCTIVITY

We have considered only the two extremes of frequency: $\omega\tau \ll 1$ (where the AC conductivity is real and approximately equal to the DC conductivity σ_{dc}, as we assumed in our treatment of the skin depth) and $\omega\tau \gg 1$ (where the AC conductivity **J**/**E** is imaginary and equal to $-\sigma_{dc}/i\omega\tau$, as we found in our treatment of the plasma frequency). For low frequencies, we used only the second term on the left side of equation (2.27), while for high frequencies, we used only the first term. For intermediate frequencies, *both terms* must be used, which yields as the general solution

$$\frac{\mathbf{J}}{\mathbf{E}} = \frac{\sigma_{dc}}{1 - i\omega\tau} = \sigma' + i\sigma'' = \sigma_{ac} \tag{2.32}$$

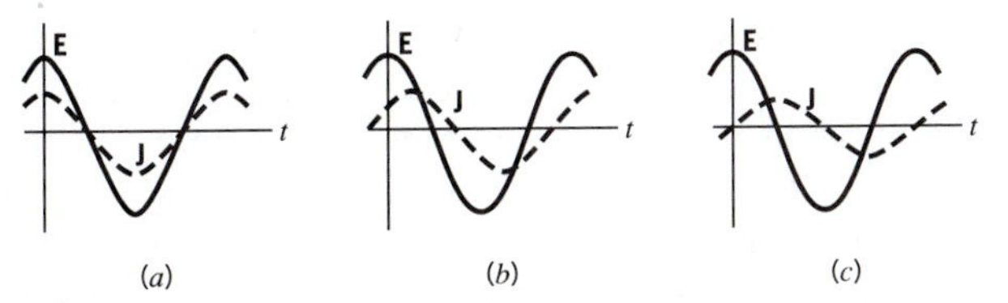

Figure 2.5. Current is (a) in phase with field at low frequencies, (b) lags by intermediate angles at intermediate frequencies, and (c) lags by 90° at high frequencies.

This AC conductivity can be seen to have the appropriate limits at both very high and very low frequencies. For $\omega\tau \ll 1$, it equals the DC conductivity σ_{dc} and is real, meaning that current and electric field are in phase (Fig. 2.5a). For $\omega\tau \gg 1$, it equals $-\sigma_{dc}/i\omega\tau$ and is purely imaginary, meaning that current and field are out of phase by 90° (Fig. 2.5c). For intermediate frequencies, the AC conductivity given by (2.32) has both real (σ') and imaginary (σ'') parts, and current lags the field by an angle between 0 and 90° (Fig. 2.5b).

Equivalently, the phase relations between current and field for the free electrons can be shown by vectors rotating clockwise in the complex plane, as in Fig. 2.6. With electric field varying as $\mathbf{E}_o e^{-i\omega t}$, current varies as $\mathbf{J}_o e^{-i(\omega t-\theta)}$, where θ is the phase angle between current and field. At low frequencies, current and field are in phase ($\theta = 0$, Fig. 2.6a), at very high frequencies, current lags field by 90° ($\theta = \pi/2$, Fig. 2.6c), and at intermediate frequencies, current lags field by intermediate angles (Fig. 2.6b). The phase angle can be determined from (2.32) and Euler's formula. For example, when $\omega\tau = 1$, the real and imaginary parts of AC conductivity are equal, meaning that $\theta = 45°$. In general,

$$\sigma_{ac} = \sigma' + i\sigma'' = \frac{\sigma_{dc}(1 + i\omega\tau)}{(1 + \omega^2\tau^2)} = \frac{\sigma_{dc}}{(1 + \omega^2\tau^2)^{1/2}} [\cos\theta + i\sin\theta] \quad \textbf{(2.33)}$$

where

$$\theta = \tan^{-1}\left(\frac{\sigma''}{\sigma'}\right) = \tan^{-1}(\omega\tau) \quad \textbf{(2.34)}$$

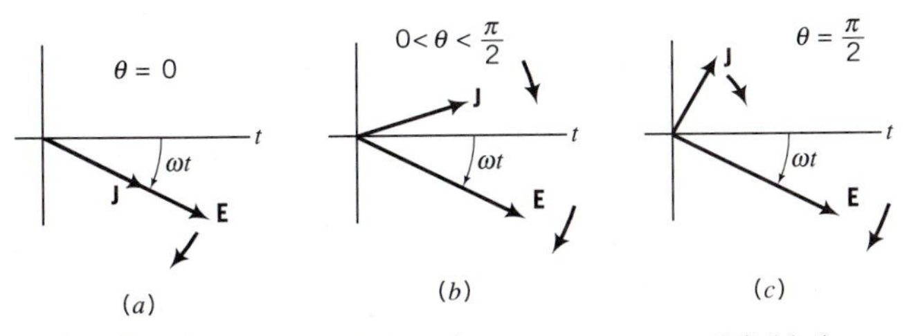

Figure 2.6. The same relations between current and field shown in Fig. 2.5 are shown here schematically as rotating vector diagrams in the complex plane, as in Fig. 2.2. (a) $\omega\tau \ll 1$, (b) intermediate frequencies, (c) $\omega\tau \gg 1$.

The mathematically complex AC conductivity of (2.32) and (2.33) therefore represents both the amplitude and the phase angle θ of the free-electron current density in a metal produced by an alternating electric field. *Similarly, other material properties described in complex form represent both amplitude and phase of a material's response to an alternating stimulus.* The real part represents the in-phase component, and the imaginary part the out-of-phase component. We chose here to describe the free-electron metal in terms of a complex conductivity, although some texts instead introduce a complex dielectric constant. In Chapter 3, we will use a *complex dielectric constant* to describe the response of an *insulator* to alternating electric fields.

Sample Problem 2.4

If the collision time is 10^{-14} seconds, at what angular frequency will the current lag an AC electric field by 30°? by 60°?

Solution

From (2.33) and (2.34), $\tan\theta = \left(\dfrac{\sigma''}{\sigma'}\right) = \omega\tau$

$\tan 30° = 0.577 = \omega\tau = 10^{-14}\omega$, so $\omega = 5.77 \times 10^{13}\ \mathrm{s}^{-1}$
$\tan 60° = 1.73 = \omega\tau = 10^{-14}\omega$, so $\omega = 1.73 \times 10^{14}\ \mathrm{s}^{-1}$

SUMMARY

Common experience tells us that most transparent materials are insulators and that metals are opaque. To examine whether conductivity and transparency are totally incompatible, we first transformed Maxwell's equations into differential form and showed that they lead to electromagnetic waves traveling in vacuum at the velocity of light. However, if the wave enters a conductor that follows Ohm's law, the wave equation becomes a *damped* wave equation. The wave vector then has an imaginary component, and the wave declines in amplitude within the metal in distances on the scale of the skin depth δ. In ordinary metals, skin depths for visible light are of the order of nanometers, and little light penetrates. This conclusion was based on inserting $\mathbf{J} = \sigma\mathbf{E}$ into the Ampere-Maxwell equation with σ equivalent to the DC conductivity, an approximation we expect to be valid as long as $\omega\tau \ll 1$.

At the other extreme, $\omega\tau \gg 1$, we found that the wave vector becomes real again, indicating no attenuation, once the frequency exceeds the plasma frequency ω_p. Although metals have plasma frequencies in the ultraviolet, conductors like indium tin oxide can have plasma frequencies in the near infrared, and therefore can be transparent to visible light despite having substantial conductivities. Such conductors can therefore be used for transparent electrodes in applications like liquid crystal displays.

In the presence of a magnetic field and with $\omega\tau \gg 1$, absorption peaks from cyclotron resonances of charge carriers have been observed in many materials, but at frequencies different from those predicted by classical physics. These measurements are interpreted in terms of carriers with an effective mass m^* different from the mass of a totally free electron, an effect that will require quantum mechanics to explain.

AC conductivity is real for $\omega\tau \ll 1$, imaginary for $\omega\tau \gg 1$, and complex $\{\sigma_{dc}/(1 - i\omega\tau)\}$ for intermediate frequencies. A complex expression for a material property represents both the amplitude and phase of a material's response to an alternating stimulus.

PROBLEMS

2-1. The divergence of a curl is zero. From that and Maxwell's equations, derive equation (2.4).

2-2. A plane wave of orange light ($\nu = 5 \times 10^{14}$ Hz), polarized with an electric field of amplitude 3×10^{-4} V/m in the x-direction, is traveling through vacuum in the positive z-direction. (a) What is the angular frequency, wavelength, wave number, velocity, and magnetic field amplitude of this light wave? (b) Write equations describing the electric and magnetic fields in sinusoidal form, with all known quantities expressed numerically. (c) Write equivalent equations in complex exponential form. (d) Calculate the maximum electric and magnetic forces exerted by this wave on an electron traveling at 10^5 m/s in the x-direction. How would those forces change if the electron were instead traveling in the y-direction? If the electron were stationary?

2-3. A plane wave of red light ($\lambda = 0.65$ μm in vacuum) enters a nonmagnetic metal ($\mu = \mu_o$) with a conductivity of 10^7 S/m. (a) By how much would the electric and magnetic fields of this wave be attenuated in traveling through 10 nm of this metal? (b) How much of the *intensity* in this wave will remain after traveling through 30 nm of this metal? (c) Recalculate (a) and (b) for the case where this metal is also ferromagnetic with a relative permeability of 300 ($\mu = 300\mu_o$).

2-4. For use in a liquid-crystal display, you want an electrode that is transparent to all visible light. You decide to use a doped semiconductor with a mobility of 0.01 m^2/V-s. What is the maximum conductivity you can use? Assume the permittivity to be that of vacuum.

2-5. A conductor is found to be transparent to ultraviolet light with wavelengths shorter than 200 nm. Assuming a permittivity equal to that of vacuum, and assuming that the effective mass equals the free-electron mass, calculate the Hall coefficient.

2-6. In vacuum, the wavelength of an electromagnetic wave varies inversely with frequency. From equation (2.24), how does the wavelength of an electromagnetic wave penetrating a conductor vary with frequency? What is this wavelength if the skin depth at this frequency is 20 nm?

2-7. From equation (2.32), what is the phase angle between the current and electric field when $\omega\tau = 1$? $\omega\tau = 0.1$? $\omega\tau = 10$? For each case, write an expression for the time dependence of current if the electric field varies as $\sin(\omega t)$.

2-8. Light entering a thin metallic film has approximately equal intensities of extreme red ($\lambda = 700$ nm in vacuum) and extreme blue ($\lambda = 400$ nm in vacuum) light. The intensity of the red light decreases by a factor of ten after penetrating 10 nm of the film. By how much has the intensity of the blue light decreased after penetrating this distance? By how much has the intensity of the red and blue light waves decreased after penetrating 20 nm?

2-9. From Fig. 1.1, estimate the skin depth for light of 400 nm wavelength (in vacuum) for (a) stainless steel, (b) doped polyacetylene, (c) sea water.

2-10. Using the Euler formula, (a) derive formulas for $\cos(A + B)$ and $\sin(A + B)$, (b) find the real and imaginary parts of the two square roots of i and represent the square roots as vectors in the complex plane. (c) Do the same for the three cube roots of i. (d) If $A_1 \cos(\omega t + \theta_1) + A_2 \cos(\omega t + \theta_2) = A_3 \cos(\omega t + \theta_3)$, find expressions for A_3 and θ_3 in terms of A_1, θ_1, A_2, and θ_2. You might find a graphical representation in the complex plane and your high-school plane geometry to be helpful.

2-11. By extensive plastic deformation, you decrease the electron mean-free path in high-purity copper by 10%. What will be the resulting changes in (a) the mobility, (b) conductivity, (c) Hall coefficient, (d) skin depth, (e) plasma frequency, and (f) the AC frequency at which the current lags the field by 45°?

2-12. Farads are units of capacitance, which we'll encounter in Chapter 3, and henries are units of inductance, which we'll encounter in Chapter 5. (a) From equation (2.15), derive the relation between farads and henries. (b) From Gauss's law, show that a farad is a coulomb per volt. (c) Given that a tesla is a weber per m^2, show from Ampere's law that a henry is a weber per ampere.

2-13. Suppose you don't like complex numbers and would like to solve equation (2.27) for AC conductivity by using the applied electric field in the form $E_o \cos(\omega t)$ rather than $E_o e^{-i\omega t}$. Since you know that the current will have both in-phase and out-of-phase components, assume that the drift velocity contains both $\cos(\omega t)$ and $\sin(\omega t)$ terms, and show that you still get the result expressed in equation (2.33).

Chapter 3

Insulators and Capacitors

3.1 POLARIZATION OF DIELECTRICS—STORING ELECTRIC ENERGY

We considered the response of a conductor to electric fields in Chapter 1 and to electromagnetic waves in Chapter 2. We now turn to the other extreme of materials: insulators, or *dielectrics*. We will consider in this chapter their response to DC and AC electric fields, and in Chapter 4 their response to electromagnetic waves.

We'll start by assuming a perfect insulator with zero DC conductivity. It has absolutely no free electrons and no mobile ions. (At elevated temperatures, ionic materials can conduct electricity by the diffusion of ions, and a few materials, known as fast-ion conductors, can have significant room-temperature ionic conductivity. However, for most practical dielectric materials, ionic conductivity at room temperature is negligible.)

Apply an electric field to a perfect insulator and nothing happens, right? Wrong! Even though a dielectric may have no free electrons, it still contains plenty of electric charge, and an electric field pushes negative charges in one direction and positive charges in the opposite direction. The dielectric *polarizes* (Fig. 3.1a), producing a net *electric dipole moment* per unit volume called the *polarization* **P**. The units for polarization are C-m per m^3, or C-m^{-2}, and it is a vector quantity, defined by convention as pointing from negative to positive charge. The application of an electric

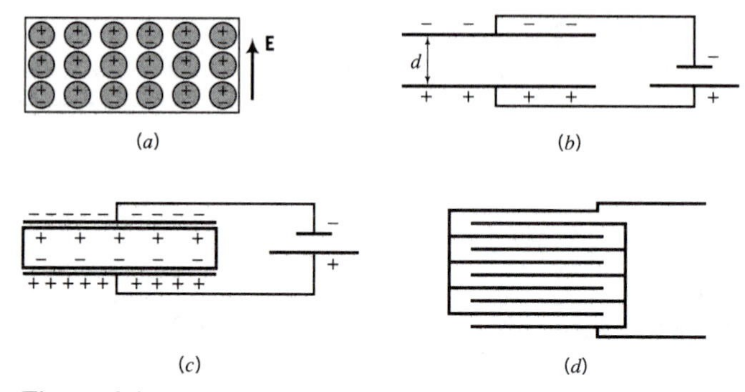

Figure 3.1. (a) Polarization of dielectric, (b) empty capacitor, (c) capacitor filled with dielectric, (d) multilayer capacitor.

field to a dielectric creates a volume polarization—and a net surface charge per unit area that is numerically equivalent to the volume polarization.

So what? Except for a transient charge motion when the field is first applied, there is no current and no conductivity. There is, however, *stored electric energy*, much as there is stored mechanical energy in a compressed spring. In DC applications, dielectrics are used to increase the stored energy in *capacitors*. Small capacitors store energy in your camera's flash unit and release it rapidly when needed. Large banks of capacitors can store many kilojoules of energy and release them rapidly into electromagnets to produce intense magnetic field pulses or into powerful lasers to produce intense light pulses.

Suppose we have a capacitor consisting of two parallel conducting plates of area A separated by a distance d (Fig. 3.1b). Assume that at first we have nothing between the plates. Applying a voltage difference V across the capacitor with a battery will produce a charge Q on each plate given by CV, where $C = \varepsilon_0 A/d$ is the capacitance in farads (one farad is one coulomb per volt). The electric field $\mathbf{E}$ between the plates will be V/d. The energy stored is $CV^2/2$.

Now insert a dielectric that fills the space between the plates (Fig. 3.1c). The electric field polarizes the dielectric, and the resulting surface charge density P on the dielectric produces an internal electric field that opposes the applied field, and also partly compensates the charge on the conducting plates. More charge flows from the battery, and the same voltage V will now produce an increased charge Q' on each plate, an increased capacitance $C' = Q'/V = \varepsilon_0 \varepsilon_r A/d$, and a corresponding increase in stored energy.

The factor of increased capacitance, charge, and stored energy is

$$\frac{C'}{C} = \frac{Q'}{Q} = \varepsilon_r \tag{3.1}$$

which is the *relative permittivity*, or *dielectric constant*, of the material (although we should warn you up front that this "constant" is not always constant). The units of the permittivity of vacuum, ε_0, and the permittivity of the dielectric, $\varepsilon = \varepsilon_0\varepsilon_r$, are farads per meter, but the dielectric constant ε_r is unitless. It can be related directly to the polarization **P**, since the surface charge on the dielectric equals the increase in charge on the plates:

$$PA = Q' - Q = C'V - CV = CV(\varepsilon_r - 1) \tag{3.2}$$

Since $C = \varepsilon_0 A/d$ and V/d is the electric field **E** between the plates of the capacitor,

$$\mathbf{P} = \varepsilon_0\mathbf{E}(\varepsilon_r - 1) \qquad \text{and} \qquad \varepsilon_r = 1 + \frac{\mathbf{P}}{\varepsilon_0\mathbf{E}} = 1 + \chi \tag{3.3}$$

where χ is called the *dielectric susceptibility* (also unitless, of course).

Another vector quantity that is sometimes useful is the electric displacement vector **D** defined by

$$\mathbf{D} = \varepsilon\mathbf{E} = \varepsilon_0\varepsilon_r\mathbf{E} = \varepsilon_0\mathbf{E} + \mathbf{P} \tag{3.4}$$

Sample Problem 3.1

If an insulator has a dielectric susceptibility of 1.32, what is the dielectric constant? In an applied electric field of 100 V/m, what is the magnitude of the polarization and the electric displacement vector?

Solution

$\varepsilon_r = 1 + \chi = 2.32$

$\mathbf{P} = \varepsilon_0\mathbf{E}\chi = (8.85 \times 10^{-12})(100)(1.32) = 1.17 \times 10^{-9}\ \text{C/m}^2$

$\mathbf{D} = \varepsilon_0\varepsilon_r\mathbf{E} = (8.85 \times 10^{-12})(2.32)(100) = 2.05 \times 10^{-9}\ \text{C/m}^2$

As materials scientists, we now want to relate these macroscopic properties—polarization and dielectric constant—to what's happening on the atomic level. Once again, we start by thinking small.

In dielectrics, there are five major microscopic mechanisms of polarization: electronic, ionic, orientational (dipolar), interfacial (space charge), and ferroelectric.

Electronic Polarization. Consider first a material with purely covalent bonding, such as diamond. In the absence of an applied electric field, the cloud of negatively charged electrons in each carbon atom is symmetrically distributed about the positively charged nucleus. However, an applied electric field pushes the electrons in one direction and the nucleus in the opposite direction. The atom distorts, and the center of the atom's negative charge no longer coincides with the position of the nucleus. If the separation between the center of negative charge and the nucleus is

d, and the total charge of the nucleus is q, each atom now has an *electric dipole moment* of

$$\mathbf{p} = qd \tag{3.5}$$

If there are N atoms per m^3, the total electric dipole moment per unit volume of the material, its *polarization* $\mathbf{P}$, will be

$$\mathbf{P} = N\mathbf{p} = Nqd \tag{3.6}$$

For moderate applied electric fields, the electric dipole moment of each atom is proportional to the field:

$$\mathbf{p} = \alpha_e \mathbf{E} \tag{3.7}$$

where the proportionality constant α_e is called the *electronic polarizability*. (For simplicity, we assume here that the electric field felt by the atom is the same as the overall electric field applied to the dielectric, an assumption we will reexamine later.) In purely covalent materials, like diamond, this is usually the *only* source of polarization. From (3.3), (3.6), and (3.7),

$$\varepsilon_r = 1 + \frac{N\alpha_e}{\varepsilon_0} = 1 + \chi \tag{3.8}$$

This equation relates a macroscopic material property, the dielectric constant, to a microscopic property, the electronic polarizability. The static dielectric constants of diamond, silicon, and germanium result entirely from electronic polarization, and are 5.9, 12, and 16, respectively, the increase resulting from increased atomic size as you descend a column of the periodic table. Larger atoms are more polarizable, that is, more easily distorted by the application of an electric field, and α_e is proportional to the atomic volume. Many polymers contain mostly small atoms (C, H, O) and therefore have relatively small electronic polarization, making them the best choice for microelectronic applications where minimum capacitance, and hence minimum dielectric constant, is desired.

Ionic Polarization. Consider instead a simple ionic material like NaCl. In addition to the distortion of the electron clouds of each ion, an electric field pushes the negatively charged chlorine ions in one direction and the positively charged sodium ions in the opposite direction, thereby stretching some Na-Cl bonds and compressing others. This leads to an additional contribution α_i to the polarizability of a dielectric, which will appear as an additional term in (3.8). However, this contribution to polarization and the dielectric constant cannot respond as rapidly to changing fields as electronic polarization, since it involves the relative displacement of entire ions. This leads to a marked frequency dependence of the dielectric constant, which we discuss later.

Orientational (Dipolar) Polarization. Some molecules, like H_2O, have permanent dipole moments, which in the absence of an electric field have no preferred orientation. An applied field tends to rotate the molecules and produce a preferred orientation and an additional contribution to polarization—another term added to (3.8). In polar liquids like water, this contribution can in fact dominate polarization at low frequencies. Although unimportant in most solids at room temperature, orientational polarization can occur in special cases where there are certain forms of crystal defects in ionic crystals and sufficient short-range ionic mobility.

Interfacial (Space Charge) Polarization. In some dielectrics, electrons or ions can move more than interatomic distances, producing the build-up of layers of charge at internal interfaces such as grain boundaries or interphase boundaries. Such internal charge layers can make large contributions to dielectric constants at low frequencies.

Ferroelectric Polarization. A few ionic crystals, called *ferroelectrics*, have symmetries that allow them to have a spontaneous ionic polarization in the absence of an electric field. In limited temperature ranges, some ferroelectrics, such as barium titanate and lead zirconium titanate (PZT), can have dielectric constants of several thousands. (These high values of dielectric susceptibility result largely from the motion of ferroelectric domain walls, much as the motion of ferromagnetic domain walls yields high magnetic susceptibilities, a topic we'll discuss in Chapter 5.)

Capacitors commonly have capacitances of microfarads or less, but modern materials and modern materials processing have produced capacitors about 10 cm^3 in volume with capacitances as high as one farad. (A 1-farad capacitor is useful for special applications where you wish to store a lot of charge and a lot of electrical energy in a small space. I use one in classroom demonstrations of generator-motor action.) This apparently is a multilayer capacitor with a large number of closely spaced plates (Fig. 3.1d), separated by a dielectric with a very high dielectric constant. Guessing $\varepsilon_r \approx 10^4$ suggests an interplate spacing of about a micron and a total plate area of about 10 m^2 (squeezed into a volume of only 10^{-5} m^3). Even if these guessed numbers are a bit off, clearly some sophisticated processing is involved! (See Problem 3-9.)

3.2 DC INSULATOR, AC CONDUCTOR

In DC circuits, capacitors are used primarily to store energy. They do not transmit direct current. In AC circuits, capacitors serve to block direct currents but *transmit alternating currents*, and offer a frequency dependence to their passage of AC. Currents enter and leave capacitors in each cycle, and currents must also flow in the dielectric. We mentioned earlier that in DC applications there is a transient charge flow immediately after the application of the voltage. In this sense, components in AC circuits always operate in a transient condition.

Suppose we apply to our ideal dielectric our by now traditional alternating electric field, $\mathbf{E} = \mathbf{E}_0 e^{-i\omega t}$. This will produce an alternating polarization

$$\mathbf{P} = \varepsilon_0(\varepsilon_r - 1)\mathbf{E}_0 e^{-i\omega t} \quad \mathbf{(3.9)}$$

A change in surface charge density **P** requires charge flow of

$$\mathbf{J} = \frac{d\mathbf{P}}{dt} = -i\omega\varepsilon_0(\varepsilon_r - 1)\mathbf{E}_0 e^{-i\omega t} \tag{3.10}$$

Thus in our dielectric $\mathbf{J}/\mathbf{E} = -i\omega\varepsilon_0(\varepsilon_r - 1)$. Note that this mathematically imaginary AC conductivity is zero for DC ($\omega = 0$) and increases with increasing frequency.

(We materials scientists and engineers focus on the current density and electric field in the dielectric. The electrical engineer focuses instead on the current I into, and the voltage V across, the capacitor, and finds that $I/V = -i\omega C' = -i\omega\varepsilon_r C$. In AC applications with constant voltage amplitude, the current through a capacitor increases linearly with increasing frequency—as long as the dielectric material can keep up with the rapidly changing fields.)

As in Chapter 2, imaginary conductivity indicates that current and field are $\pi/2$ out of phase. But in the dielectric the current *leads* the field by $\pi/2$ (Fig. 3.2a), whereas in the case of free electrons and $\omega\tau \gg 1$ (Fig. 2.5b) the current *lags* the field by $\pi/2$. Figure 3.2b represents this phase relationship of current and voltage in a capacitor in the form of vectors rotating clockwise in the complex plane.

If you still feel uncomfortable with complex numbers, suppose instead that the electric field varies as $\cos(\omega t)$, the real part of our traditional $e^{-i\omega t}$. In our capacitor, the current will then vary as $\cos(\omega t + \pi/2) = -\sin(\omega t)$, the real part of $-ie^{-i\omega t}$. For free electrons at high frequencies, current will vary as $\cos(\omega t - \pi/2) = +\sin(\omega t)$, the real part of $ie^{-i\omega t}$. Check it out. If you remember Euler's formula, you can always move between complex exponential notation and sines and cosines.

Whenever current and field are *exactly* 90° out of phase, there is no energy loss; energy flows into and out of the dielectric reversibly. (The instantaneous power is the product of current and voltage, and the product $\sin(\omega t)\cos(\omega t)$ integrates to exactly zero.) This is a result of our assumption of an *ideal* dielectric, with only a real part to the dielectric constant ε_r. We have assumed that the polarization can respond instantaneously and losslessly and remain in phase with the alternating electric field (making its derivative, the current, exactly $\pi/2$ out of phase).

That's asking a lot of our material, particularly as we go to higher and higher

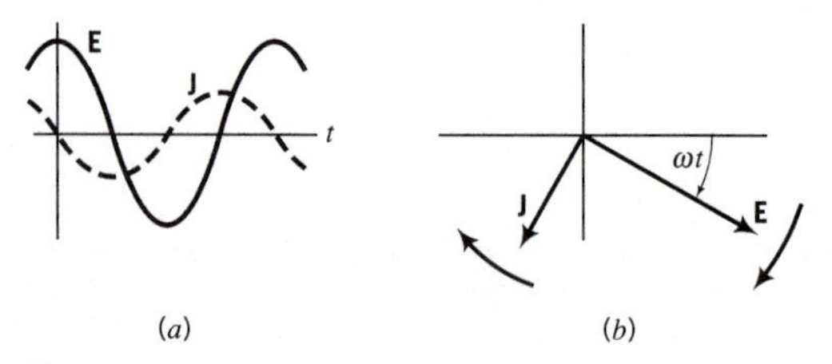

Figure 3.2. (a) In a capacitor, current *leads* field by 90° (compare to Fig. 2.5). (b) The same relationship shown as vectors rotating clockwise in the complex plane (compare to Fig. 2.6).

frequencies. *Eventually, the polarization will have a hard time keeping up, and the current will no longer be exactly 90° out of phase with the field.* In other words, in a *real* dielectric, ε_r will eventually have an *imaginary* part. **P** and **D** will no longer be exactly in phase with **E**, **J** will no longer be exactly $\pi/2$ out of phase with **E**, and integration of the instantaneous power over a cycle will no longer be zero. I heat my tea each morning with the imaginary part of its dielectric constant—yet it's not imaginary tea!

In the next two sections, we consider two basic models—one based on relaxation, one based on resonance—that yield predictions of how the real and imaginary parts of the dielectric constant may vary with frequency. We'll find that each mechanism of polarization has a characteristic range of frequencies where **P** can no longer keep up with **E**, and, as a result, the real component of ε_r changes markedly and an imaginary component of ε_r appears.

3.3 RELAXATION AND HEATING TEA

Let's consider orientational polarization and those water molecules in my cup of tea. With no applied field, their intrinsic electric dipoles will be randomly directed. Once a field is applied, there will be a tendency for the dipoles to align with the field, but that tendency will be opposed by thermal disorder. For any given field and temperature, an equilibrium polarization will be determined by a balance between the electrical forces and thermal disorder, but a finite time will be required for the H_2O molecules to rotate into this equilibrium distribution. According to Debye's model of orientation polarization, if the equilibrium polarization in a given field is $\mathbf{P}_0$, the polarization a time t after application of the field is

$$\mathbf{P}(t) = \mathbf{P}_0\{1 - \exp(-t/\tau_r)\} \tag{3.11}$$

where τ_r is the *relaxation time*.

If the period of an AC electric field is large compared to τ_r, that is, $\omega\tau_r \ll 1$, the polarization will be able to keep up with the changing field with negligible phase lag. In this limit, the dielectric constant will have little or no imaginary component. If the period is instead short compared to τ_r, that is, $\omega\tau_r \gg 1$, the dipoles will have no time to rotate in response to the rapidly changing field. There will be no orientational polarization and no contribution to *either* the real or the imaginary component of the dielectric constant.

It is at intermediate frequencies, where the period and relaxation time are not very different, that phase lags will be significant and the dielectric constant will be complex. If we define $\varepsilon_d(\omega)$ as the frequency-dependent contribution to the dielectric constant resulting from orientational polarization and $\varepsilon_d(0)$ as the DC ($\omega = 0$) contribution, the Debye analysis predicts

$$\varepsilon_d(\omega) = \frac{\varepsilon_d(0)}{1 - i\omega\tau_r} = \varepsilon_d' + i\varepsilon_d'' \tag{3.12}$$

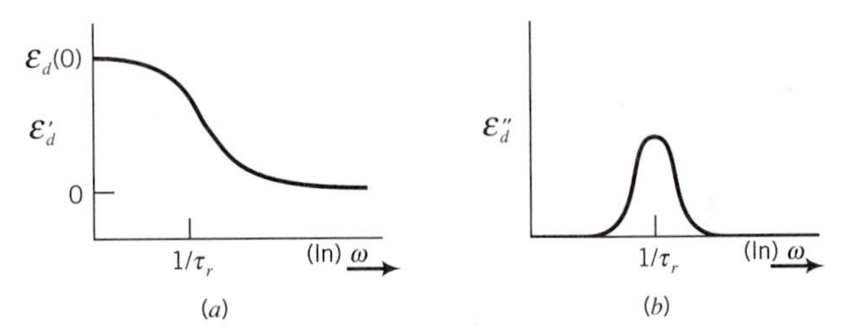

Figure 3.3. Frequency dependence of the real and imaginary parts of the contribution of the orientational polarization to the dielectric constant, from Debye's relaxation model.

In this intermediate frequency region, we have a *complex dielectric constant*, with the real and imaginary parts of the contribution from orientational polarization given by:

$$\varepsilon_d' = \frac{\varepsilon_d(0)}{1 + \omega^2\tau_r^2} \quad \text{and} \quad \varepsilon_d'' = \frac{\varepsilon_d(0)}{1 + \omega^2\tau_r^2}\,\omega\tau_r \tag{3.13}$$

Note that both the complex dielectric constant in (3.12) and the complex conductivity of a free-electron metal given by equation (2.32) vary as $1/(1 - i\omega\tau)$. Although the two cases involve very different physics—one dealing with the collisions of free electrons and the other with the collisions of rotating molecules—both contain a relaxation time, and similar mathematics is involved.

As shown in Fig. 3.3, the real part of this contribution to the dielectric constant ε_d' decreases gradually with increasing frequency from its static value $\varepsilon_d(0)$ to 0, while the imaginary part ε_d'' is zero at very low and at very high frequencies, but goes through a maximum at $\omega\tau_r = 1$. This represents a frequency region where the dielectric is no longer lossless, and some electrical energy is dissipated as heat. It is convenient but not accidental that for the H_2O dipoles in my tea, microwave frequencies (about 10^9 Hz) are in this region of optimum heating. I heat my tea in a microwave oven.

Sample Problem 3.2
In a strongly polar liquid like water, the low-frequency dielectric constant is dominated by orientational polarization. In terms of the relaxation time τ_r, at what frequency does the imaginary part of the dielectric constant grow to become 1% of the real part? Assume $\varepsilon_d(0) \gg 1$.

Solution
From (3.13), $(\varepsilon_d''/\varepsilon_d') = \omega\tau_r$, so it equals 0.01 when $\omega = 0.01/\tau_r$

3.4 ELECTRONIC AND IONIC RESONANCE

Next we consider the frequency dependence of the *electronic* polarizability α_e. When an electric field displaces the bound electrons of an atom, creating an electric dipole moment, there is an elastic restoring force that can be assumed to be proportional to the displacement. This differs from the case of free electrons in metals, where there is no restoring force. So the general equation for the dynamics of bound electrons in alternating fields includes not only the two terms in equation (2.27) but *also a restoring force term Kx*:

$$m\frac{d^2x}{dt^2} + m\gamma\frac{dx}{dt} + Kx = -eE = -eE_0e^{-i\omega t} \quad \textbf{(3.14)}$$

Here the various terms have been expressed in terms of the electron displacement x (so electron velocity is dx/dt). The restoring force is represented by a "spring constant" K, and the viscous damping by the parameter γ (replacing $1/\tau$). A charge q displaced by a distance x produces an electric dipole moment qx, so the frequency dependence of polarizability α_e (and therefore of the dielectric constant) will be reflected directly in the frequency dependence of x.

From left to right, the various terms in (3.14) represent the inertia of the electron (ma), a viscous damping force proportional to velocity, the restoring force Kx, and the force produced by an alternating electric field. As we noted in Section 2.4, for the free electrons of conductors (where $Kx = 0$) the damping term dominated at low frequencies (giving us Ohm's law) and the inertial term dominated at very high frequencies (giving us the plasma frequency). *For the bound electrons of dielectrics, the restoring force term Kx is nonzero, and it plays a very important role.*

The Kx term makes the dynamics of bound electrons in alternating fields more complicated than the dynamics of free electrons, which gave us (2.32), and the dynamics of rotating dipoles, which gave us (3.12). Rather than tackling (3.14) directly, we'll consider first the effect of the Kx term without the existence of the damping and the alternating field terms. We'll then add the field and then add the damping, approaching a solution to (3.14) in steps.

Any such system has a natural resonant frequency determined by the mass m and the spring constant K. In the absence of an external force ($\mathbf{E} = 0$) and in the limit of no damping ($\gamma = 0$), equation (3.14) simplifies to

$$m\frac{d^2x}{dt^2} + Kx = 0 \quad \textbf{(3.15)}$$

which you may recognize as the equation for a *harmonic oscillator*. The function $x = x_0e^{-i\omega t}$ is a solution to this differential equation if $\omega^2 = K/m$. The mass m and the spring constant K of this simple undamped harmonic oscillator determine its *resonant frequency*

$$\omega_r = \left(\frac{K}{m}\right)^{1/2} \quad \textbf{(3.16)}$$

If we apply an alternating electric field $\mathbf{E} = \mathbf{E}_0 e^{-i\omega t}$ (this problem is called the *forced* harmonic oscillator), our differential equation becomes

$$-eE_0 e^{-i\omega t} - Kx = m\frac{d^2x}{dt^2} \tag{3.17}$$

Plugging in $x = x_0 e^{-i\omega t}$, we find that the amplitude of oscillation is

$$x_0 = \frac{-eE_0}{m(\omega_r^2 - \omega^2)} \tag{3.18}$$

This function approaches a constant value of $-eE_0/K$ as $\omega \rightarrow 0$ (it is negative because the electron moves opposite to the field), is positive for $\omega > \omega_r$, approaches zero as $\omega \rightarrow \infty$, but blows up to infinity as $\omega \rightarrow \omega_r$ from either direction (Fig. 3.4a). When the frequency of an applied AC field matches a natural resonant frequency of a system, things can get out of hand.

For a more accurate model for the physical phenomenon we are trying to understand, electronic polarization, we must include the damping force $\gamma m(dx/dt)$ and solve the full equation (3.14). Our solution of course becomes a bit more complicated. Our field $\mathbf{E} = \mathbf{E}_0 e^{-i\omega t}$ will now produce an amplitude of oscillation given by

$$x_0 = \frac{-eE_0}{m(\omega_r^2 - \omega^2 - i\omega\gamma)} = \frac{-eE_0[(\omega_r^2 - \omega^2) + i\omega\gamma]}{m[(\omega_r^2 - \omega^2)^2 + \omega^2\gamma^2]} \tag{3.19}$$

As seen in Fig. 3.4b, the real part of this function has the same limits for $\omega \rightarrow 0$ and $\omega \rightarrow \infty$ as equation (3.18) for the undamped oscillator. However, rather than going to infinity as $\omega \rightarrow \omega_r$, it goes through a finite maximum and a finite minimum before decaying to zero at high frequencies. Equation (3.19) also has an *imaginary part*, which goes through a maximum near $\omega = \omega_r$ (Fig. 3.4c). Our displacement x_0 (and therefore our polarizability α_e and dielectric constant ε_r) are *complex*, with both in-phase and out-of-phase components.

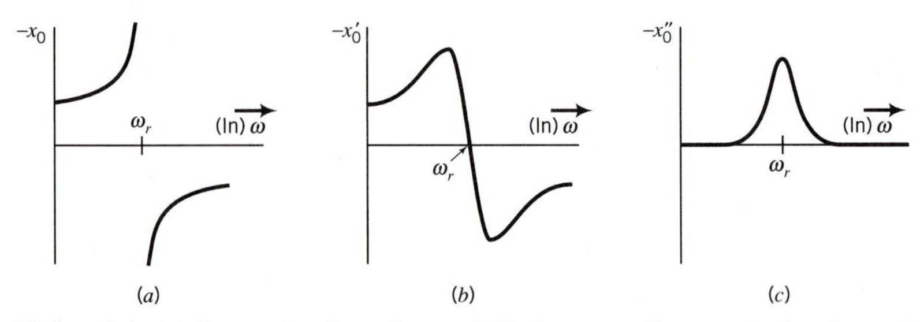

Figure 3.4. (a) Frequency dependence of displacement (hence α_e) of undamped oscillator—equation (3.18), where ω_r is the resonance frequency. (b) and (c) Frequency dependence of real (in-phase) and imaginary (out-of-phase) components of displacement for damped oscillator—equation (3.19). (b) and (c) represent, schematically, the expected frequency dependence of the real and imaginary parts of the contribution to the dielectric constant of electronic (or ionic) polarization.

Since the amplitude of the induced electric dipole moment is qx_0, we expect the resulting polarization **P** and dielectric constant to vary with frequency as in Figs. 3.4b and 3.4c. *The imaginary component of the displacement will correspond to imaginary components of polarization and dielectric constant and to a current within the dielectric that is no longer exactly $\pi/2$ out of phase with the electric field.*

At very low frequencies ($\omega \ll \omega_r$), the displacement x_0 (and therefore the polarization) can keep up with the alternating field, so there is little lag and little energy loss and the dielectric constant is real. At very high frequencies ($\omega \gg \omega_r$), the electrons simply cannot respond to the rapidly changing field; there is no polarization ($x_0 \rightarrow 0$), no contribution to the dielectric constant, and no energy loss. It is in the intermediate frequency range, where ω is not that different from ω_r, that the response of the charge to the alternating field is literally complex. It is in this frequency region that the imaginary component ε_r'' is substantial, and a portion of the electrical energy is transformed into heat. The resonant frequency for electronic polarization is generally in the *ultraviolet* ($\approx 10^{15}$ Hz).

Although electronic polarization requires only the motion of electrons, *ionic* polarization involves the displacement of entire ions, positive ions in one direction, negative ions in the other. The differential equation governing the dynamics of ionic polarization is equivalent to (3.14), with force terms corresponding to acceleration and damping and a restoring-force term. Thus the frequency dependence of ionic polarization is also qualitatively represented by Fig. 3.4. However, since the movement of entire ions is involved, *the masses involved are much greater*, and the resonant frequencies given by (3.14) are much lower, generally in the *infrared* ($\approx 10^{13}$ Hz). Ionic polarization therefore cannot keep up with the higher frequencies of visible light, and the dielectric constant at the frequencies of visible light ($> 10^{14}$ Hz, frequencies higher than the infrared but lower than the ultraviolet) results almost exclusively from electronic polarization.

Sample Problem 3.3

The dielectric constant of NaCl is 5.9 at low frequencies, but the effective dielectric constant at the frequencies of visible light is only 2.25. What are the relative contributions of electronic and ionic polarization to dc polarization?

Solution

The frequencies of visible light are above the resonant frequencies for ionic polarization (which are in the infrared), so that the dielectric susceptibility ($\chi = \varepsilon_r - 1$) of $2.25 - 1 = 1.25$ comes only from electronic polarization. The DC susceptibility of $5.9 - 1 = 4.9$ includes contributions from both, 1.25 from electronic polarization, and $4.9 - 1.25 = 3.65$ from ionic polarization. Thus the ionic contribution to DC polarization in NaCl is nearly three times as large (3.65 vs. 1.25) as the contribution from electronic polarization. (The low-frequency dielectric constant is $1 + 1.25 + 3.65 = 5.9$.)

3.5 THE DIELECTRIC "CONSTANT"

Electronic, ionic, and orientational polarization each have characteristic frequency ranges in which the imaginary component of the dielectric constant ε_r'' and associated "dielectric heating" go through a maximum. Each involves an eventual decrease in the real part ε_r' with increasing frequency. However, the variation of ε_r' with frequency is different, because electronic and ionic polarization each involve a restoring force and a resonance, whereas orientational polarization does not. Figure 3.5 shows the schematic variation with frequency of the real and imaginary parts of the dielectric constant expected for a material with contributions from all three—orientational (up to $\approx 10^9$ Hz for water), ionic (up to $\approx 10^{13}$ Hz), and electronic polarization (up to $\approx 10^{15}$ Hz). Materials with important contributions to the dielectric constant from interfacial or ferroelectric polarization will have other frequency-dependent terms to further complicate things. Since these mechanisms involve the long-range motion of ions or ferroelectric domain walls, their critical frequencies can be as low as MHz or even kHz.

In choosing a dielectric for electronic applications, criteria include the real and imaginary parts of the dielectric constant and their variation with frequency and temperature. The energy absorption related to ε_r'' is often reported in terms of the *loss tangent*

$$\tan \delta = \varepsilon_r''/\varepsilon_r' \tag{3.20}$$

where the angle δ represents the lag angle between electric displacement **D** and electric field. At 25° C and a frequency of 1 MHz, $\tan \delta$ is about 0.0001 for pure silica glass, 0.001 for Al_2O_3, and 0.01 for soda-lime glass.

Materials with only electronic polarization generally have less variation of dielectric constant with frequency and lower losses. Materials that also have substantial ionic polarization can have higher dielectric constants, but at the price of more

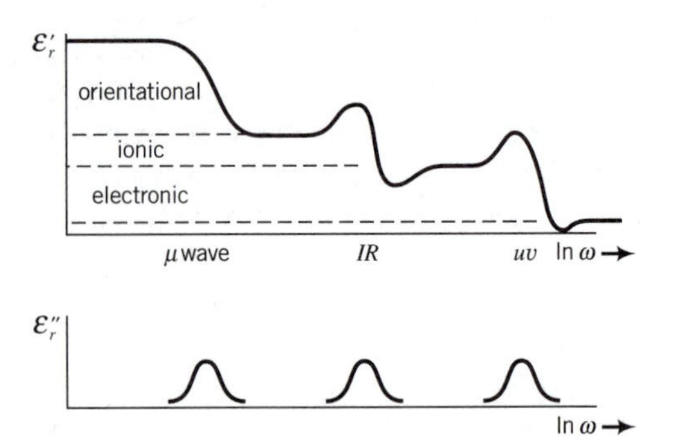

Figure 3.5. Schematic variation of real and imaginary parts of total dielectric constant with frequency for a material with contributions from orientational, ionic, and electronic polarization.

variation with frequency and higher losses at some frequencies. This is especially true of ferroelectrics, which have huge dielectric constants but can have high losses, a strong dependence on both frequency and temperature, and even a variation with electric field, that is, a nonlinear variation of polarization with field. Thus the proper choice of dielectric will depend on which factors are of importance for the specific application. In many applications of capacitors in electronic circuits, low dielectric losses and small variation with temperature are important. Often, a very low dielectric constant is also desired to minimize capacitance; as noted earlier, polymers (which have small atoms and hence low electronic polarizability) are usually the best choice in this case. In other applications, high capacitance (high charge storage) is desired, and a high dielectric constant is important. Which polarization mechanisms can be used depends on the frequency range encountered in the application. Although a low loss tangent is desirable in most cases, sometimes dielectric heating can be useful. It certainly helps me with my morning tea.

One other important property of dielectrics should be mentioned: *dielectric strength*, the maximum electric field that a dielectric can sustain before "dielectric breakdown" and the loss of its insulating properties occurs. This property is often highly sensitive to microstructure but typically is in the range of 10^6 to 10^8 V/m. The dielectric strength limits how much energy can be stored in a capacitor of fixed dimensions, and even plays a role in limiting the optical energy density in solid lasers. (High light intensity corresponds to high electric fields.)

Finally, we made the simplifying assumption earlier that the local electric field at each location within the dielectric was equivalent to the electric field applied to the sample. More detailed treatments of dielectrics recognize that the electric field creating an electric dipole at each point is affected also by all the other electric dipoles within the material, which results in a relation between dielectric constant and polarizability more complicated than equation (3.8). This relation, called the *Clausius-Mosotti relation*, is

$$\frac{(\varepsilon_r - 1)}{(\varepsilon_r + 2)} = \frac{N\alpha}{3\varepsilon_0} \tag{3.21}$$

(Note that this expression approaches equation (3.8) for dielectric constants close to one.) Although this changes the quantitative connection between microscopic polarizability and the macroscopic dielectric constant, it does not change any of the qualitative arguments presented in this chapter.

SUMMARY

Application of an electric field to a dielectric creates a polarization—an electric dipole moment per unit volume and a surface charge per unit area—by pushing negative and positive charges in opposite directions. The most important mechanisms in most dielectrics are electronic and ionic polarization. Polar liquids and some solids also have a contribution from orientation of permanent dipoles. Other mechanisms of polarization, including interfacial (space charge) and ferroelectric polarization, are important in some high-ε_r solids.

Polarization in a dielectric is directly related to the dielectric constant, which can be measured by the increase in charge stored in a capacitor when the dielectric is inserted between the plates. Although capacitors do not transmit direct current, they transmit alternating currents, their effective conductivity increasing linearly with increasing frequency. If an alternating electric field is applied to an *ideal dielectric* (ε_r real), polarization **P** remains in phase with the field, the alternating current ($d\mathbf{P}/dt$) leads the field by exactly 90°, and there is no energy loss in the dielectric.

In a real dielectric, however, polarization cannot remain exactly in phase with rapidly changing electric fields. This leads to a variation of dielectric constant with frequency, and, in certain frequency ranges, an imaginary component of the dielectric constant that corresponds to energy loss and heating of the dielectric. For each mechanism of polarization there is a frequency range in which there is maximum heating and beyond which the contribution to polarization rapidly decreases.

A relaxation model of orientational polarization and a harmonic-oscillator model of ionic and electronic polarization yield the frequency dependence of the real and imaginary parts of the dielectric constant for each case. The orientational polarization associated with water molecules in liquids decreases in the range of microwave frequencies, and the associated dielectric loss produces heating of liquids and foods in microwave ovens.

Dielectric loss associated with ionic and electronic polarization peaks near the corresponding resonant frequencies, which vary as $(K/m)^{1/2}$, where K is the spring constant, which describes the restoring force that opposes displacement, and m is the displaced mass. Resonant frequencies are typically in the infrared for ionic polarization and in the ultraviolet for electronic polarization (thus the dielectric constant for visible light is determined only by electronic polarization). The choice of dielectrics for various applications depends on the relative importance of high (or low) dielectric constant, low dielectric loss (often expressed in terms of the loss tangent), dielectric strength, and the dependence of the dielectric constant on frequency, temperature, and other variables.

PROBLEMS

3-1. (a) What are the units of electronic polarizability in terms of farads? (b) Find the relation between farads and ohms.

3-2. At $t = 0$, a voltage V is applied across a resistance R and a capacitance C connected in series. Derive the time dependence of (a) the current, (b) the voltage across the capacitance, and (c) the voltage across the resistance. (Hint: current $I = dQ/dt$)

3-3. (a) A voltage of 9 volts is applied across an empty capacitor with plates of 10 cm^2 area separated by a distance of 3 mm. What is the electric charge on the capacitor plates? (b) A dielectric with a relative permittivity of 25 is inserted between the capacitor plates. What is now the electric charge on the capacitor plates? What is the charge on the surface of the dielectric? What is the total electric dipole moment per unit volume in the dielectric? (c) Suppose instead that the 9-volt battery was disconnected from the capacitor before the dielectric was inserted. After insertion of the dielectric, what is the electric charge on the capacitor plates and on the surface of the dielectric? What is the voltage across the capacitor? What is the total electric dipole moment per unit volume in the dielectric?

3-4. The electronic polarizability of inert gases increases rapidly with increasing atomic number: He-0.18, Ne-0.35, Ar-1.74 ($\times 10^{-40}$ farad-m^2). How do you expect the electronic polarizability of the ions F^- and Na^+ to compare with the polarizability of Ne? Why?

3-5. Derive a relationship between the electronic polarizability α_e and the spring constant K for the hydrogen atom.

3-6. The dielectric constant of LiF is 9.27 at low frequencies and approaches 1.9 at high frequencies. What are the percentage contributions of electronic and ionic polarization to the low-frequency polarization? Which ion contributes most to the electronic polarization?

3-7. Assuming $x = x_0 e^{-i\omega t}$, solve differential equation (3.14) for x_0 and show that (3.18) is the solution for $\gamma = 0$ and (3.19) is the solution for $\gamma \neq 0$.

3-8. By comparing Figs. 3.3 and 3.4, it can be seen that Debye's relaxation model and the damped oscillator model yield similar qualitative behavior for the frequency dependence of the imaginary part of the dielectric constant but qualitatively different behavior for the frequency dependence of the real part. Briefly explain this difference.

3-9. You construct a multilayer capacitor with N metallic plates of area A separated from neighboring plates by a thickness d of a dielectric with dielectric constant ε_r. Assuming the plates are of negligible thickness, derive a formula for the total capacitance per unit volume in terms of d and ε_r. Check this formula against the numerical estimates given in Section 3.1 for a 1-farad capacitor.

3-10. What is the maximum value of ε_d'' predicted by Debye's relaxation model? At what frequency does that occur? What is the loss tangent at that frequency? Assume $\varepsilon_d(0) \gg 1$.

3-11. Diamond has a dielectric constant of 5.68. Calculate the polarization, electric displacement, and dielectric susceptibility when diamond is exposed to an electric field of 1 V/mm.

3-12. From the polarizability given in Problem 3-4, calculate the dielectric susceptibility and dielectric constant of argon gas under standard temperature (273 K) and pressure (1 atmosphere $= 1.01 \times 10^5$ Pascals). (Assume an ideal gas, which has 2.69×10^{25} molecules per m^3 at standard temperature and pressure.)

3-13. For the case of orientational polarization, the equilibrium polarization as a function of applied field and temperature can be derived from classical statistics. The energy of an electric dipole of strength **p** in an electric field **E** equals $-\mathbf{p} \cdot \mathbf{E} = -pE \cos \theta$. There will thus be a torque of $\mathbf{p} \times \mathbf{E} = pE \sin \theta$ attempting to rotate each dipole into its lowest energy orientation, parallel with the field, but that torque will be opposed by thermal disorder. In thermal equilibrium, the probability of each dipolar orientation will be given by $\exp(\mathbf{p} \cdot \mathbf{E}/k_B T)$. (a) By integrating this equilibrium distribution over all solid angles $(2\pi \sin \theta d\theta)$, show that the net dipole moment in the field direction, as a fraction of full alignment (all dipoles parallel to the field), is given by $\{\coth(\mathrm{a}) - (1/a)\}$, where $a = pE/k_B T$. (b) Derive an expression for the dielectric susceptibility in the limit of low applied fields $(pE \ll k_B T)$, assuming the material contains N dipoles per unit volume.

Chapter 4

Lenses and Optical Fibers (Optical Properties of Insulators)

4.1 REFRACTION: SLOWING THE WAVES

We saw in Chapter 3 that the bound electrons in insulators, by shifting only tiny fractions of a nanometer in response to electric fields, are able to do quite a lot. By increasing the dielectric constant (thereby increasing capacitance), they increase the ability of capacitors to store electrical energy and they increase the ability of capacitors to transmit alternating currents. In some frequency ranges, via the imaginary part of the dielectric constant, they produce dielectric heating (which is often undesirable but sometimes can be useful). It turns out that the response of bound electrons to the high-frequency alternating fields of light and other electromagnetic waves can also be put to good use. By bending light as it enters the lenses of our eyeglasses, they help us to see, and by confining light within optical fibers, they help us to send telephone conversations and other information across long distances.

Let's consider an electromagnetic wave with an electric field in the x-direction traveling in vacuum in the positive z-direction. We can write the alternating electric and magnetic fields as

$$\mathbf{E} = E_0\mathbf{i}e^{-i(\omega t - kz)} \qquad \text{and} \qquad \mathbf{B} = B_0\mathbf{j}e^{-i(\omega t - kz)} \tag{4.1}$$

where $B_0 = E_0/c$, and $c = \omega/k = (\mu_0\varepsilon_0)^{-1/2}$ (as shown in Chapter 2). How is this wave changed when it enters an ideal (ε_r real) dielectric? In the dielectric, the wave equations are

$$\nabla^2\mathbf{E} = \mu\varepsilon \frac{\partial^2\mathbf{E}}{\partial t^2} \quad \text{and} \quad \nabla^2\mathbf{B} = \mu\varepsilon \frac{\partial^2\mathbf{B}}{\partial t^2} \tag{4.2}$$

solutions to which are waves with velocity $\omega/k = (\mu\varepsilon)^{-1/2}$. If we assume our dielectric is nonmagnetic, $\mu = \mu_0$. Substituting $\varepsilon = \varepsilon_0\varepsilon_r$, the velocity within a dielectric becomes

$$c' = \omega/k = (\mu_0\varepsilon_0\varepsilon_r)^{-1/2} = c\varepsilon_r^{-1/2} \tag{4.3}$$

Since $\varepsilon_r > 1$, we can see that the velocity of the wave inside the dielectric has been reduced from its velocity in vacuum by the factor $\varepsilon_r^{-1/2}$. It is traditional in optics to define the light velocity within a material as $c' = c/n$, where n is called the *index of refraction.* We therefore have demonstrated an important relationship between the dielectric constant ε_r, an electrical property, and the index of refraction n, an optical property:

$$n = \varepsilon_r^{1/2} \qquad \varepsilon_r = n^2 \tag{4.4}$$

When the electromagnetic wave entered the dielectric, some properties of the wave changed, and some didn't. *The angular frequency ω didn't change.* We can convince ourselves of this by considering the boundary conditions at the surface between vacuum and dielectric, where the time dependence of the fields must be identical on both sides of the interface. But the velocity ω/k *did* change. Since it decreased by the factor n, the wave number k must have *increased* by the factor n. And since $k = 2\pi/\lambda$, *the wavelength λ must have decreased by the factor n.* One other thing: Faraday's equation (2.7) tells us that the ratio of the electric and magnetic field amplitudes (2.18) is ω/k, so that also decreased by a factor n.

Sample Problem 4.1

A beam of red light with a wavelength in vacuum of 650 nm enters an insulator with an effective dielectric constant ε_r of 1.7 at the frequencies of visible light. What will be the frequency, angular frequency, velocity, and wavelength of this light when traveling within the insulator?

Solution

$n = \varepsilon_r^{1/2} = 1.3$,
$\nu = 3 \times 10^8/6.5 \times 10^{-7} = 4.62 \times 10^{14}$ Hz (in vacuum *and* in the insulator)
$\omega = 2\pi\nu = 2.90 \times 10^{15}\ \text{s}^{-1}$ (also unchanged)
$c' = c/n = 3 \times 10^8/1.3 = 2.3 \times 10^8$ m/s (decreased in the insulator)
$\lambda = 6.5 \times 10^{-7}/1.3 = 5 \times 10^{-7}$ m (also decreased!)

All those changes result from the AC polarization of the dielectric and the resulting ε_r. We can see this from the mathematics, but what's physically happening inside the dielectric? Considered in detail on the atomic level, it's really pretty complicated. The alternating electric fields of the light wave set all the bound charges into oscillatory motion. Oscillating electric charges act as tiny antennas sending out electromagnetic waves that interfere with the incoming wave and the waves from all the other oscillating charges. The net effect of the interference is a slowing of the incoming wave, represented by the index of refraction. Sometimes it's easier to focus on the mathematics.

My eyeglass lenses help me see because the light changes direction when it enters the glass—it is refracted. When light moves from a material with index n_1 into a material with a different index n_2, its change of direction is defined by *Snell's law*:

$$n_1 \sin\theta_1 = n_2 \sin\theta_2 \tag{4.5}$$

where the angles in each material represent the angle between the direction the light wave is traveling and the normal to the interface between the two materials. (These angles are usually called the *angle of incidence* and the *angle of refraction*. There's also a reflected wave, with an angle of reflection equal to the angle of incidence.) According to (4.5), when light moves from a material of lower index of refraction into a material of higher index of refraction, for example, from air into glass (Fig. 4.1a), it is bent *toward* the surface normal ($n_2 > n_1$, so $\theta_2 < \theta_1$). Conversely, when moving from glass or water into air (Fig. 4.1b), it is bent away from the surface normal.

In this latter case, the maximum angle of refraction is 90°. This corresponds to an angle of incidence called the *critical angle* θ_c, beyond which there can be no refracted wave. We then have *total internal reflection* (Fig. 4.1c). From (4.5), $n_2 \sin\theta_c = n_1 \sin 90°$, so

$$\theta_c = \sin^{-1}\frac{n_1}{n_2} \tag{4.6}$$

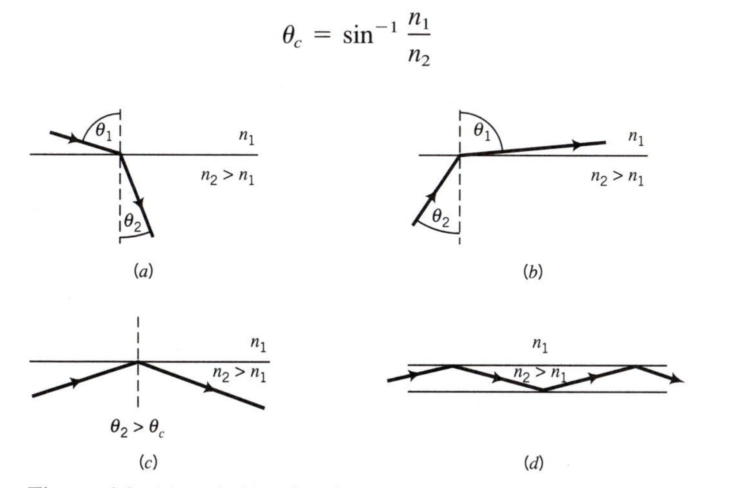

Figure 4.1. (a) and (b) refraction, (c) total internal reflection, (d) light beam confined to an optical fiber by total internal reflection.

If we have a light wave traveling within a long fiber of a material with an index of refraction higher than that of the material surrounding it, and the angle of incidence with the surface is maintained greater than θ_c, total internal reflection will keep the wave from escaping the fiber (Fig. 4.1d). It is this phenomenon that now allows long-distance communication by lightwaves in "optical fibers," which can carry much more information than can be carried by electrical signals in copper wires of comparable size. A bundle of optical fibers (a "light pipe") can be used to transmit images from inside our bodies, allowing physicians to study internal structures and functions from outside our bodies. All these wonders result from the AC polarization of the dielectric—tiny oscillations of the bound electrons that slow the speed of light. The most common optical fibers are hair-thin silica-based glass, surrounded by a plastic cladding to protect the glass from abrasion.

When light moves from one material into another at normal incidence, it has no change of direction ($\theta_2 = \theta_1 = 0$). However, the indices of refraction determine how much of the light is transmitted, and how much is reflected. For light, "how much" is measured in terms of *intensity*, which is proportional to the *square* of the field amplitude and has units of watts/m^2. In vacuum,

$$I = c\varepsilon_0 \frac{E_0^2}{2} \tag{4.7}$$

The intensity can be seen to equal the energy density in J/m^3 ($\varepsilon_0 E_0^2/2 = B_0^2/2\mu_0$) of the wave in vacuum times c, the wave velocity in vacuum. Traveling within a dielectric, the expression for energy density in terms of electric field will increase by the factor ε_r and the wave velocity will decrease by the factor $n = \varepsilon_r^{1/2}$, so equation (4.7) should be multiplied by the factor n for the intensity within a dielectric.

If light traveling in vacuum (or air, which has $n \approx 1$) is in normal incidence on a dielectric with index of refraction n, the *reflectivity* (fraction of the incident intensity that is reflected) of the dielectric is given by

$$R = \frac{(n - 1)^2}{(n + 1)^2} \tag{4.8}$$

For a glass with $n = 1.5$, only 0.25/6.25 (4%) will be reflected, and 96% will be transmitted. Although it's possible to see your reflection in window glass, it's a weak reflection. Increasing polarizability of the dielectric (increasing ε_r and n) will lead to increased reflection. Formula (4.8) can be derived from boundary conditions at the interface that require continuity of the total electric and magnetic fields, considering incident, reflected, and transmitted waves. See Problem 4-13.

All of the above is for an ideal dielectric with a real dielectric constant. If the dielectric constant has an imaginary component ε_r'', the index of refraction will also have an imaginary component, the *extinction coefficent* κ. Then the equation for reflectivity will be changed to

$$R = \frac{(n - 1)^2 + \kappa^2}{(n + 1)^2 + \kappa^2} \tag{4.9}$$

Clearly a nonzero extinction coefficient will increase the reflectivity. In a conductor, the κ^2 term dominates, and reflectivity approaches one (for $\omega < \omega_p$).

Sample Problem 4.2

At the frequencies of visible light, a transparent plastic has an effective dielectric constant ε_r of 1.7. If surrounded only by air, what will be the critical angle for internal reflection? What will be the reflectivity for normal incidence?

Solution

$n = \varepsilon_r^{1/2} = 1.30$, so (since air has $n \approx 1$) $\theta_c = \sin^{-1}(1/1.3) = 50°$,
$R = (0.3/2.3)^2 = 0.017$ (only 1.7% reflected)

4.2 DISPERSING THE WAVES

It's time for a reality check on equation (4.4), the relation between the dielectric constant and the index of refraction. The dielectric constant (measured at low frequencies) and the square of the index of refraction (measured at the frequencies of visible light) are given in Table 4.1 for several representative materials.

For diamond and liquid hydrogen, which are purely covalent materials, the agreement is very good. Equation (4.4) is confirmed. And agreement is good for polyethylene, a polymer that consists of a long chain of CH_2 units. But the agreement is not quite that good for teflon, a polymer chain of CF_2 units. The discrepancy is more substantial for SiO_2 and NaCl, and huge for acetone and water. What's the problem?

This is a clear demonstration of an effect discussed in Chapter 3, *the frequency-dependence of the dielectric constant (Fig. 3.5).* Where the dielectric constant is

Table 4.1. Low-frequency Dielectric Constant and the Square of the Index of Refraction for Visible Light for Various Materials

Material	ε_r	n^2
Diamond	5.68	5.66
H_2 (liquid)	1.23	1.23
Polyethylene	2.30	2.28
Teflon	2.10	1.89
SiO_2 (quartz)	3.85	2.13
NaCl	5.90	2.25
Acetone	20.7	1.85
Water	80.4	1.77

determined only by electronic polarization, as in covalent diamond and hydrogen, polarization can keep up with the frequencies of visible light (between 10^{14} and 10^{15} Hz) pretty well and the dielectric constant at optical frequencies is nearly the same as that measured at low frequencies. We remember, however, that ionic and orientational polarization cannot keep up with optical frequencies, making the effective dielectric constant at visible light frequencies less than that at low frequencies. Let's see if that explains the data for the other materials.

In polyethylene, although the C-C bonds along the chain are purely covalent, the C-H bonds are slightly polar. However, from the agreement seen in the table, ionic polarization makes only a minor contribution to the dielectric constant. Polarization is largely electronic, and equation (4.4) is satisfied, as in diamond and hydrogen. The structure of teflon is identical to that of polyethylene, but with the C-H bonds replaced by C-F bonds. The C-F bond is more polar than the C-H bond (more electron transfer), and apparently ionic polarization makes a modest contribution to the dielectric constant at low frequencies (about $2.10 - 1.89 = 0.21$) but none at optical frequencies.

Ionic compounds like SiO_2 and NaCl would be expected to have a significant component of ionic polarization at low frequencies, and that expectation is confirmed by the data. The dielectric constant of NaCl of 5.90 at low frequencies decreases to an effective dielectric constant of only 2.25 at light frequencies, where ionic polarization can no longer contribute. Apparently, the contribution of ionic polarization to the low-frequency dielectric constant is 3.65 ($5.90 - 2.25$), while that of electronic polarization is only 1.25 ($2.25 - 1$). Similarly, the contributions of ionic and electronic polarization to the low-frequency dielectric constant of SiO_2 are 1.72 ($3.85 - 2.13$) and 1.13 ($2.13 - 1$), respectively.

In liquids with molecular dipole moments, like water and acetone, orientational polarization is largely responsible for their high dielectric constants at low frequencies, but orientational polarization begins to drop out at microwave frequencies (with which I heat my tea). Their effective dielectric constants at light frequencies, where only electronic polarization contributes, are much smaller. The data in Table 4.1 do not disprove equation (4.4); they result only from the fact that the dielectric constant is frequency dependent.

Sample Problem 4.3

What percentage of the low-frequency dielectric susceptibility of acetone is contributed by orientational polarization? Assume the contribution of ionic polarization is negligible.

Solution

The low-frequency dielectric susceptibility ($\chi = \varepsilon_r - 1$) of acetone is $20.7 - 1 = 19.7$. At visible light frequencies, it is $1.85 - 1 = 0.85$, contributed by electronic polarization. Apparently $20.7 - 1.85 = 18.9$ is contributed by orientational polarization. $18.9/19.7 = 96\%$

When the dielectric constant varies with frequency, so does the index of refraction, which means that the velocity of electromagnetic waves varies with frequency. *This variation of wave velocity with frequency is called dispersion.* Electromagnetic waves in vacuum always travel with a velocity of c, independent of frequency, so electromagnetic waves in vacuum are nondispersive. In a dielectric, on the other hand, there is always some dispersion. It is mild in some materials but strong in others, especially in certain frequency ranges. As we learned in Chapter 3, the variation of wave velocity with frequency is accompanied by an imaginary part of the dielectric constant and by energy loss. Dielectric energy loss can be useful in heating tea, but it is not desirable in optical fibers used for telecommunication because it results in attenuation, which we will discuss in the next section.

In optical fibers, dispersion directly deteriorates the *form* of signal pulses, since the pulses will in general contain light with a range of frequencies. With dispersion, the various component waves of different frequencies travel at different velocities, resulting in a change of shape of the pulse as it travels down the fiber (Fig. 4.2a). Thus it is desirable to choose the material and the communication wavelength to minimize this *material dispersion.*

There is another factor that contributes to distortion of signal pulses in optical fibers that depends simply on geometry, not on material properties. It is called *modal dispersion.* A light wave traveling directly down the central axis of an optical fiber has a shorter travel distance than one traveling at an angle to the axis and undergoing multiple internal reflections (Fig. 4.2b). These two paths, and any intermediate ones, are different "modes" of transmission of the light. Even if the material itself is nondispersive, the different path lengths mean that light that started out as a sharp pulse will broaden with distance traveled down the fiber. Fortunately, modal dispersion can be combated by clever control of material properties.

The index of refraction of the glass can be changed by changing its chemical composition. Suppose we produced the fiber so that its core was not uniform in composition but had a composition that varied slightly with distance from the axis, designed so that the index of refraction gradually decreased with distance from the axis (Fig. 4.2c). This is called a *graded-index* fiber. Then the waves traveling at an angle to the axis will on average travel faster than the wave traveling directly down

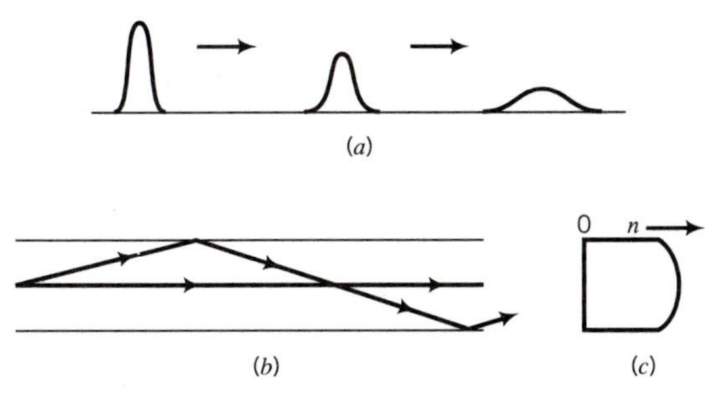

Figure 4.2. (a) Broadening of signal pulse caused by dispersion, (b) geometric source of modal dispersion, (c) graded index of refraction to combat modal dispersion.

the axis, compensating for their increased path length. If perfectly designed, all the waves will take the same amount of time to travel down a mile of fiber, and modal dispersion will disappear. In practice, it has not been found possible to completely eliminate modal dispersion, but with graded-index fibers, it can be greatly reduced. By reducing the spreading of signal pulses, minimizing both material and modal dispersion allows the transmission of more pulses per second, thereby enhancing the possible communication rate of the fiber.

Although material dispersion can create problems in optical fibers, it has a positive side as well. A slight variation of light velocity with frequency in water across the range of visible light (n = 1.342 for violet light, 1.330 for red light) produces one of the wonders of nature: the rainbow. Dispersion in glass also produces a variation of n from red to violet, allowing a glass prism to "disperse" a beam of white light into a beautiful spectrum of colors. Real dielectrics are better than ideal dielectrics not only for heating tea, but also in creating beauty.

4.3 ATTENUATING THE WAVES

Another factor limiting fiber-optic communications is the *attenuation* that light experiences when passing through many miles of fiber. Attenuation can occur both by absorption processes (associated with ε_r'') and by scattering (deflection) processes. Light intensity I is found to decrease exponentially with distance z traveled down the fiber, and this attenuation in optical fibers is usually defined in terms of *decibels per kilometer* (dB/km), where

$$\text{dB} = 10 \log_{10} \frac{I_0}{I} \tag{4.10}$$

Thus an attenuation of 60 dB corresponds to a decrease in light intensity from its initial value I_0 by a factor of 10^6 (or, equivalently, a decrease in electric field amplitude by a factor of 10^3).

Sample Problem 4.4

After traveling within an optical fiber for 15 km, the intensity of light is found to have decreased to half its original intensity. What is the attenuation in dB/km of this fiber? How much of the original intensity will be left after traveling a total of 30 km?

Solution

$(I_0/I) = 2$, and $10 \log_{10} (2) = 3.01$ dB

3.01 dB/15 km = 0.20 dB/km

The light intensity will again be halved in traveling the second 15 km, so $(I/I_0) = 0.25$, a total loss of $2 \times 3.01 = 6.02$ dB

Much of the attenuation can be attributed to energy loss associated with ε_r'', the imaginary part of the dielectric constant (which corresponds, you'll recall, with the part of the current density $d\mathbf{P}/dt$ that is in phase with the electric field). Our wave with $\mathbf{E} = \mathbf{E}_0 e^{-i(\omega t - kz)}$ is a solution of the wave equation for the dielectric if

$$k = (\omega^2 \mu_0 \varepsilon)^{1/2} = (\omega^2 \mu_0 \varepsilon_0)^{1/2} (\varepsilon_r' + i\varepsilon_r'')^{1/2} \tag{4.11}$$

If we assume that $\varepsilon_r'' \ll \varepsilon_r'$, which is generally the case, the imaginary part of the wave vector k becomes

$$k_{imag} = \frac{\omega \varepsilon_r''}{2c\sqrt{\varepsilon_r'}} = \frac{\omega \sqrt{\varepsilon_r'}}{2c} \tan \delta \tag{4.12}$$

As we saw in Chapter 2, the imaginary part of k yields an exponential decay of $\mathbf{E}$ of the form $\exp(-k_{imag}\, z)$. Thus the attenuation in dB is directly proportional to $\tan \delta$, the dielectric loss tangent that we defined in (3.20). Remembering that $\log_{10}(e) = 0.434$, if k_{imag} is expressed in m^{-1}, the attenuation in dB per meter can be seen from (4.10) to equal 8.69 times k_{imag}.

Once again, I remind you that "imaginary" is merely an unfortunate mathematical term. Attenuation of light in dielectrics is very real, and attenuation of light in glass is of great technological importance in optical fiber applications. The higher the attenuation, the shorter the length of fiber before an optical amplifier must be introduced to restore the signal amplitude. One of the triumphs of materials science and engineering in recent decades has been a decrease in the optical attenuation of glass by several orders of magnitude (Fig. 4.3), produced by improving its chemical and structural perfection. Of course, attenuation is a function of frequency, and it is found to be a minimum in the near-infrared, at a wavelength between one and two microns. In addition to impurity effects (especially OH^- ions), attenuation in the infrared is related to ionic polarization of the glass (ε_r'' from ionic polarization peaks in the infrared; see Fig. 3.5).

Although the effects of ionic polarization are decreased at shorter wavelengths, decreasing wavelength leads to a rapid increase in attenuation in the visible range

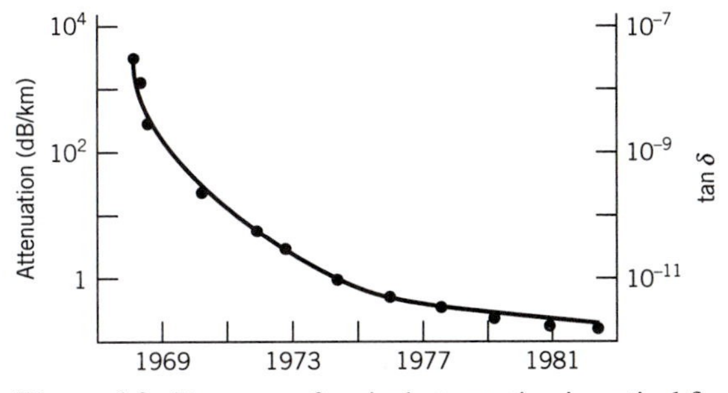

Figure 4.3. Decrease of optical attenuation in optical fibers produced by improved materials.

associated with another effect—*Rayleigh scattering*. By its very nature, the amorphous structure of glass produces variations in local density that scatter (deflect) light. As with scattering of light by air molecules, Rayleigh scattering is proportional to $\lambda^{-1/4}$, so that scattering of blue light is much greater than that of red light. Thus the sky is blue, sunsets are red (light that has traveled through a lot of air, which removes blue light by preferential scattering), and optical fibers transmit best in the near-infrared.

The effects of Rayleigh scattering on color can be easily demonstrated with a glass rod that is intentionally produced with many small pores or other fine light-scattering centers. If you shine white light into one end of the rod, the sides of the glass rod will appear bluish from the scattered light, but the light coming out the other end will be notably reddened because of the preferential removal of blue. The scattering centers produce an artificial sky and an artificial sunset inside a glass rod. (You can see the same effect by putting a few drops of milk in a glass of water and shining light through it.)

4.4 DIFFRACTING THE WAVES

Until now, in our consideration of various optical properties—skin depth, plasma frequency, refraction, dispersion, attenuation—we have implicitly considered our solids to be continuous media. We have not specifically taken into account the crystalline structure of solids, the periodic array of atoms. We can get away with that as long as the wavelength of our electromagnetic wave is long compared to interatomic spacings. However, it's worth remembering that *special things happen when a wave interacts with a periodic structure with spacings comparable to the wavelength.* In the case of crystal lattices, which have periodic spacings on the order of 1 nm or less, no diffraction effects would be expected for visible or ultraviolet light. An electromagnetic wave with a wavelength of 1 nm has a frequency of 3×10^{17} Hz. Electromagnetic waves in this wavelength and frequency range are called *x-rays*. As you probably know, x-ray diffraction is a common technique for characterizing crystal structures.

Bragg's law states that diffraction occurs from a given crystal plane when

$$2d \sin \theta = n\lambda \tag{4.13}$$

where d is the spacing between crystal planes, θ is the angle between the incident beam and the plane (Fig. 4.4), n is an integer, and λ is the wavelength. Clearly *diffraction can occur if* $\lambda \leq 2d$*, or, equivalently, if the wave number* $k \geq \pi/d$.

The most common x-ray diffraction techniques determine a series of interplanar spacings by employing "monochromatic" (single wavelength) x-rays and varying θ either by rotating a single crystal or by using a powder sample that contains many crystals of different orientations. With diffraction data from a sufficient number of different crystal planes, the crystal structure of a material can be determined. If instead you expose a stationary single crystal to an x-ray beam that contains a wide *range* of wavelengths ("white" x-rays), the angle θ between the incident beam and

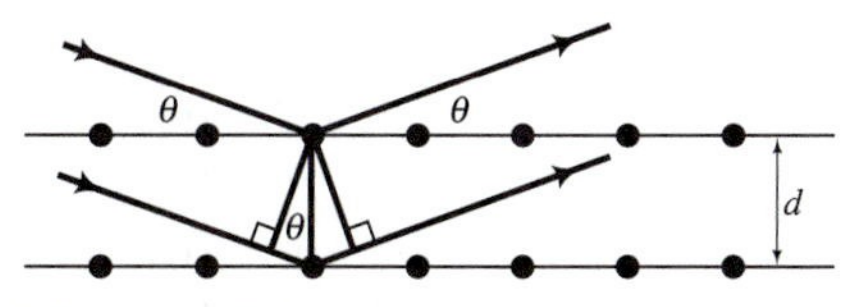

Figure 4.4. Bragg diffraction. The two diffracted rays interfere constructively when the difference in path length ($2\,d \sin\theta$) equals an integral number of wavelengths ($n\lambda$).

each crystal plane remains fixed, but for each interplanar spacing d, a different wavelength λ satisfying (4.13) will produce diffraction. The resulting *Laue pattern* of diffraction spots from many planes can be used to determine the structure and orientation of the crystal.

To get diffraction effects with visible light, which has vacuum wavelengths ranging from 400 nm to 700 nm, requires much larger periodic spacings—on the order of 1 μm. You may have looked through those novelty eyeglasses constructed from scribed transparent plastics, which turn any white light source into a colorful spectacle, and you've probably noticed the separation of colors in light reflected from the surface of a compact disc. Such transmission or reflection "gratings" with a fixed spacing d will, it is clear from (4.13), diffract the different components of white light at different values of θ, thereby separating white light into the full spectrum of colors.

There are also a few diffraction gratings in the natural world that separate colors via (4.13). The skin of the gopher snake and the wing cases of some beetles have periodic structures of the right spacing to serve as reflection gratings for visible light. The gemstone opal contains patches of periodic arrangements of tiny SiO_2 spheres that produce complex colored patterns that change dramatically as the opal is turned. Several metallic alloys undergo phase transformations (called *eutectic* and *eutectoid* transformations) that yield regular patterns of two alternating lamellar phases, with periods on the order of microns (micrometers). Polished surfaces of such alloys act as diffraction gratings, producing beautiful colored reflections. But the wavelengths of visible light are much too long for diffraction effects from periodic crystal lattices; for that we need x-rays, or other waves with wavelengths on the order of nanometers, not micrometers. We will later encounter diffraction effects in crystals with sound waves (Chapter 7) and with electron waves (Chapters 8 and 13).

SUMMARY

Through AC polarization, bound charges in dielectrics slow the velocity of electromagnetic waves by a factor called the *index of refraction* $n = \varepsilon_r^{1/2}$. Refraction also leads to a reduction in wavelength and in the ratio of electric to magnetic fields, as well as to an increase in the energy density of the electromagnetic wave.

Snell's law shows that light traveling in a material with an index of refraction greater than that of its surroundings will be held within that material by the phenomenon of total internal reflection as long as the angle of incidence with the normal to the interface is greater than a critical angle θ_c determined by the two indices of refraction. This allows the long-distance transmission of light signals in optical fibers. The index of refraction also determines the reflectivity of a dielectric.

In materials in which electronic polarization is dominant, dielectric constants measured at low frequencies and indices of refraction measured at the frequencies of visible light are in good agreement with the equation $n = \varepsilon_r^{1/2}$. In materials in which ionic or orientational polarization make substantial contributions, the low-frequency dielectric constant will be higher than that calculated from the square of the index of refraction at visible light frequencies, which is determined only by electronic polarization. Such data can be used to calculate the relative contributions of different mechanisms of polarization.

Variation of the dielectric constant, index of refraction, and wave velocity with frequency is called *dispersion.* Material dispersion in an optical fiber causes broadening of signal pulses. Pulses can also be broadened by a geometric effect called *modal dispersion*, which can be minimized by creating a composition gradient across the fiber, thereby producing a graded index of refraction.

The imaginary part of the dielectric constant leads to exponential attenuation of light in an optical fiber (commonly expresssed in dB/km), but modern processing has greatly decreased attenuation by increasing the chemical purity and structural perfection of glass fibers. Minimum attenuation is in the near-infrared, increasing at longer wavelengths due to impurity effects and ionic polarization in the glass and increasing at shorter wavelengths due to Rayleigh scattering from density fluctuations.

In discussing the interaction of electromagnetic and other waves with solids, crystalline structure can be ignored and the solid considered continuous only if wavelengths are large compared to interatomic spacings. However, when wavelengths are sufficiently small (i.e., wave numbers are sufficiently high), diffraction effects must be taken into account. With electromagnetic waves, diffraction in crystals becomes important at x-ray frequencies.

PROBLEMS

4-1. The plane wave of Problem 2-2 enters a crystal of NaCl. (a) What are the frequency, angular frequency, wavelength, wave number, and velocity of this wave inside the NaCl? (b) What is the ratio of magnetic and electric field amplitudes inside the NaCl? (c) What is the critical angle for total internal reflection? (d) Sketch the expected qualitative variation of the real part of the dielectric constant as a function of frequency from low frequencies to visible light frequencies.

4-2. A light wave in air is incident on a water surface. (a) At what angle of incidence are the reflected and refracted waves at right angles to each other?

(b) What is the reflectivity of the water for normal incidence? (c) At what angle of incidence is this wave totally reflected by the water?

4-3. A light wave is in normal incidence on a diamond crystal. (a) What is the reflectivity? (b) What is the ratio of the electric field in the reflected wave to that in the incident wave? (c) What percentage of the intensity of the incident wave is transmitted? (d) What is the ratio of the electric field in the transmitted wave to that in the incident wave?

4-4. The dielectric constant of quartz is 3.85 at low frequencies, and its index of refraction for visible light is 1.46. What fraction of its polarization is ionic (a) at low frequencies (b) at frequencies of visible light?

4-5. From data given in Table 4.1, what fraction of the polarization of water at low frequencies is electronic polarization?

4-6. At low frequencies, the majority of the dielectric constant of water and of acetone derives from orientational polarization. Do you expect the same to be true for liquid carbon tetrachloride (CCl_4)? Explain.

4-7. A glass fiber of circular cross section, with a uniform index of refraction of 1.5, is coated with a plastic with an index of refraction of 1.2. The intensity of light entering one end is uniformly distributed over all angles. (a) What fraction of the light intensity will be transmitted down the fiber by total internal reflection? (b) After traveling a length of fiber equal to ten fiber diameters, what will be the maximum difference in path length traveled by different light beams?

4-8. (a) Derive equation (4.12) from equation (4.11). (b) Confirm that, as claimed in the text, the attenuation in dB/m equals 8.69 k_{imag}.

4-9. For light traveling through a weakly conducting material with a skin depth of 1 cm, what is the attenuation in dB/cm? in dB/km?

4-10. Engineers designing a fiber-optic communications link decide that as soon as the transmitted light intensity falls to 0.1% of its original value, an amplifier must be inserted in the line to restore the intensity to its original value. How many amplifiers are required in a 1200 km line if the attenuation in dB/km of the glass is (a) 100 (b) 10 (c) 1 (d) 0.1?

4-11. You decide that the attenuation in the near-infrared of your silica glass fibers associated with ionic polarization is too high, and you wish to explore the effect of compositional changes. To decrease attenuation, would it be better to substitute ions with lower mass or with higher mass than those in pure silica? Explain.

4-12. An electromagnetic wave with $\nu = 10^{14}$ Hz is traveling through diamond in the negative x-direction. (a) Calculate the wave number and angular frequency and write an expression in complex exponential form for the electric field of this wave. (b) What is the velocity of this wave? (c) What is the ratio of magnetic to electric field amplitudes?

4-13. Derive equation (4.8) from the boundary conditions at the interface requiring that the total electric and magnetic fields be continuous across the interface. (Hint: Reflection affects electric and magnetic fields differently, and recall that the ratio of electric to magnetic fields in the dielectric will be different from that in vacuum.)

4-14. Write an expression for the intensity of an electromagnetic wave (a) in vacuum and (b) in a dielectric in terms of the magnetic field amplitude B_0.

Chapter 5

Inductors, Electromagnets, and Permanent Magnets

5.1 INDUCTORS—STORING MAGNETIC ENERGY

Chapters 1 and 3 focused on the response of materials to electric fields, and Chapters 2 and 4 on their response to electromagnetic waves. We turn now to their response to magnetic fields. All conductors respond to *changing* magnetic fields, because, via Faraday's law of induction, changing magnetic fields induce electric fields. In contrast, most materials have little response to *steady* magnetic fields; they are essentially nonmagnetic. But the few materials that *are* magnetic (some are conductors, some are insulators) are very important to technology, and they're worth a chapter.

Magnets provide the forces that convert electrical energy into motion in motors, and into sound in loudspeakers, making them the "movers and shakers" of modern technology. They also play many other roles. In operating a computer, for example, magnetic materials are important not only in motors and speakers but also in the hard and floppy disks that store the data, in the heads that write and read the data, in the monitor that presents the data to you on the screen, in the transformers that convert input voltages into voltages appropriate for the electronic circuits, and, of course, in the generators and transformers that deliver the electric power to your wall socket. But we'll introduce magnetic materials with their use in a basic component of AC electric circuits, the *inductor*.

AC electric circuits have three categories of passive elements—resistors, capacitors, and inductors. A capacitor stores energy via charges and electric fields. In Chapter 3, we introduced dielectric materials by describing how they increase the stored energy of capacitors (by the factor ε_r) through their polarization, that is, their electric dipole moment per unit volume. An inductor stores energy via currents and magnetic fields. Here we'll introduce magnetic materials by describing how they increase the stored energy of inductors (by the factor μ_r) through their magnetization—their magnetic dipole moment per unit volume.

We'll soon see that there are major differences between the response of dielectric materials to electric fields and the response of magnetic materials to magnetic fields, but there are many instructive parallels. So we'll start by reminding you of the stored energy in an empty (no dielectric) capacitor with plates of area A and separation d and an applied voltage V.

EMPTY CAPACITOR

Electric field between the plates	$E = \dfrac{V}{d}$
Energy per unit volume	$\dfrac{\varepsilon_0 E^2}{2} = \dfrac{\varepsilon_0 V^2}{2d^2}$
Total stored energy	$\dfrac{\varepsilon_0 V^2}{2d^2}$ multiplied by Ad, or $\dfrac{\varepsilon_0 A V^2}{2d} = \dfrac{CV^2}{2}$

The capacitance $C = \varepsilon_0 A/d$ incorporates the dimensions of the capacitor. The units for C are farads, and the units for ε_0, the permittivity of vacuum, are farads/meter.

Now consider instead a simple inductor in the form of a solenoid (Fig. 5.1a) of area A, length l, and n turns per unit length, carrying a current I. (This is another time when you might want to have your introductory physics textbook within reach.)

EMPTY INDUCTOR

Magnetic field inside coil	$B = \mu_0 nI$
Energy per unit volume	$\dfrac{B^2}{2\mu_0} = \dfrac{\mu_0 n^2 I^2}{2}$
Total stored energy	$\dfrac{\mu_0 n^2 I^2}{2}$ multiplied by Al, or $\dfrac{\mu_0 n^2 AlI^2}{2} = \dfrac{LI^2}{2}$

The *inductance* $L = \mu_0 n^2 Al$ incorporates the dimensions of the inductor. The units for L are henries, and the units for μ_0, the permeability of vacuum, are henries/meter. (The units for inductance commemorate Joseph Henry, an American physicist. All the other units in introductory physics derive from European scientists.)

As we noted in equation (3.1), inserting a dielectric within the plates of the

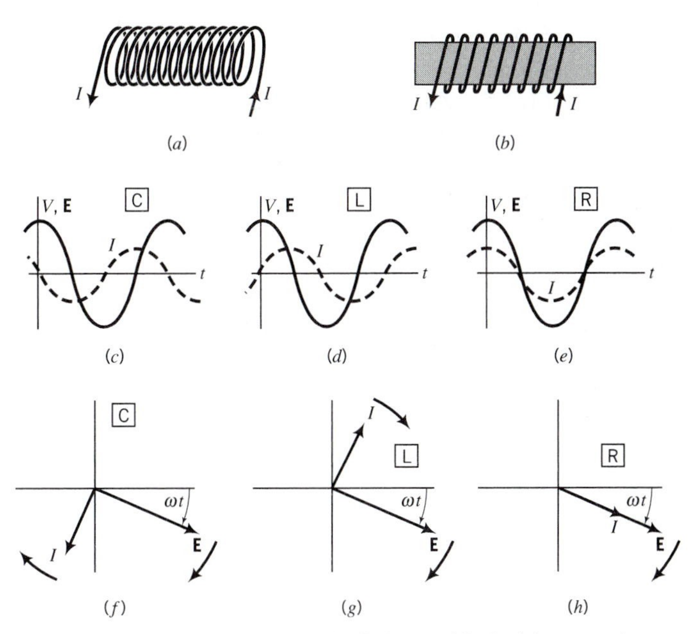

Figure 5.1. (a) empty inductor, (b) inductor filled with magnetic material, and current-field phase relations in (c) an ideal capacitor, (d) an ideal inductor, and (e) an ideal resistor. (f), (g), and (h) show these same phase relations with rotating vectors in the complex plane.

capacitor increases its capacitance, and its stored electric energy, by the factor ε_r, the relative permittivity or dielectric constant. Most materials have dielectric constants that are significantly different from one, resulting from their polarization in response to an electric field.

Similarly, inserting an appropriate magnetic material within the coils of the inductor (Fig. 5.1b) increases its inductance, and its stored magnetic energy, by the factor μ_r, the relative permeability:

$$\frac{L'}{L} = \mu_r \tag{5.1}$$

As we'll discuss in the next section, only a few materials have relative permeabilities significantly different from one; most materials are essentially nonmagnetic and develop little magnetization in response to a magnetic field. But those few magnetic materials with $\mu_r \gg 1$ were used as compasses many centuries ago, allowed the development of the electrical industry in the late 19th century, and remain crucial to modern technology. Before considering the magnetic properties of these materials, a few more words about inductance and its role in AC electric circuits are appropriate.

Whenever the current I through an inductor is changing, the magnetic field within it is changing, and that produces, via Faraday's law of induction, a voltage across the inductor given by $L(dI/dt)$. As with a capacitor, the voltage across an inductor in an AC circuit is 90° out of phase with the current. However, with a capacitor, the current leads the voltage (Figs. 3.2, 5.1c, and 5.1f). With an inductor, the current *lags* the voltage (Figs. 5.1d and 5.1g). (The time-honored way to remember this distinction is "E*L*I the *I*CEman.")

The voltage across a pure resistor is proportional to current I, across a pure inductor it is proportional to dI/dt, and across a pure capacitor it is proportional to the charge Q, the time integral of current. The resulting phase relations between current and voltage for resistors, capacitors, and inductors are shown together in Fig. 5.1c–e, and in 5.1f–h with vectors in the complex plane. (The figure is drawn for ideal capacitors and ideal inductors. Real components also have some resistance.) If voltage varies as $V_0 e^{-i\omega t}$, the three currents are:

Resistor $\quad I(t) = \frac{V(t)}{R} = \frac{V_0}{R} e^{-i\omega t}$ $\quad$ In phase with voltage

Capacitor $\quad I(t) = C\frac{dV(t)}{dt} = -i\omega C V_0 e^{-i\omega t} = \omega C V_0 e^{-i(\omega t + \pi/2)}$ $\quad$ Leads voltage

Inductor $\quad I(t) = \frac{1}{L}\int V(t)dt = i\frac{V_0}{\omega L} e^{-i\omega t} = \frac{V_0}{\omega L} e^{-i(\omega t - \pi/2)}$ $\quad$ Lags voltage

(For those who prefer sines and cosines, if the voltage varies as $\cos \omega t$, the current varies through a resistor as $\cos \omega t$, through a capacitor as $-\sin \omega t$, and through an inductor as $+\sin \omega t$.)

Whereas I/V increases linearly with frequency for an ideal capacitor, across an ideal inductor it is V/I that increases linearly with frequency (since dI/dt increases linearly with frequency). So the *impedance* V/I to alternating currents increases with frequency for inductors and decreases with frequency for capacitors. This opposite response of capacitors and inductors to changes in frequency is used in tuning circuits, which combine an inductor and a capacitor to produce a preferential response to a particular frequency. For more about AC circuits, we refer you to your introductory physics text or an electrical engineering text. Our focus in this chapter is on the magnetic material that we put inside the solenoid to raise the inductance by the factor μ_r.

More generally, in any electromagnet, the magnetic material in the "core" of the electromagnet increases the magnetic field inside the coil by the factor μ_r. Suppose that you tried to produce a magnetic field of about one tesla with an empty solenoid. If your coil had one turn per mm ($n = 10^3$ m^{-1}), you'd need a current of about a thousand amps ($nI = 10^6$ A/m, so $B = \mu_0 nI \approx 1$). But if you inserted an iron core with a relative permeability μ_r of 10^4, you'd need only 100 milliamps. That's why most electromagnets have iron cores.

5.2 MAGNETIZATION—IN A SPIN

Inside the empty solenoid, the magnetic field was $B = \mu_0 nI$. How does inserting a magnetic material enhance that field? For simplicity, we'll assume the sample is long and thin or in the shape of a toroid, to avoid the complications of end effects (i.e., demagnetizing fields associated with magnetic poles). The magnetic field is increased by the factor μ_r and now has two parts, a contribution from the current through the coil (an *applied* field), and a contribution from the material itself:

$$B = \mu_0 nI + \mu_0 M = \mu_0(H + M) = \mu_0 \mu_r H \quad \textbf{(5.2)}$$

We have two new terms in this equation: M and H. The *magnetization M* represents the material's response to the applied magnetic field, and equals the net magnetic dipole moment per unit volume. A *magnetic dipole* can be thought of as a current loop—a current I circulating around an area A—with units of current times area: amp-m^2. Thus the units for magnetization are amp-m^2 per m^3, or A/m.

Since n is the number of turns per meter of the solenoid, the units for $H = nI$ are also A/m. But what do we call **H**? Most introductory physics texts call **B** the *magnetic field* and don't discuss **H**. Many other books call **H** the magnetic field and call **B** either the *magnetic induction* or the *magnetic flux density*, because the name "magnetic field" was historically preempted by **H**. I agree with Nobel laureate Edward Purcell, who in his classic text *Electricity and Magnetism* calls this common practice "clumsy and pedantic." So, not wishing to be clumsy and pedantic, I'll go along with Purcell and continue to call **B** the magnetic field. If you insist on a name for **H**, call it the *H-field* or the *applied field.*

Whatever you choose to call **H**, equation (5.2) defines it in terms of the magnetic field $\mu_0\mathbf{H}$ applied to the material from its surroundings, whereas the total magnetic field **B** inside the sample also includes the magnetic field $\mu_0\mathbf{M}$ resulting from the material's magnetization **M**. The relative permeability μ_r of the material represents the ratio of the total field in the material to the applied field:

$$\mu_r = \frac{\mathbf{B}}{\mu_0\mathbf{H}} = 1 + \frac{\mathbf{M}}{\mathbf{H}} = 1 + \chi_m \quad \textbf{(5.3)}$$

where χ_m is the (unitless) *magnetic susceptibility.*

Equation (3.3) was the corresponding equation for dielectric constant, with $(\varepsilon_r - 1)$ proportional to polarization **P**, the net *electric* dipole moment per unit volume; in (5.3), $(\mu_r - 1)$ is proportional to the magnetization **M**, the net *magnetic* dipole moment per unit volume. The factor ε_r represented the increase in capacitance (or increased charge on the capacitor plates for a given voltage) produced by inserting the dielectric. The factor μ_r represents the increase in inductance (or increased magnetic field in the solenoid for a given current) produced by inserting the magnetic material.

So much for the parallels. Now for the differences. For most solids, ε_r is significantly different from one, but μ_r is not. The magnetic susceptibility χ_m of most solids falls somewhere between -10^{-5} and $+10^{-3}$ (thus $\mu_r \approx 1$). Materials with small positive magnetic susceptibilities are called *paramagnetic*, and those with small negative magnetic susceptibilities (the field from **M** opposes the field from **H**) are called *diamagnetic*. For most practical purposes, they can be considered non-magnetic. A few materials, however, have magnetic susceptibilities of many thousands, so that susceptibilities and relative permeabilities, from (5.3), are essentially equal. And, like the dielectric constants of ferroelectrics, susceptibilities and permeabilities can vary with field, that is, **M** can vary nonlinearly with **H**. To understand a bit about these materials, we have to look at the physical origin of the magnetization **M**.

Sample Problem 5.1

The magnetic susceptibilities of Ag, W, and Fe are -2×10^{-5}, 6×10^{-5}, and $\approx 10^4$, respectively. What is the relative permeability of each of these metals?

Solution

$\mu_r = 1 + \chi_m = 0.99998$ for Ag (diamagnetic), 1.00006 for W (paramagnetic), and $\approx 10^4$ for Fe (ferromagnetic, see following).

The origin of magnetism is electricity in motion. The magnetism of an empty solenoid results from electric current, that is, long-distance motion of electrons. However, electrons in magnetic materials can produce magnetization and magnetic fields without ever leaving their home atoms. Simplified models of the atom describe it as a tiny solar system, with the electrons revolving about the nucleus like planets about the sun, and also rotating about their own axes. (We'll become more sophisticated about atomic structure in Chapter 9.) Both of these electron motions produce magnetism, but the first component, called *orbital magnetism*, is not important in most materials. In iron and most strongly magnetic materials, it is the second component of electron motion, its *spin* about its own axis, that creates most of the magnetism.

The magnetic dipole moment associated with the spin of a single electron is called the *Bohr magneton*, and it equals $\mu_B = 9.27 \times 10^{-24}$ amp-m^2. The very existence of spin is deeply rooted in quantum mechanics, and was first demonstrated by Dirac by applying the principles of relativity to the quantum mechanics of an electron. In fact, the equation defining the Bohr magneton includes Planck's constant h, the basic constant of quantum mechanics:

$$\mu_B = \frac{eh}{4\pi m} \tag{5.4}$$

In the spirit of our semiclassical treatment, for now we simply take spin as a basic property of electrons and delay our discussion of quantum mechanics and the origin of Planck's constant until Chapter 8.

If an atom with one spinning electron is magnetic, shouldn't an atom or molecule with *two* spinning electrons be more magnetic? No! Consider an atom of helium and a diatomic molecule of hydrogen (H_2). Each has two electrons, but an interaction force of quantum-mechanical origin between the spins of neighboring electrons (called *exchange*) makes the two electrons in the helium atom and in the hydrogen molecule spin in opposite directions. Net spin magnetism: zero. A similar cancellation of spin magnetism by the pairing of opposite spins occurs for most electrons of most atoms. For example, an atom of radon contains 86 electrons, with 43 spins pointing one way, and 43 pointing the opposite way. Net spin magnetism: zero. Such materials have only a tiny diamagnetic susceptibility. Luckily for modern technology, not all atoms are so unbiased.

In many materials, the electron spins on each atom do not completely cancel; each atom has a net spin and a net magnetic dipole moment. Yet most such materials still remain only very weakly magnetic because the individual atomic magnets are uncoupled and point in random directions. If an external magnetic field is applied, the individual magnets, like tiny compass needles, attempt to align with the field. The energy of an atomic magnetic dipole of strength $\mathbf{p}_m$ in a field $\mathbf{H}$, and the related torque on the dipole, are given by dot and cross products, respectively:

$$E = -\mu_0 \mathbf{p}_m \cdot \mathbf{H} \qquad \mathbf{T} = \mu_0 \mathbf{p}_m \times \mathbf{H} \tag{5.5}$$

(These equations are analogous to those for the energy of and torque on an electric dipole in an electric field; see Problem 3-13.) This torque will be opposed by thermal disorder, with the probability of each dipolar orientation given by $\exp(\mu_0 \mathbf{p}_m \cdot \mathbf{H}/k_B T)$. At room temperature and usual magnetic fields, the energy of an individual atomic dipole is small compared to $k_B T$, and this leads to only a tiny paramagnetic susceptibility. The relative permeability μ_r of paramagnetic materials is therefore approximately one (except at very high fields and very low temperatures).

Sample Problem 5.2

In an H-field of 10^6 A/m, if an atom has a magnetic dipole moment of three Bohr magnetons and the moment rotates from antiparallel to the field to parallel to the field, by how much does its energy decrease? At $T = 300$ K, what fraction of $k_B T$ is this energy?

Solution

$\Delta \mathbf{E} = 2\mu_0(3\mu_B)\mathbf{H} = (2.52 \times 10^{-6})(2.78 \times 10^{-23})(10^6) = 7.01 \times 10^{-23}$ Joules

$(7.01 \times 10^{-23})/(4.14 \times 10^{-21}) = 0.0169$

(Thus the orientation of isolated atomic spins will be dominated by thermal disordering except at very high fields and very low temperatures.)

In a few materials, the net spins of neighboring atoms are instead coupled strongly by the exchange force. This interatomic exchange force can be either positive or negative, producing either parallel or antiparallel spins. In chromium and manganese, each atom is strongly magnetic, but neighboring atomic magnets are antiparallel and cancel each other out. Such materials are called *antiferromagnetic*, and usually have only a small paramagnetic susceptibility. In *ferromagnetic* iron, cobalt, and nickel, however, the exchange force between neighboring atomic magnets makes them point in the *same* direction. *Rather than cancelling each other out, all the atomic magnets add up to produce a substantial macroscopic magnetization* **M**. Ferromagnetism (from the Latin word for *iron*) is thus a phenomenon of large-scale interatomic cooperation.

In some oxides of iron known as *ferrites*, exchange coupling is antiparallel, but opposing dipole moments are not equal, and a net magnetization remains. Such materials are called *ferrimagnetic*. Magnetizations are generally less than in ferromagnetic materials, but otherwise their macroscopic magnetic behavior is similar.

Thermal disorder also opposes the exchange force, leading to an eventual disappearance of ferromagnetism above a critical temperature T_c called the *Curie temperature*. Above their Curie temperatures, ferromagnetic materials behave like strong paramagnets, with a paramagnetic susceptibility proportional to $(T - T_c)^{-1}$, the so-called Curie-Weiss law.

If there are N atoms per unit volume, each with a magnetic dipole moment of $\mathbf{p}_m$, and all are coupled by exchange to point in the same direction, this will produce a *saturation magnetization* M_s (total magnetic dipole moment per unit volume) given by

$$\mathbf{M}_s = N\mathbf{p}_m \tag{5.6}$$

To get a feel for the numbers involved, let us assume that each atom has a magnetic moment corresponding to a net unbalanced spin of one electron spin, so that $\mathbf{p}_m = 9.27 \times 10^{-24}$ A-m^2 (one Bohr magneton, about 10^{-23} A-m^2). A metal has about 10^{29} atoms/m^3. From (5.6), we'd then get a saturation magnetization of about 10^6 A/m, and the corresponding magnetic field $\mu_0\mathbf{M}_s$ would be of the order of one tesla (10 kilogauss). That in fact is the right order of magnitude (iron's saturation magnetic field is about twice that).

With all those atomic magnets in a ferromagnet coupled together by the exchange force, the saturation magnetization interacting with an applied field **H** will yield a very large energy per unit volume. Combining (5.6) and (5.5), this energy density (in J/m^3) is:

$$E = -\mu_0\mathbf{M}_s \cdot \mathbf{H} \tag{5.7}$$

With even a small applied field **H**, there will be a strong tendency for the saturation magnetization of the ferromagnet to align with the field. The energy of individual atomic magnetic moments of a paramagnet in a field **H** was small compared to k_BT, but in a bulk ferromagnet the exchange forces couple all the atomic spins together (a phenomenon some call "electronic fascism"), and the energy represented by equation (5.7) is very large compared to $k_B T$. (An intermediate case is a small ferromagnetic particle, containing only a thousand atoms or less. Here thermal fluctuations can yield an intermediate behavior dubbed *superparamagnetism*.)

Sample Problem 5.3
Consider a ferromagnetic particle consisting of 10^6 atoms, each of which has a magnetic dipole moment of two Bohr magnetons. In an H-field of 10^6 A/m, if the total magnetic dipole moment of the particle rotates from antiparallel to the field to parallel to the field, by how much does its energy decrease? For what temperature will this equal $k_B T$?

Solution
$\Delta E = 2\mu_0(2 \times 10^6 \mu_B)H = 4.67 \times 10^{-17}$ joules
$T = (4.67 \times 10^{-17})/(1.38 \times 10^{-23}) = 3.38 \times 10^6$ K
(The exchange coupling of atomic spins in ferromagnets makes them much less susceptible to thermal effects than isolated atomic spins.)

5.3 MAGNETIC DOMAINS AND M(H)

We still have a bit of explaining to do. We know that each iron atom is a tiny magnet because the spins of its 26 electrons do not completely cancel, and iron is ferromagnetic because the net spins of neighboring atoms are coupled by exchange to point in the same direction. But what controls μ_r and the variation of **M** with **H**? And how do we explain that some materials, like the ones we use to hold notes to refrigerators, are *permanent magnets*, while others, like the refrigerator itself, lose their macroscopic magnetization as soon as an external magnetic field is removed? We must introduce the concept of *magnetic domains*.

Within ordinary iron or steel, like the walls of your refrigerator, each iron atom has a net spin, and that spin is coupled to the spin of its neighboring atoms, so that the iron is locally magnetized to its saturation value $\mathbf{M}_s$. However, it is divided into domains magnetized in different directions, as shown schematically in Fig. 5.2c. In this figure, two domains are magnetized up, two down. Net macroscopic magnetization: zero. Each end of the material has a mix of north and south poles, and external magnetic fields are very limited in extent. (It is to reduce external magnetic fields that domains form.) The only unhappy atoms (atoms with spins not parallel to their neighbors, the exchange force therefore not satisfied) are the few in the boundaries between domains, across which the direction of magnetization changes. These boundaries are called *domain walls*, and controlling the behavior of magnetic domain walls is the major task of materials scientists who work with magnetic materials. But let's focus on the domains first, and consider the domain walls a bit later.

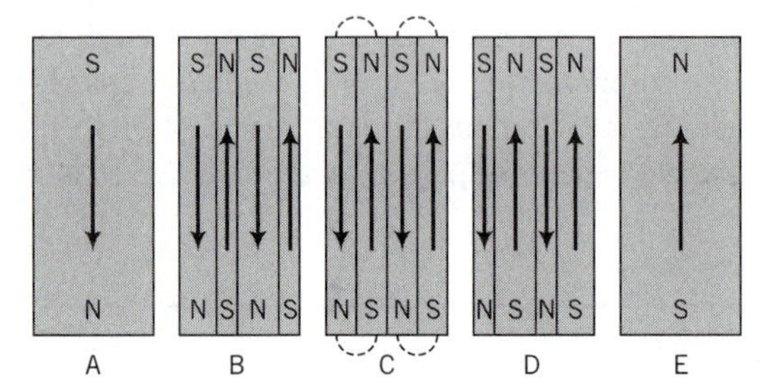

Figure 5.2. Magnetic domain structures. (a) and (e): in a saturated ferromagnet; (b) and (d): in a partly-magnetized ferromagnet; and (c) in an unmagnetized ferromagnet.

Now suppose we apply an upward magnetic field $\mu_0\mathbf{H}$ to the multidomain sample in Fig. 5.2c. From (5.7), the energy per unit volume of the upward-pointing domains is now $-M_s\mu_0H$, while the energy per unit volume of the downward-pointing domains is $+M_s\mu_0H$. Energy will be lowered by $2M_s\mu_0H$ per unit volume if domain walls move so as to increase the volume of the favorably oriented domains at the expense of unfavorably oriented domains. Thus the applied magnetic field produces a force on a unit area of domain wall—a "magnetic pressure"—of $2M_s\mu_0H$.

Suppose that in Fig. 5.2d, walls have moved so that 75% of the sample consists of upward-pointing domains, and 25% of downward-pointing domains. There is now a net macroscopic magnetization of $0.5\mathbf{M}_s$ parallel to the field. If the walls continue to move under the influence of the applied field, eventually the downward-pointing domains will completely disappear (Fig. 5.2e) and the net sample magnetization will be $\mathbf{M}_s$. No further increase in macroscopic magnetization can occur; the magnetization is saturated.

Sample Problem 5.4

An H-field of 40 A/m is applied to a ferromagnet as in Fig. 5.2 and the total magnetic field B within the sample is measured to be one tesla. If the saturation magnetization is 1.4×10^6 A/m, what percentage of the sample consists of domains magnetized parallel to the field? What is the magnetic susceptibility?

Solution

$B = 1 = (1.26 \times 10^{-6})(40 + M)$, so $M \approx 7.94 \times 10^5$ A/m $= 57\%$ of M_s.
If $x\%$ is aligned parallel with the field, $(100 - x)\%$ is aligned antiparallel, and the net percentage parallel to the field is $x - (100 - x) = 2x - 100 = 57$, so $x = 79\%$
$\chi_m = M/H = (7.94 \times 10^5)/40 = 1.99 \times 10^4$

If the applied field is now removed, and if domain walls can form and move easily, the reverse process will occur: (e) to (d) to (c). At (c), the sample is once again unmagnetized—there is no net macroscopic magnetization. If we apply a field in the reverse direction, we go from (c) to (b) to (a)—we saturate the magnet in the opposite direction. Figure 5.2 shows schematically what happens inside the iron core of an electromagnet when current is applied to the coil (or what happens in the refrigerator door when you bring a permanent magnet up to it). Macroscopic magnetization occurs by the motion of domain walls. Of course in real materials hundreds or thousands of domain walls may be moving rather than just three, and the field is not always parallel to the axis of magnetization. (We'll deal with the case of a nonparallel field in Section 5.4.)

Figure 5.3 shows how the *net* macroscopic magnetization (magnetization up minus magnetization down) varies with the strength of the applied field. This is a somewhat idealized "magnetization curve" of a reversible ferromagnetic material. The slope dM/dH (magnetic susceptibility) at low fields can be many thousands—only a few A/m of applied field can produce many kA/m of net magnetization—because most of the work of aligning the atomic magnets has already been done by the exchange force between neighboring atoms. All the applied field has to do is to favor some domains over the others via equation (5.7). We should note, however, that dM/dH will be high only in low fields and will, of course, decrease to zero in high fields as the material saturates.

When domains grow or shrink, all the action is taking place within the moving domain walls, where the atomic magnets are flipping over from one direction to another. It's easier to flip over a few atomic magnets at a time rather than to flip them all simultaneously. When you zip up your jacket, all the action takes place inside the zipper—it's easier to engage a few teeth at a time than to engage them all simultaneously. (Dislocations, if you are familiar with them, represent another analogy to magnetic domain walls—it's easier to shear an atomic plane a few atoms at a time than to shear them all simultaneously.)

We've assumed so far that it's easy to form and move magnetic domain walls. Suppose that instead we had a material in which that was difficult. The magnetization curve would now look more like Fig. 5.4. After you've applied and removed an

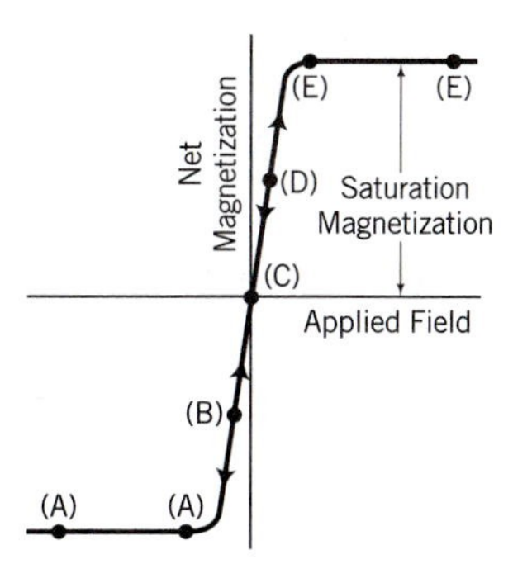

Figure 5.3. Reversible magnetization curve of an ideal soft magnetic material. Letters (A)–(E) correspond to domain pictures in Fig. 5.2.

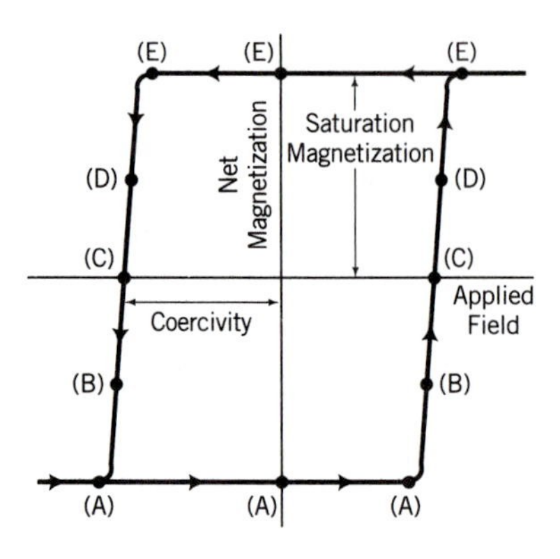

Figure 5.4. Magnetization curve (hysteresis loop) of a hard (permanent) magnet. Letters (A)–(E) correspond to domain structures in Fig. 5.2.

applied field pointing upwards, the material would remain magnetized upward (a "permanent" magnet). You would have what is called a *remanent magnetization*. It would then take a substantial applied field in the opposite direction to form and move domain walls and eventually reverse the magnetization. Remove that field, and you'd have a remanent magnetization in the opposite direction. It now takes a substantial upwards field to reverse the magnetization again. We get irreversible **M**(**H**) behavior, a *hysteresis loop*. The reverse field required to reduce the net magnetization to zero is called the *coercivity*.

Materials that behave like Fig. 5.3, with reversible or nearly reversible **M**(**H**) curves, are called *soft* magnetic materials. They are useful for the cores of electromagnets, for AC applications like inductors and transformers, and for refrigerator doors. Materials that behave like Fig. 5.4, with hysteretic **M**(**H**) curves, are called *hard* magnetic materials, and they would not be suitable for inductors. They are useful instead for applications calling for a permanent magnetic field, such as loudspeakers, permanent-magnet motors, data storage media (disks, tapes, credit and ATM cards, etc.), and for holding notes to refrigerator doors.

Within good soft (low coercivity, approaching Fig. 5.3) magnetic materials, magnetic domain walls can form and move easily. Within good hard (high coercivity, like Fig. 5.4) magnetic materials, they cannot. The ease with which domain walls can move depends both on the microstructure of the magnet and on some basic properties of the material. One of these basic properties is magnetic anisotropy.

5.4 ROTATING AGAINST ANISOTROPY

In the preceding section, we assumed that the saturation magnetization $\mathbf{M}_s$ was pointed either up or down, and that the applied field **H** was parallel (or antiparallel) to $\mathbf{M}_s$. The domain structure we assumed is actually very common, because ferromagnetic materials have *magnetic anisotropy* and prefer to be magnetized in particular crystalline directions called their *easy axes*. The easy axes are $\langle 1\,0\,0 \rangle$ in bcc iron, $\langle 1\,1\,1 \rangle$ in fcc nickel, and the hexagonal [0 0 0 1] axis in hcp cobalt. (To differentiate magnetic anisotropy resulting from crystal structure alone from other possible sources of anisotropy, this is often called *magnetocrystalline* anisotropy.)

To say that a material "prefers" to be magnetized along an easy axis means that magnetization along that axis corresponds to a minimum in energy, and any departure from that direction of magnetization will lead to an increase in energy. For simplicity, we consider a "uniaxial" material like cobalt that has a *single easy axis.* If ϕ is the angle between the magnetization vector and the direction of the easy axis, the *anisotropy energy* per unit volume in such a material is given approximately by:

$$E_{an} = K \sin^2 \phi \tag{5.8}$$

where K is the *anisotropy constant*, expressed in units of J/m^3.

Note that this energy is minimum both for $\phi = 0$ and for $\phi = 180°$, that is, for magnetization pointing in either direction along the easy axis. Magnetic anisotropy can be measured by applying a magnetic field $\mu_0 \mathbf{H}$ *perpendicular to the easy axis* (Fig. 5.5). This field exerts a torque on the magnetization in each domain that attempts to rotate it toward the field direction, since the energy of interaction with the field, from (5.7), will be:

$$E_H = -M_s \mu_0 H \cos(90° - \phi) = -M_s \mu_0 H \sin \phi \tag{5.9}$$

per unit volume. This rotation will be opposed by the anisotropy energy $E_{an} = K \sin^2 \phi$, and the equilibrium orientation of the magnetization will be determined by minimizing the sum of these two energy terms, i.e., by setting $d(E_H + E_{an})/d\phi = 0$. (The saturated magnetization in each domain must compromise between its desire to remain along the easy axis, represented by E_{an}, and its desire to be parallel to the field, represented by E_H.) The resulting net component of the magnetization parallel to the field will be:

$$\frac{M}{M_s} = \sin \phi = \frac{M_s \mu_0 H}{2K} = \frac{H}{H_{an}} \tag{5.10}$$

where

$$H_{an} = \frac{2K}{\mu_0 M_s} \tag{5.11}$$

is called the *anisotropy field.* So with field applied perpendicular to the easy axis, the gradual rotation of the magnetization gives a linear **M**(**H**) curve (Fig. 5.5) that reaches saturation ($M = M_s$) at $H = H_{an}$, allowing measurement of K. The magnetic susceptibility $\chi_m = M/H = M_s/H_{an}$ for $H < H_{an}$ and the differential susceptibility $dM/dH = 0$ for $H > H_{an}$ (once the rotation is completed).

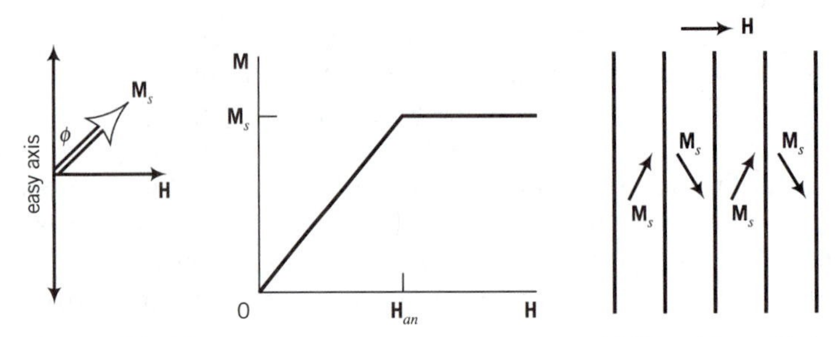

Figure 5.5. Measuring anisotropy constant from **M**(**H**) with applied field **H** perpendicular to easy axis. Here **M** is the net component of magnetization in the field direction.

Sample Problem 5.5

Measuring **M**(**H**) of a uniaxial ferromagnet with the applied field perpendicular to the single easy axis (as in Fig. 5.5) reveals that the anisotropy field is 6×10^5 A/m and the saturation magnetization is 1.4×10^6 A/m. (a) What is the magnetic anisotropy constant K? (b) What is the component of magnetization parallel to the applied field when the field strength is 6×10^4 A/m? 6×10^5 A/m? 6×10^6 A/m?

Solution

(a) From (5.11), $K = \mu_0 M_s H_{an}/2 = (1.26 \times 10^{-6})(1.4 \times 10^6)(6 \times 10^5)/2 = 5.29 \times 10^5$ J/m^3

(b) From (5.10), the component of magnetization parallel to the field is $M = M_s \sin\phi = M_s(H/H_{an})$ for $H \leq H_{an}$

so $M = M_s \sin\phi = (1.4 \times 10^6)(6 \times 10^4)/(6 \times 10^5) = 1.4 \times 10^5$ A/m at $H = 6 \times 10^4$ A/m

and $M = M_s \sin\phi = (1.4 \times 10^6)(6 \times 10^5)/(6 \times 10^5) = 1.4 \times 10^6$ A/m at $H = 6 \times 10^5$ A/m

For $H > H_{an}$, $M = M_s$ (saturation), so $M = 1.4 \times 10^6$ A/m at $H = 6 \times 10^6$ A/m

We have separately considered an applied field *parallel* to the easy axis (Section 5.3) where net magnetization increases by domain-wall motion, and an applied field *perpendicular* to the easy axis (Section 5.4), where net magnetization increases by rotation of the direction of magnetization in each domain away from the easy axis. If the applied field is at an intermediate angle, as will of course be the general case for an unaligned polycrystal, *the* **M**(**H**) *curve will result from both domain-wall motion and magnetization rotation.* The field component parallel to the easy axis in each grain will produce net magnetization by domain-wall motion and the field component perpendicular to the easy axis will produce net magnetization by rotation. Usually domain-wall motion proceeds first, followed by rotation.

Sample Problem 5.6

(a) If the applied field is at an angle of 25° to the easy axis, what fraction of saturation (in the field direction) can be produced by domain-wall motion alone? (b) What if the angle between the applied field and easy axis was 85°?

Solution

(a) Once domain walls have moved and disappeared, the magnetization has been saturated in the direction of the easy axis, and the fraction along the field direction is cos(25°) = 0.906. To further increase the magnetization towards saturation will require rotation of the magnetization out of the easy axis.
(b) Here the fraction is only cos(85°) = 0.087.

5.5 THE DOMAIN WALL

Although the formation of magnetic domains lowers the energy of external magnetic fields, there is an increase in energy associated with the domain walls themselves, which have a finite width across which the spin orientation gradually changes from the magnetization direction of one domain to that of the other (Fig. 5.6). One contribution to the energy of domain walls is the magnetic anisotropy K, because the spins in the wall deviate from the easy axis; see (5.8). Another contribution comes from the fact that the spins in the wall are not exactly parallel to their neighbors, that is, from the *exchange force* mentioned earlier. The contribution to wall energy from exchange can be expressed in terms of the *exchange constant* A, a measure of the increase in energy of neighboring spins when they are not perfectly parallel. It is the exchange force, of course, that creates the phenomenon of ferromagnetism, and A is proportional to the Curie temperature, the temperature above which thermal effects overcome exchange and ferromagnetism is lost. The units for A are J/m.

From the point of view of exchange energy alone, domain walls would prefer to be very wide, thereby minimizing the angle between neighboring spins. From the point of view of anisotropy energy alone, a domain wall should be infinitely thin, so that all spins would point in one direction or the other along the easy axis. Exchange favors a wide wall, anisotropy a narrow wall, and the equilibrium domain-wall width is determined by a balance between these two competing energy terms

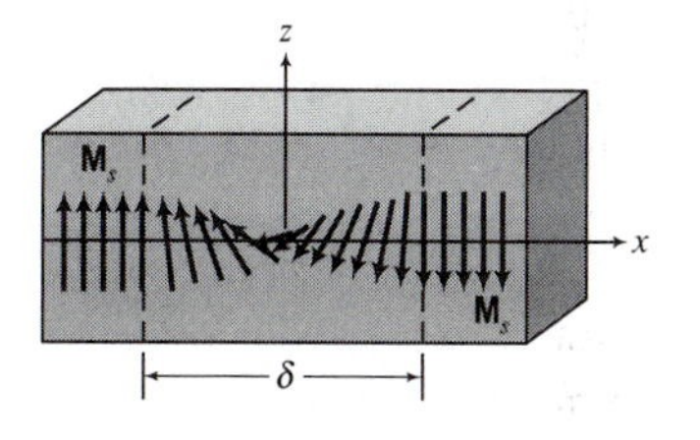

Figure 5.6. Change of spin orientation across wall between two magnetic domains. Within the domain wall, spins are neither in the easy-axis direction nor exactly parallel to their nearest-neighbor spins. Both these deviations contribute to the surface energy of the domain wall.

(see Problem 5-7). Minimizing the sum of the exchange and anisotropy energies of the wall yields a total domain-wall surface energy (energy per unit area) of

$$\gamma = \pi(AK)^{1/2} \tag{5.12}$$

and a domain-wall width of

$$\delta = 4(A/K)^{1/2} \tag{5.13}$$

The exchange constant A does not vary much among practical magnetic materials, which all have fairly high Curie temperatures. *However, the anisotropy K can vary by many orders of magnitude.* In very low-K materials (e.g., amorphous alloys used for transformer cores), the domain-wall width δ can be more than 10 μm, whereas in very high-K materials (e.g., rare-earth compounds used for permanent magnets), δ can be less than 10 nm. The domain-wall energy γ can also vary by as much as a factor of 1000 between high-K and low-K materials.

Sample Problem 5.7
Within a magnetic domain wall, what is the average energy per unit volume?

Solution
A unit area of domain wall has an energy of γ and extends over a volume equal to $1 \times \delta = \delta$. Thus $\gamma/\delta = \pi K/4$.

5.6 HARDER AND SOFTER

For AC applications, like inductors, transformers, and magnetic recording heads, engineers want soft (low-coercivity) materials capable of easily changing their state of magnetization many times each second. For others, like permanent-magnet motors, loudspeakers, and magnetic data storage, engineers want hard (high-coercivity) materials that retain their magnetization once magnetized. As noted earlier, whether a magnetic material is soft or hard depends on whether domain-wall motion within the material is easy or difficult.

The ease of magnetic domain-wall motion depends sensitively on a factor we have not yet considered—the homogeneity of the material's microstructure. Until now, we have implicitly assumed that the magnetic material was chemically and physically homogeneous, that every volume element was the same as every other volume element. But real materials contain inhomogeneities, and these inhomogeneities can have an important effect on the ease of domain-wall motion and hence on coercivity.

Consider, for example, the boundaries between grains (crystals) in a polycrystalline material. Within grain boundaries, the arrangement of atoms is less regular than within the grains, and the local chemical constitution is often different. Impurity atoms tend to congregate there. Thus properties like A and K, and therefore the magnetic domain-wall energy γ, may be very different in the region of the grain boundary. There may be other inhomogeneities in the microstructure—perhaps precipitates or inclusions of a second phase, or accumulations of crystal defects (e.g., dislocations). Any of these departures from chemical and physical homogeneity might produce local variations in γ, shown schematically in Fig. 5.7.

Now consider a magnetic domain wall attempting to move under the influence of an applied field. Those energy dips and energy bumps in Fig. 5.7 will make it difficult. Any inhomogeneity producing a local decrease in domain-wall energy would produce attractive forces on the wall (given by the gradient of the wall energy, $d\gamma/dx$) that would make it hard to escape. Any inhomogeneity producing a local *increase* in wall energy would produce instead *repulsive* forces that would make it hard to pass. By obstructing the motion of magnetic domain walls, inhomogeneities increase coercivity.

Some magnetic materials would be expected to be much more sensitive to such microstructural effects than others. Materials with a high anisotropy K have narrow walls and high surface energy, and inhomogeneities with dimensions comparable to the domain-wall width δ could produce large forces on the wall (large $d\gamma/dx$). But materials with a low anisotropy have wide walls of low surface energy, so the forces on the walls produced by the inhomogeneities would be low (low $d\gamma/dx$).

This reasoning suggests that engineers seeking hard (hysteretic, high-coercivity) magnetic behavior should seek high-K materials with inhomogeneous microstructures. Providing a "bumpy road" will obstruct wall motion, and if anisotropy is high, it's relatively "easy to be hard." Engineers seeking soft (reversible, low-coercivity) magnetic behavior should instead seek low-K materials with homogeneous microstructures, offering a smooth highway to facilitate wall motion.

Today's best soft magnetic materials indeed have low anisotropy and few grain boundaries. They are mostly either large-grained (with a cubic crystal symmetry, like Fe-Si and Fe-Ni) or noncrystalline (Fe-rich or Co-rich amorphous metals, produced by rapid quenching from the melt). Today's best *hard* magnetic materials (barium-iron oxides and rare-earth compounds like $Fe_{14}Nd_2B$) instead have *high*

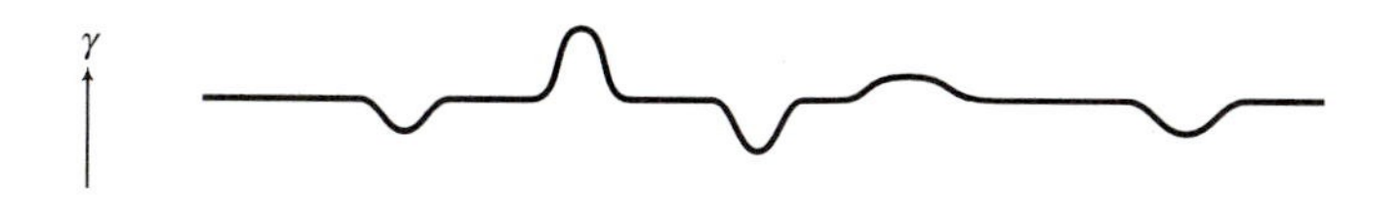

Figure 5.7. Schematic of variation of domain-wall energy γ in the vicinity of chemical or physical inhomogeneities, e.g., grain boundaries. Such variations obstruct easy domain-wall motion and thereby increase coercivity. Microstructural inhomogeneities can produce similar variations of energy and a resulting obstruction of motion of other entities whose mobility influences the properties of solids; e.g., fluxoids (Chapter 6) and dislocations (Chapter 7).

anisotropies (associated with hexagonal and other low-symmetry crystal structures) and very fine grains (many barriers to domain-wall motion).

In recent decades, materials scientists and engineers have developed both soft and hard magnetic materials far superior to those that were available in the past. Figure 5.8 shows, on a log scale, the range of coercivities available today, from coercivities of over 10^6 A/m in rare-earth compounds like Co_5Sm and $Fe_{14}Nd_2B$ to coercivities less than 1 A/m in amorphous cobalt-based alloys. Improvements in hard magnetic materials have led to remarkable decreases in the size of electric motors and speakers. Improvements in soft magnetic materials have led to substantial reductions of the energy losses in transformers.

Summarizing the differences between soft and hard magnetic materials:

Soft Ferromagnetic Materials	Hard Ferromagnetic Materials
(Temporary magnets)	(Permanent magnets)
For AC-field and on/off applications	For DC-field applications
Reversible **M**(**H**), like Fig. 5.3	Hysteretic **M**(**H**), like Fig. 5.4
Low coercivity	High coercivity
Low anisotropy K	High anisotropy K
Wide, low-energy domain walls	Narrow, high-energy domain walls
Easy domain-wall motion	Hard domain-wall motion
Homogeneous microstructure	Inhomogeneous microstructure

Progress also continues in hard magnetic materials for data recording on computer disks, where storage density has increased by many orders of magnitude in recent decades. For recording media, intermediate coercivities, typically between 20 and

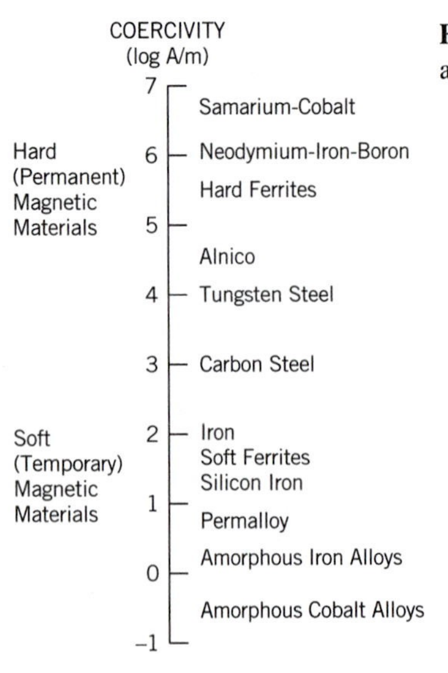

Figure 5.8. Range of coercivities (note log scale) available with magnetic materials today.

100 kA/m, are desired, since you want to enter and change data with moderate fields. Iron oxide particles remain the dominant recording medium for audio and video tapes, but cobalt-rich metallic thin films (which have higher magnetizations than oxides) are the major current choice for hard disks. Also contributing to progress in this field are improvements in soft magnetic materials for write and read heads. The latest read heads are based on the phenomenon of *magnetoresistance*, the change of electrical resistivity with magnetic field.

Although most properties of solids depend to a degree on microstructure, a few properties, like magnetic coercivity, are *especially* sensitive to microstructure because they depend on the interaction of microstructure with a moving entity, in this case the magnetic domain wall. Such properties are called *structure-sensitive*. Another example of a structure-sensitive property is the critical current density of superconductors, as we'll see in Chapter 6.

SUMMARY

Inductance is increased by the factor μ_r (the relative permeability) if a soft magnetic material is inserted within the coils of a solenoid. Magnetization **M** (magnetic dipole moment per unit volume) is the source of the increase. Most materials are either diamagnetic or paramagnetic, with relative permeabilities little different from one. However, a few remarkable materials called *ferromagnets* can have relative permeabilities of many thousands. The primary source of magnetization **M** is unbalanced electron spins, and ferromagnetism results from an *exchange force* between net spins of neighboring atoms that couples them to remain parallel, allowing the magnetic dipole moments of many atomic magnets to act in unison.

When macroscopically unmagnetized, ferromagnets are divided into *magnetic domains*. Within each domain, the material is locally magnetized, but different domains are magnetized in different directions. Within the domain walls that separate domains, magnetization rotates gradually from the orientation of one domain to the orientation of the other. Domains form to lower the energy of external magnetic fields, and the price paid is the energy of the *domain walls*, which depends both on magnetic anisotropy K and exchange constant A.

The magnetization curve **M**(**H**) of a ferromagnet is determined primarily by the formation and motion of magnetic domain walls when the applied field is *parallel* to the easy axis. If domain walls can form and move easily, we have "soft" (reversible, low-coercivity) magnetic behavior with high permeabilities (desirable for AC applications such as inductors, transformers, and recording heads). If formation and motion of domain walls is difficult, we have "hard" (hysteretic, high-coercivity) magnetic behavior (suitable for permanent-magnet applications such as motors, speakers, and magnetic recording media).

If the field **H** is instead applied *perpendicular* to the easy axis of a uniaxial ferromagnet (single easy axis), the magnetization will rotate out of the easy axis and the anisotropy field $H_{an} = 2K/\mu_0 M_s$ can be measured from the resulting **M**(**H**). If the field **H** is applied at a *general* angle to the easy axis, magnetization in the field direction will usually occur first by domain-wall motion and later by magnetization rotation.

Low magnetic anisotropy K and a homogeneous microstructure are favorable for soft magnetic behavior, while high magnetic anisotropy and an inhomogeneous microstructure are favorable for hard magnetic behavior. Coercivity depends on the interaction of magnetic domain walls with microstructure, and is an example of a highly *structure-sensitive* material property.

PROBLEMS

5-1. By comparing expressions for the voltages across a resistor and an inductor, find the relation between henries and ohms. Also express henries and farads in terms of only the basic SI units—meters, kilograms, seconds, and amperes.

5-2. At $t = 0$, a voltage V is applied across a resistance R and an inductance L connected in series. Derive the time dependence of (a) the current, (b) the voltage across the inductance, and (c) the voltage across the resistance.

5-3. A pure capacitance C with charge Q stored on its plates is connected at $t = 0$ to a pure inductance L. Derive the time dependence of (a) the charge on the capacitance and (b) the current. (Hint: You may notice that the differential equation describing this circuit is that of a harmonic oscillator.) What is the resonant frequency of this circuit?

5-4. If the circuit of Problem 5-3 also includes a resistance R, derive the time dependence of the current.

5-5. Cobalt has a saturation magnetization of 1.4×10^6 A/m, and an atomic volume of 6.7 cm^3/mole. (a) What is the magnetic dipole moment of each cobalt atom in Bohr magnetons? (b) You apply a magnetic field in the easy direction with a coil of 10 turns per meter carrying a current of 0.3 milliamps, and determine that the total field **B** within the sample is one tesla. Calculate **M** and determine what percentage of the volume now consists of domains with magnetization parallel to the field.

5-6. A solenoid with one turn per millimeter carrying a current of one milliamp produces a magnetization of 10^5 A/m of a tested material. The susceptibility of the material remains unchanged up to a current of 10 milliamps, beyond which the magnetization of the material remains unchanged. (a) What is the saturation magnetization of this material? (b) What is the susceptibility at low applied fields? (c) What is the permeability of this material at low fields?

5-7. For a general domain-wall width of w, the contribution to the wall's surface energy from exchange is proportional to $1/w$ ($E_A = C_A A/w$), and the contribution from anisotropy is proportional to w ($E_K = C_K K w$). The equilibrium domain-wall width δ is determined by minimizing the sum of these two terms. From equations (5.12) and (5.13), determine the proportionality constants C_A and C_K.

5-8. (a) In an applied field **H**, what is the energy per unit volume in a domain in which the saturation magnetization is parallel to the field? (b) In an applied field **H**, what is the energy per unit volume in a domain in which the saturation magnetization is antiparallel to the field? (c) These two magnetic domains are separated by a domain wall. The applied field **H** tends to expand one domain and contract the other, in essence producing a force or pressure on the wall. Calculate this pressure (in pascals) for the case of a saturation magnetization of 10^6 A/m and $H = 100$ A/m.

5-9. A field **H** is applied parallel to the magnetic easy axis. Most of the material is magnetized anti-parallel to the field (a memory of an earlier applied **H**), but one cylindrical domain of radius r with magnetization parallel to the field has formed. Although the magnetic pressure from **H** (see Problem 5-8) tries to expand this domain, the surface energy γ of the domain wall tries to resist expansion. Derive a formula for the total energy as a function of r (setting $E = 0$ at $r = 0$), and calculate the radius at which the magnetic pressure overcomes the wall energy.

5-10. Calculate the magnetic pressure on a wall between oppositely directed domains if the field **H** is applied at a general angle θ to the easy axis. Assume that **H** is not sufficiently large to produce significant magnetization rotation.

5-11. A field **H** is applied at an angle of 45° to the magnetic easy axis of a material with a single easy axis. Magnetization occurs first by domain-wall motion (leaving just one domain) and then by rotation of the magnetization. Calculate the form of the "approach to saturation," that is, the H-dependence of M/M_s for the limiting case of $H \gg H_{an}$.

5-12. Two ferromagnetic materials each have a saturation magnetization of 10^6 A/m and an exchange constant A of 10^{-11} J/m. However, they differ greatly in their magnetic anisotropies. One (an amorphous metal) has $K = 5$ J/m^3, while the other (a rare-earth compound) has $K = 10^6$ J/m^3. For each of these two materials, calculate the anisotropy field, domain-wall energy, and domain-wall width. In which of these materials do you expect the ease of domain-wall motion to be more sensitive to microstructure? Why?

5-13. In an applied magnetic field $\mu_0\mathbf{H}$, the energy per unit volume of a material with constant susceptibility is $E = \int_0^H \mu_0 M dH = -\mu_0\chi_m H^2/2$. (a) If **H** and the x-axis are vertical, what is the force per unit volume on a paramagnetic or diamagnetic material with magnetic susceptibility χ_m in a vertical field gradient dH/dx? (b) If the mass density of the material is ρ, what value of $H(dH/dx)$ is required to counter the gravitational force and produce magnetic levitation? (c) Frogs and humans have a mass density of roughly 1 g/cm^3 and a magnetic susceptibility of about -10^{-5} (they are diamagnetic). What value of $H(dH/dx)$ will be required to levitate a frog or a freshman? (The magnetic levitation of small frogs was accomplished in 1997 in a Dutch laboratory. Magnets large enough and strong enough to levitate college students have not yet been constructed.)

5-14. To simplify the text, we assumed sample shapes in which the "demagnetizing fields" associated with the poles at sample ends (where the magnetic field enters or exits the material) could be neglected. In such cases, the only H-field acting on the magnetization is the external applied field. However, demagnetizing H-fields from poles often are important. Consider a uniformly magnetized permanent magnet of spherical shape with no applied field. For this case, sketch the patterns of **B**, **H**, and **M** vectors both inside and outside the magnet. (Hint: **B** is always divergenceless, but at the poles, **M** clearly is not.)

Chapter 6

Superconductors and Superconducting Magnets

6.1 THE BIG CHILL AND SUPERCONDUCTIVITY ($R = 0$)

In scientific research, as in space exploration, you maximize your chances of discovering something new if you, as in *Star Trek*, "boldly go where no one has gone before." Captains Kirk and Picard comb the galaxies, and Dutch physicist Heike Kamerlingh Onnes measured the properties of materials at temperatures colder than anyone had reached before.

In the early years of this century, Onnes and other physicists were exploring lower and lower temperatures. As materials get colder, gases turn to liquids, liquids turn to solids, and, within solids, the random vibrational motion of atoms becomes less and less (when you chill, the atoms still). In 1908, Onnes and his research group at the University of Leiden became the first to reach a temperature so low that even helium, the most stubborn of the gases, became liquid.

With the production of liquid helium, the laboratory at Leiden became "the coldest spot on earth." Under atmospheric pressure, the boiling point of liquid helium is only 4.2 K (i.e., 4.2 Kelvin degrees above absolute zero), and Onnes could reach even lower temperatures by lowering the pressure above the liquid. At these record-breaking cryogenic temperatures, Onnes decided to measure, among other things, the electrical conductivity of metals.

Onnes found that as temperature decreased, the conductivity of various metals steadily increased. This was no surprise, because it was already known that electron mobility and conductivity in solids were limited by the thermal vibration of atoms, which of course decreased as temperature decreased (see Fig. 1.3). The surprise came in 1911, when he cooled a wire of mercury below 4.2 K. The electrical conductivity abruptly increased to *infinity*! Current continued to flow through the wire, but there was no voltage—no electrical resistance. Onnes had discovered a remarkable phenomenon that he named *superconductivity*.

To realize just how remarkable superconductivity is, consider the most common commercial superconducting composite, which consists of many fine filaments of a niobium-titanium (Nb-Ti) alloy imbedded in a matrix of nearly pure copper. Such composites are the primary materials used today in the windings of superconducting electromagnets for applications such as magnetic resonance imaging and particle accelerators. These multifilamentary composite wires are formed from large billets by extrusion, swaging, and extensive wire drawing. In the process, diameters of the Nb-Ti filaments are reduced from centimeters to microns. At room temperature, the conductivity of the copper is about 10^8 S/m, and that of the Nb-Ti alloy is nearly two orders of magnitude lower. Any current that you pass through this composite at room temperature will almost entirely choose the high-conductivity path, the copper matrix.

If you cool the composite to 9.1 K (Fig. 6.1), the conductivity of the copper increases by more than two orders of magnitude because of reduced thermal vibrations in the copper lattice (which results in an increase in the electron mean free path, collision time, and mobility). The conductivity of the Nb-Ti also increases, but only slightly, because the resistance in the alloy comes mostly from the presence of two different atomic species, not from lattice vibrations. At 9.1 K, the conductivity of copper is more than 10,000 times the conductivity of Nb-Ti. Any current passing through the composite will certainly avoid the high-resistivity filaments.

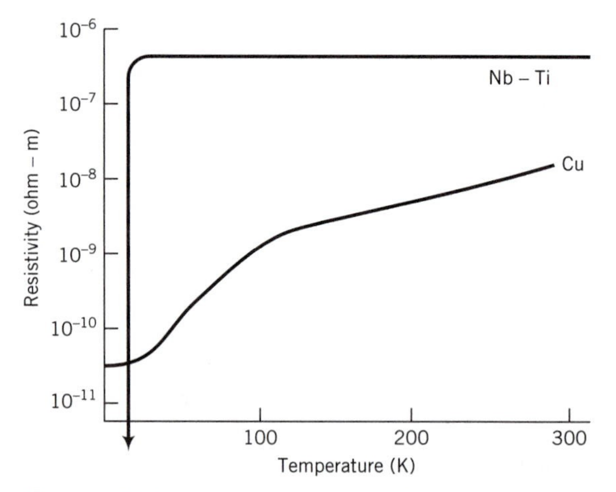

Figure 6.1. Resistivity (note log scale) vs. temperature for copper and Nb-Ti.

Now cool a bit further to 9.0 K. At this temperature, Nb-Ti becomes superconducting; its conductivity becomes suddenly infinite. Compared to that, the pure copper matrix looks like an insulator! Electrical current will now totally pass through the Nb-Ti filaments, which essentially short-circuit the comparatively high-resistance copper. Compared to infinity, even a conductivity of 10^{10} S/m looks pretty low!

Sample Problem 6.1

You have a composite in which the Nb-Ti filaments comprise 25% of the cross section, the copper matrix 75%, and you use it to carry a current of 100 amps. If the resistivities of copper and Nb-Ti are 2×10^{-8} and 1×10^{-6} ohm-m, respectively, at room temperature, and 4×10^{-11} and 8×10^{-7} at 9.1 K, how much current will each phase carry at each of those temperatures? How much current will each carry at 9.0 K, when the Nb-Ti becomes superconducting?

Solution

At room temperature, the conductivity of copper (5×10^{7} S/m) is 50 times that of Nb-Ti (1×10^{6} S/m), so the current density must also be 50 times greater than that in the Nb-Ti (since the electric field in each phase is the same; see the sample problem in Chapter 1 about conductivity in composites). Copper also has three times the cross-sectional area of the Nb-Ti, so the total current carried by copper is 150 times that carried by Nb-Ti. Thus the Nb-Ti carries 100/151 = 0.66 amps and the copper carries 99.34 amps.

At 9.1 K, the conductivity of copper (2.5×10^{10} S/m) is 20,000 times that of Nb-Ti (1.25×10^{6} S/m), so the current carried by Nb-Ti is only 100/60,001 = 0.0017 amps, and the copper carries 99.9983 amps.

At 9.0 K, the Nb-Ti is superconducting and carries the full 100 amps (shorting out the copper). Compared to a superconductor, even pure copper at cryogenic temperatures is an insulator. See also Problem 1-7.

Onnes did not have multifilamentary Cu-Nb-Ti composites. In fact, he limited his studies to pure metals. He found that mercury was not the only metal that lost all electrical resistance at cryogenic temperatures. Although some metals, like copper, never became superconducting, tin, lead, indium, and many other metals each became superconducting when cooled below its *critical temperature* T_c. The critical temperatures were different for each metal but remained very low, only a few degrees above absolute zero.

What can you do with metals that have no electrical resistance—that allow frictionless flow of electrons? Onnes realized that the electrical resistance of metals at room temperature and the associated heating and energy loss was what limited the magnetic fields obtainable with ordinary electromagnets. But if you built an electromagnet with wire that became superconducting when cooled below its critical temperature T_c, you should be able to achieve huge currents, and produce huge magnetic fields. He wrote in 1913:

> *The extraordinary character of this state* [*superconductivity*] *can be well elucidated by its bearing on the problem of producing intense magnetic fields with the aid of coils without iron cores. Theoretically it will be possible to obtain a field as intense as we wish by arranging a sufficient number of ampere windings* [*i.e., sufficient* $H = nI$] *round the space where the field has to be established . . . a field of 100,000 gauss could then be obtained by a coil of say 30 centimeters in diameter and the cooling with helium would require a plant which could be realized in Leiden with a relatively modest financial support.*

Onnes received the financial support he was fishing for, but never attained his goal of a 100-kilogauss (10-tesla) electromagnet. Unlike many scientists looking for financial support, Onnes had openly foreseen the possibility of failure: "There remains of course the possibility that a resistance is developed in the superconductor by the magnetic field."

Unfortunately for him, that's how it turned out. The pure metals he studied all lost their superconductivity at magnetic fields of only a few hundred gauss. He learned that superconductivity disappeared not only when the temperature exceeded T_c, but also when the applied field exceeded a *critical magnetic field* $\mu_0 H_c$, and when the current through the material exceeded a *critical current* I_c. The phenomenon of superconductivity was remarkable, but it was also delicate. It disappeared if the temperature, magnetic field, or electric current were too high. All three quantities had to be kept below critical values that were characteristic for each material, and an increase in any one of the three variables decreased the critical values of the other two. The delicate phenomenon of superconductivity existed only if temperature, field, and current were all kept below a critical surface in *T-H-I* space (Fig. 6.2a).

Specifically, in the absence of applied current, the critical field H_c varied parabolically with temperature (Fig. 6.2b) as

$$H_c(T) = H_0(1 - t^2) \tag{6.1}$$

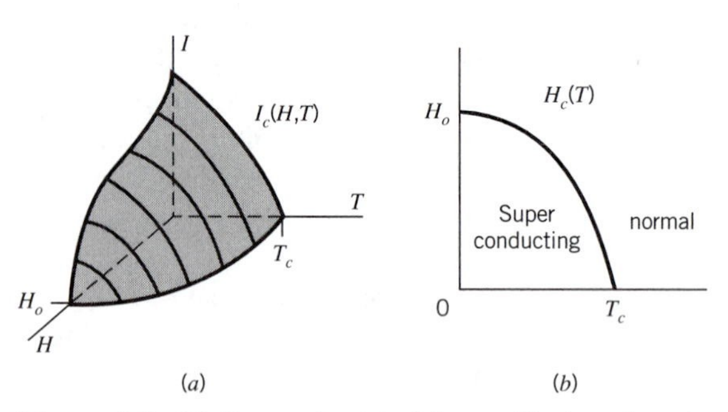

Figure 6.2. (a) For each material, a critical surface in *T-H-I* space defines the limits of superconductivity. (b) Critical field vs. temperature.

where $t = T/T_c$, the *reduced temperature*, and H_0 is the critical field extrapolated to absolute zero.

Sample Problem 6.2

Tin has a critical temperature of 3.7 K, and a critical field at 1.5 K of 20 kA/m. What is the critical field at 2.5 K?

Solution

At 1.5 K, $t = T/T_c = 0.405$. From (6.1), $H_0 = 20/(1 - t^2) = 24$ kA/m
At 2.5 K, $t = T/T_c = 0.676$, so $H_c = 24(1 - 0.457) = 13$ kA/m

Even in the absence of an applied field, a current creates a magnetic field that can exceed the critical field. From Ampere's law, a wire of radius r carrying a current I produces at its surface a circumferential magnetic field of $B = \mu_0 I/2\pi r$. The wire will begin to lose superconductivity when this field exceeds $\mu_0 H_c$, giving a critical current I_c of

$$I_c = 2\pi r H_c \tag{6.2}$$

If *both* a current and a field are applied to the superconductor, the applied field and the current-produced field will add at some points to produce a larger field, so that the applied field will lower the critical current below the value given by (6.2). The detailed form of the decrease will depend on the relative orientation of the field and current.

To achieve Onnes's dream of a 10-tesla electromagnet, one must have a material that not only has a critical magnetic field $\mu_0 H_c$ over 10 tesla, but also has a substantial critical current I_c in the presence of a 10-tesla magnetic field, since the inner wires of the solenoid would be exposed to the full field. Half a century passed before his dream was achieved. Before discussing the 1961 breakthrough that moved superconductors from scientific curiosities to useful engineering materials, we'll discuss the effect of magnetic fields on superconductors in terms of a 1933 discovery—the *Meissner effect.*

6.2 MEISSNER EFFECT—SUPERDIAMAGNETISM ($B = 0$)

Onnes was aware that a material with infinite conductivity would shield its interior from magnetic field. Apply a magnetic field to copper and, by Faraday's law, currents are induced (eddy currents) that attempt to keep the field from penetrating. In copper, electrical resistance rapidly damps those currents and magnetic field soon penetrates. In a superconductor, there is no electrical resistance, and surface supercurrents will continue to keep the applied magnetic field $\mu_0\mathbf{H}$ from entering the material. Within the superconductor, the magnetic field $\mathbf{B}$ remains zero. So, from (5.2) and (5.3),

$$B = \mu_0 nI + \mu_0 M = \mu_0(H + M) = \mu_0\mu_r H = \mu_0(1 + \chi_m)H = 0 \tag{6.3}$$

We recall that materials with negative magnetic susceptibilities are called *diamagnets*. As long as the superconductor completely shields its interior from the applied magnetic field $\mu_0\mathbf{H}$, it is a perfect diamagnet, a *superdiamagnet*:

$$\mathbf{M} = -\mathbf{H} \qquad \mu_r = 0 \qquad \chi_m = -1 \tag{6.4}$$

The magnetization **M** of the superconductor comes not from electron spins as in ferromagnets, but from surface supercurrents that create a field that opposes and exactly cancels the applied field, so that $\mathbf{B} = 0$ inside the superconductor. (Strictly speaking, the surface supercurrents flow to a small but finite "penetration depth" below the surface, so that **B** is not zero within this thin surface layer.) If we now increase the applied field until it exceeds the critical field, superconductivity will be lost, the magnetization **M** and the magnetic susceptibility χ_m will vanish, magnetic field will penetrate the material, **B** will equal $\mu_0\mathbf{H}$, and the relative permeability μ_r will equal one. (We ignore here any weak diamagnetism or paramagnetism the material may have in the absence of superconductivity.)

Sample Problem 6.3
At 1.5 K, tin has a critical field of 20 kA/m. Calculate **B** and **M** for tin in applied fields of 19 and 21 kA/m.

Solution
At 19 kA/m (below H_c), $B = 0$ and $M = -H = -19$ kA/m
At 21 ka/m (above H_c), $M \approx 0$ and $B \approx \mu_0 H = 2.65 \times 10^{-5}$ tesla

But suppose we now *decrease* the applied field so that it once again is less than the critical field. Onnes and his colleagues expected from Faraday's law that currents would be induced to oppose any $d\mathbf{B}/dt$. They assumed that once the material developed infinite conductivity, it would trap within it, with surface supercurrents, the field that was in it when it became superconducting, that is, $\mu_0 H_c$ (Fig. 6.3a). The results of that experiment were so "obvious" that nobody did the experiment until 1933, but the results were quite different from the expectations—once the applied field was reduced below the critical field, rather than trapping the magnetic field, the superconductor *expelled* the magnetic field (Fig. 6.3b).

This expulsion of magnetic field with decreasing applied field is called the *Meissner effect*. Infinite conductivity alone predicted trapping the field, not expulsion. It predicted $d\mathbf{B}/dt = 0$ (since $\mathbf{E} = 0$), but not $\mathbf{B} = 0$. Apparently, the superdiamagnetism described in (6.3) and (6.4) is *reversible*, showing that it is *a separate and distinct property of superconductors, and not just a result of infinite conductivity.* Figure 6.3b shows the reversible **M(H)** and **B(H)** curves experimentally observed,

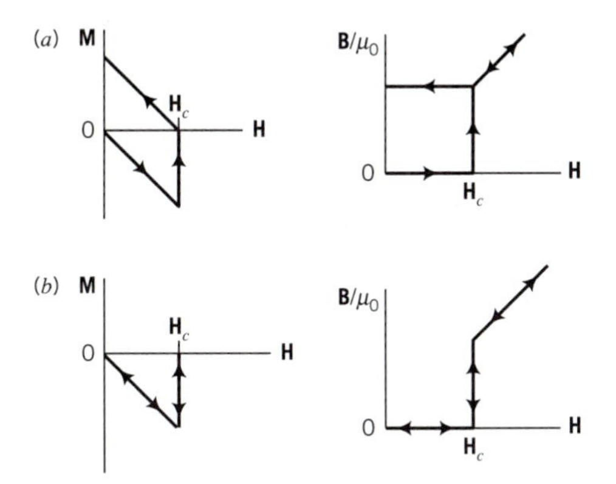

Figure 6.3. (a) **M**(**H**) and **B**(**H**) behavior expected simply from infinite conductivity, (b) observed **M**(**H**) and **B**(**H**) behavior of Type I superconductor, showing flux expulsion in decreasing field (the Meissner effect).

and Fig. 6.3a those that would be expected simply from infinite conductivity. Reversibility allowed the application of thermodynamics to superconductivity, with the "superconducting state" of the material the equilibrium state for $H < H_c$ and the "normal (nonsuperconducting) state" the equilibrium state for $H > H_c$.

Most elemental superconductors behave as in Fig. 6.3b, exhibiting perfect and reversible superdiamagnetism for $H < H_c$, and losing diamagnetism and superconductivity for $H > H_c$. Such materials are called *Type I superconductors*. It is the increase in energy associated with diamagnetic surface supercurrents that makes such materials lose their superconductivity in applied magnetic fields, usually at fields less than 0.1 tesla, destroying Onnes's dreams of building a 10-tesla electromagnet.

There were hints in the literature, however, that not all superconductors behaved that way. In particular, in the 1930s a Russian physicist named Schubnikow reported clear evidence that several alloys of lead maintained evidence of superconductivity to magnetic fields much higher than the critical field of lead and had reversible magnetization behavior more complex than that in Fig. 6.3. However, his paper in a Russian journal apparently was not noticed in the West, and Schubnikow himself was killed in one of Stalin's purges. It was not until the 1960s that the existence of *Type II superconductors* became widely recognized and Onnes's dream was finally realized.

6.3 HIGH-FIELD SUPERCONDUCTORS—THE MIXED STATE

Type I superconductors lose their superconductivity in modest magnetic fields because of the increase in energy associated with complete exclusion of magnetic field. Research in the early 1960s showed that some superconductors, called Type II, instead allow *partial* penetration of magnetic field, thus reducing the energy associated with complete diamagnetism and allowing some of the superconducting state to extend to higher magnetic fields.

The reversible **M**(**H**) and **B**(**H**) curves of an ideal Type II superconductor are shown in Fig. 6.4a. The material is fully in the superconducting state, and superdiamagnetic, for fields lower than the *lower critical field* H_{c1}. However, it does not lose all trace of superconductivity and diamagnetism and complete the transition to the normal state until a much higher *upper critical field* H_{c2}. Between H_{c1} and H_{c2}, it is neither fully superconducting nor fully normal; it is in what is called the *mixed state*.

Alloying lead, for example, converts it from a Type I superconductor into a Type II superconductor, with increasing alloying producing a steady decrease in H_{c1} and a steady increase in H_{c2} (Fig. 6.4b). Although most superconducting elements are Type I superconductors, most alloys and compounds are Type II superconductors. It took 50 years from the discovery of superconductivity to firmly establish the existence of Type II superconductors because the few physicists who studied superconductivity in the early years limited their studies primarily to pure elements. In 1961, however, scientists at Bell Laboratories reported that the intermetallic compound Nb_3Sn could carry large supercurrents in the presence of large magnetic fields. The first superconducting electromagnet to achieve Onnes's dream of 10 tesla was soon built at General Electric from Nb_3Sn, and within a few years high-field superconducting electromagnets had been constructed with Nb_3Sn, Nb-Ti, and other Type II superconductors at several laboratories. To understand how these materials are capable of carrying large supercurrents in the presence of large magnetic fields requires some knowledge of the mixed state.

The magnetic structure of the mixed state, shown schematically in Fig. 6.5, has been revealed by various experimental techniques. It consists of a magnetic "flux lattice," a regular array of linear features called *fluxoids*. Each fluxoid has at its

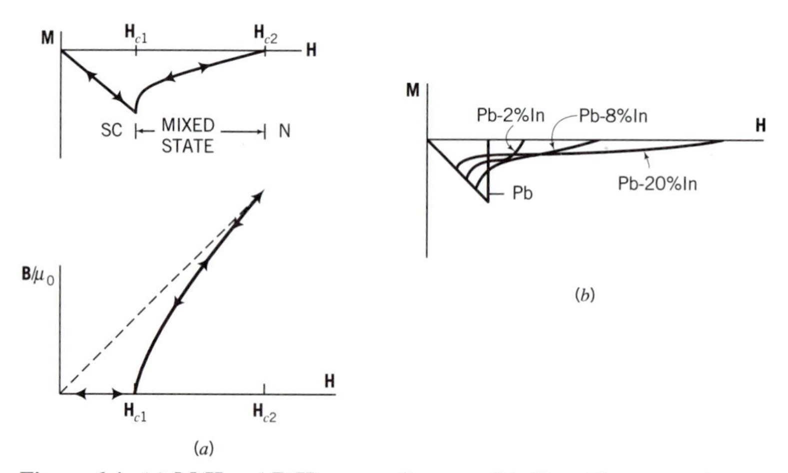

Figure 6.4. (a) **M**(**H** and **B**(**H**) curves for reversible Type II superconductor, (b) showing the increasingly Type II behavior resulting from increasing additions of In in solid solution in Pb.

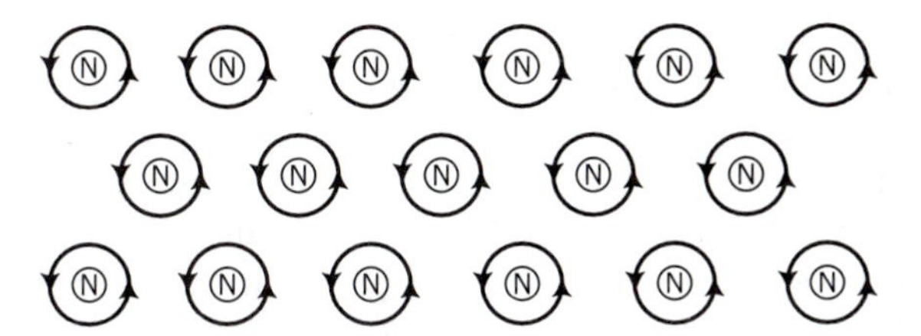

Figure 6.5. Schematic of magnetic flux lattice in mixed state of Type II superconductor (magnetic field normal to page). Each fluxoid contains 2.07×10^{-15} webers (tesla-m^2) of magnetic flux, contains a normal (nonsuperconducting) core, and is surrounded by a vortex of supercurrent.

center a narrow cylinder of material within which the superconducting state has been replaced by the normal (nonsuperconducting) state, and magnetic field has penetrated. In the superconducting matrix around each thin normal cylinder, a vortex of supercurrent confines the magnetic field to each fluxoid. Fluxoids repel each other and form a regular two-dimensional lattice. Each one carries a fixed amount of magnetic flux (the "flux quantum") equal to $\phi_0 = 2.07 \times 10^{-15}$ webers or T-m^2. If there are N fluxoids per m^2, the magnetic field in the mixed state is $B = N\phi_0$.

Sample Problem 6.4
A Type II superconductor has a lower critical field of 5 kA/m and an upper critical field of 100 kA/m. In an applied field of 10 kA/m, the sample magnetization **M** is −4 kA/m. How many fluxoids per unit area are present in the sample?

Solution
$B = \mu_0(H + M) = (1.26 \times 10^{-6})(10 \times 10^3 - 4 \times 10^3) = 7.56 \times 10^{-3}$ T
$N = B/\phi_0 = (7.56 \times 10^{-3})/(2.07 \times 10^{-15}) = 3.65 \times 10^{12}$ fluxoids/m^2

Superconductivity, like ferromagnetism, is a collective property of electrons deeply imbedded within quantum mechanics. This is clear from the formula for the superconducting flux quantum, which, like equation (5.4) for the Bohr magneton, includes Planck's constant h:

$$\phi_0 = \frac{h}{2e} \tag{6.5}$$

By being able to form the mixed state, Type II superconductors are able to allow significant field penetration and still maintain most of their volume in the thermodynamically favorable superconducting state. (Type I superconductors are unable to form a mixed state because boundaries between normal and superconducting regions in those materials have a high surface energy, making the formation of fluxoids energetically unfavorable.) But the mixed state in Type II superconductors is neither fully superconducting nor fully normal. Can a material in the mixed state really carry lossless supercurrents? Well, no and yes. It depends on microstructure.

Suppose our goal, like Onnes's, is to wind a solenoid with superconducting wire and create large magnetic fields with large lossless currents in the mixed state. In the geometry of a solenoid, the parts of the wire adjacent to the inside of the solenoid are exposed to magnetic fields perpendicular to the direction of the current. In Fig. 6.5, for example, where the magnetic field is normal to the figure, the current would be traveling, let's say, from left to right. If the current avoids the nonsuperconducting cylindrical fluxoid cores, it can find a superconducting path across the figure. But when currents flow perpendicular to magnetic fields, Lorentz forces arise. The resulting force is perpendicular to both the field and the current (thus vertical in the figure) and is exerted on the fluxoid.

In a Type II superconductor with ideal homogeneous microstructure, the fluxoids will respond to the Lorentz force produced by the current, and will move (at right angles to the current). A good skier might be able to ski through a dense forest if the trees all stayed in place, but not if they were moving. Moving fluxoids will lead, in effect, to some current passing through their normal cores, which will result in some energy loss and some electrical resistance. Thus in the mixed state between H_{c1} and H_{c2}, a Type II superconductor with a homogeneous microstructure (no grain boundaries, no dislocations, no chemical inhomogeneities to block fluxoid motion) cannot carry lossless currents. Its critical current is zero (Fig. 6.6).

But if we introduce microstructural inhomogeneities (e.g., grain boundaries, dislocations, particles of other phases, etc.) those "bumps in the road" can produce local variations in the energy of the fluxoids (as in Fig. 5.7) and thereby provide "pinning" forces on the fluxoids that will keep them from moving in response to the Lorentz force produced by the current. Experiments show (Fig. 6.6) that the introduction of

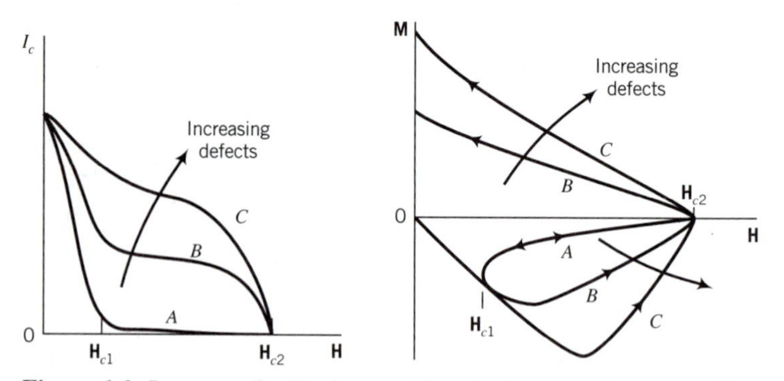

Figure 6.6. Increase of critical current in mixed state and corresponding increase in magnetic hysteresis of Type II superconductor resulting from fluxoid-pinning defects.

microstructural inhomogeneities allows Type II superconductors to carry essentially lossless currents in the mixed state. Critical current in the mixed state of Type II superconductors, much like ferromagnetic coercivity, which we discussed in Chapter 5, is a very *structure-sensitive* property. Microstructural inhomogeneities increase coercivity by pinning magnetic domain walls and increase critical current in the mixed state by pinning fluxoids. As we'll mention in Chapter 7, they also increase the mechanical strength of materials by pinning dislocations.

The similarity to ferromagnetic behavior becomes clearer if we consider the effects of fluxoid pinning on $\mathbf{M}(\mathbf{H})$ or $\mathbf{B}(\mathbf{H})$ behavior of Type II superconductors. Without pinning, magnetization curves are nearly reversible (Fig. 6.4a), as were magnetization curves of soft ferromagnets (Fig. 5.3). With fluxoid pinning, the entry of magnetic field in increasing field is delayed, and the exit of magnetic field in decreasing field is delayed. Magnetization curves are hysteretic (Fig. 6.6), as were magnetization curves of hard ferromagnets (Fig. 5.4). In fact, measurements of magnetic hysteresis curves can be used to estimate the pinning forces exerted on fluxoids by the microstructure, and to predict critical currents.

The Nb-Ti alloy filaments in the commercial superconducting composites mentioned earlier contain extremely high densities of dislocations, produced by the mechanical forming operations that transformed them from large rods to fine wires. The dislocations, plus fine precipitates of Ti-rich particles produced by intermediate heat treatments, provide the fluxoid pinning forces that give Nb-Ti large mixed-state critical currents in fields as high as 9 tesla at 4.2 K. The high-field superconducting compound Nb_3Sn, used for superconducting magnets capable of producing higher fields than Nb-Ti magnets because of a higher $\mathbf{H}_{c2}$, is too brittle to allow introduction of high dislocation densities by plastic deformation. However, it is processed to produce extremely fine grain sizes, so that in Nb_3Sn it is the Sn-rich grain boundaries that provide most of the pinning forces on fluxoids. As noted earlier, it was with a magnet based on Nb_3Sn that Onnes's dream of a 10-tesla electromagnet was first accomplished—a half-century late.

6.4 SUPERCONDUCTIVITY AT "HIGH" TEMPERATURES

The development of high-field Type II superconductors in the 1960s transformed superconductivity from an interesting scientific curiosity into a phenomenon of engineering utility. Superconducting electromagnets found several important applications, most notably in magnetic-resonance-imaging (MRI) magnets, now the basis of a standard (but expensive) diagnostic instrument in most major hospitals. The ability to carry lossless electric currents could have many other important applications, but the number of practical devices that use superconductivity remains limited, largely because superconductivity is a low-temperature phenomenon. Before 1986, the highest critical temperatures found were only a few degrees above 20 K. Most superconducting devices use liquid helium, which is expensive and difficult to work with and requires elaborate thermal shielding to separate the supercold device from its room-temperature surroundings.

The next materials breakthrough came in 1986, when Müller and Bednorz of the

IBM Zurich Laboratories found a complex oxide, La-Ba-Cu-O, that appeared to be superconducting above 30 K. The result was confirmed in other laboratories, and early in 1987, another complex oxide, Y-Ba-Cu-O, was found to have a critical temperature above 90 K. This was well above the boiling point of liquid nitrogen (77 K). Since liquid nitrogen is much cheaper and much easier to work with than liquid helium, this raised hopes of finding many applications, including but not limited to electromagnets, of these new "high-temperature" superconductors. The media suddenly discovered superconductivity, which became the subject of numerous magazine cover stories and front-page articles in *The New York Times*.

The ensuing years have seen discoveries of oxides with yet higher critical temperatures, the current (1998) record under ambient pressures being about 130 K. But these superconducting oxides have been found to be very difficult materials to work with. Although they are all Type II superconductors with very high upper critical fields, their major drawback to date has been limited critical currents in the mixed state. Whereas grain boundaries are useful fluxoid-pinning features in low-temperature superconductors like Nb_3Sn, they appear to be "weak links" in high-temperature superconductors and are a major cause of the limited critical currents. In addition, thermal effects make fluxoid pinning at 77 K much more difficult than fluxoid pinning at 4.2 K. The materials are also brittle multielement ceramics, and are difficult to process economically into long lengths of superconducting wire.

Although many billions of dollars have been poured into research on superconducting oxides since 1987 and thousands of researchers have studied these exciting new materials, the dreams of widespread applications have not yet been fulfilled. Progress continues, and some applications have begun to appear. However, from the current viewpoint, it still appears that the most significant technological breakthrough in superconducting materials was the discovery of high-field superconductors (Nb-Ti, Nb_3Sn) in the 1960s, rather than the discovery of high-temperature superconducting oxides in the 1980s.

The unpredicted discovery of high superconducting critical temperatures in oxides has revealed how limited our understanding of the basic origin of superconductivity still is. Superconductivity remained almost a total mystery until 1957, when Bardeen, Cooper, and Schrieffer produced a theory that explained most of the experimental facts known at that time. The *BCS theory* explains superconductivity as arising from an electron-lattice interaction that produces an attractive force between electrons that is strong enough to overcome their Coulomb repulsion. This attractive force allows electrons to form into "Cooper pairs" that can carry electric current without losing energy in collisions, thereby achieving infinite mobility. The $2e$ charge of a Cooper pair appears in the denominator of equation (6.5).

The BCS theory involves highly complex quantum mechanics. A simplified explanation of the attractive force between electrons created by the lattice starts with the attraction of one electron for neighboring positive ions, which displace slightly. However, because of their much higher mass, they move much more slowly than the electron. By the time they have displaced slightly, increasing the local positive charge density, the first electron has moved on. However, the increase in local positive charge density creates an attractive force on a second electron, which effectively acts as an attractive force between the first and second electron.

An analogy commonly used is that of bowling balls on a mattress. If you put one bowling ball on a mattress, it depresses the mattress locally. A second bowling ball placed on the mattress will then roll downward towards the first, an apparent "attraction" between the two bowling balls that is created by the response of an intermediary—the mattress. In superconductivity, the positive ions of the lattice act like the mattress, their response creating an effective attractive force between electrons.

Although the BCS theory was capable of explaining many experimental observations on low-temperature superconductors, it is generally believed to be inadequate to explain high-temperature superconducting oxides. Evidence indicates that Cooper pairs are the carriers of lossless current in these materials (the flux quantum remains equal to $h/2e$), but the mechanism of attractive interaction that forms the pairs remains a matter of controversy. Optimists continue to hope that improved understanding of the remarkable phenomenon of superconductivity will eventually lead to room-temperature superconductors with the ability to transmit electricity over long distances, and to create large magnetic fields in electromagnets, without electrical resistance and associated energy loss. After 75 years during which maximum critical temperatures increased to only 20 K, the next 5 years saw further increases of over 100 K. Although there's a still a long way to go to reach room temperature, recent progress does offer some reason for hope.

SUMMARY

Many materials, when cooled below a critical temperature T_c, abruptly lose their electrical resistance and develop infinite conductivity. This phenomenon of superconductivity exists as long as the temperature remains below T_c, the applied field remains below the critical field H_c, and the current remains below the critical current I_c. These critical values are characteristic for each material, and interrelated in that an increase in any one of the three variables—temperature, field, or current—decreases the critical values of the other two. A critical surface in T-H-I space defines the range of superconductivity for each material.

An ideal Type I superconductor is also a superdiamagnet ($\mathbf{M} = -\mathbf{H}$, $\mu_r = 0$, $\chi_m = -1$), excluding magnetic field from its interior with surface supercurrents until losing superconductivity and transforming to the normal (nonsuperconducting) state abruptly at the critical field H_c. If the applied field is decreased from above H_c, magnetic field is spontaneously expelled (the Meissner effect) at the critical field. This reversibility allows a thermodynamic treatment of the superconducting-normal transition and identifies the transition as resulting from the increase in energy of the superconducting state associated with the diamagnetic surface supercurrents. Most elemental superconductors are Type I superconductors.

In contrast, most superconducting alloys and compounds are Type II superconductors, in which the transition from the superconducting to the normal state in increasing field is gradual rather than abrupt. By allowing partial penetration of the magnetic field, these materials decrease the energy associated with flux exclusion, and can maintain large portions of their volume in the superconducting state up to high magnetic fields. Full superconductivity and full superdiamagnetism are maintained only up to a lower critical field H_{c1}, but the fully normal state is not attained until an upper critical field H_{c2}. Between H_{c1} and H_{c2}, these materials are in a mixed state that is neither fully superconducting nor fully normal. Magnetic field penetrates in the form of quantized fluxoids, each carrying 2.07×10^{-15} T-m^2 of magnetic flux (one "flux quantum"). Each fluxoid has a cylindrical core in the normal state, surrounded by a vortex of supercurrent in the surrounding superconducting matrix, and fluxoids repel each other to form a two-dimensional magnetic flux lattice.

A current traversing the mixed state in a direction perpendicular to the magnetic field (and therefore to the fluxoids) exerts a Lorentz force on the fluxoids. If the superconductor has an ideally homogeneous microstructure, the fluxoids will move in response to this current-produced force, and some energy loss and electrical resistance will appear. Thus a Type II superconductor with homogeneous microstructure will not be able to carry lossless currents in the mixed state. However, if microstructural defects (like grain boundaries, dislocations, or second-phase particles) are introduced, they provide pinning forces on the fluxoids. By resisting fluxoid motion, inhomogeneities allow lossless current flow and substantial critical currents in the mixed state. Alloys and compounds with high upper critical fields and inhomogeneous microstructures can therefore carry high supercurrents in the presence of high magnetic fields and have been used to produce high-field electromagnets.

Until 1986, most research had focused on metallic superconductors, with critical temperatures that remained below 25 K. In recent years, however, many oxide superconductors have been found with much higher critical temperatures, some as high as 130 K. Although these oxides are Type II superconductors with very high upper critical fields, critical currents at high temperatures and high fields remain limited. Eventually these materials may become capable of producing superconducting electromagnets that operate at higher temperatures or produce higher fields than present low-temperature metallic superconductors. The BCS theory of superconductivity, based on the formation of Cooper pairs via an electron-lattice interaction, appears inadequate to explain high-temperature superconductivity.

PROBLEMS

6-1. Lead has a critical temperature of 7.2 K and a critical field at 4.2 K of 430 A/cm. (a) What is the critical field at $T = 1$ K? (b) At 4.2 K, how much current can a 1-mm-diameter lead wire carry without resistance? (c) If a field of 200 A/cm is applied to a sample of lead at 4.2 K, what will be the sample magnetization? The relative permeability? The susceptibility? The average magnetic field inside the lead? (d) What will be the values of those four quantities if the applied field is instead 500 A/cm?

6-2. Figure 6.3 shows the difference between the magnetic behavior expected simply from infinite electrical conductivity and the actual behavior observed with an ideal Type I superconductor. Briefly explain the difference.

6-3. An ideal (reversible) Type II superconductor has a lower critical field of 10^4 A/m and an upper critical field of 10^6 A/m. In an applied field of 10^5 A/m, the magnetization is -5000 A/m. (a) What was the magnetization in an applied field of 10^4 A/m? What will it be in an applied field of 10^6 A/m? (b) In an applied field of 10^5 A/m, what will be the average magnetic field B within the material?

6-4. If the fluxoids in the mixed state form a triangular lattice like that in Fig. 6.5 and the average magnetic field in the material is 0.1 T, what is the spacing between neighboring fluxoids?

6-5. In an applied field of 630 A/cm, a Type II superconductor has a magnetization of -100 A/cm. What is the average magnetic field inside the material? How many fluxoids does it contain per unit area?

6-6. You wind a superconducting solenoid with a composite conductor in which the superconductor occupies about 20% of the total cross section, which is 1 cm $\times$ 4 mm. You wish to produce a solenoidal field of 10^7 A/m, but in the presence of such a large field the superconductor has a critical current density of only 10^4 A/cm^2. (Unlike a Type I superconductor, where current flows only on the surface, the critical current of a Type II superconductor is typically proportional to its cross-sectional area.) How many turns per meter must your solenoid have?

6-7. In the preceding problems, we have considered only solid superconductors. Suppose instead you have a thin-walled hollow tube of Type I superconductor. Would you expect the **M**(**H**) and **B**(**H**) behavior of this sample to behave more like the curves in Fig. 6.3a or like those in Fig. 6.3b? Explain.

6-8. One method of measuring magnetization is to place the sample in a field gradient dH/dx, and measure the resulting force in the x-direction. (a) From the magnetic energy per unit volume ($E = -\mu_0 \chi_m H^2/2$—see Problem 5-13), calculate the force per unit volume in an applied field of 1000 A/cm and a field gradient of 1000 A/cm^2 on (a) a Type I superconductor (assume that the critical field is not exceeded), (b) a soft ferromagnet with a relative permeability of 10^6, and (c) a paramagnet with a susceptibility of 10^{-4}. For each of these materials—superconductor, ferromagnet, paramagnet—is the force exerted in the direction of increasing or decreasing field?

6-9. You wind a solenoid with a wire of Type I superconductor. You notice that although there is no voltage across the windings when the current is held constant, a nonzero voltage appears whenever you change the current. What is the cause of this voltage? (Hint: See Chapter 5.)

6-10. As seen in Fig. 6.4b, lead-20% indium is a Type II superconductor. When well annealed, the **M**(**H**) and **B**(**H**) curves are nearly reversible. Sketch what you expect these curves to look like after the alloy is heavily cold-worked to introduce a high density of dislocations.

6-11. If **H** and the *x*-axis are in the vertical direction, what value of $H(dH/dx)$ will be required to levitate a superconductor with a mass density of 7 g/cc? How does this compare with the value of $H(dH/dx)$ required to levitate a frog or a freshman? (See Problem 5-13.)

6-12. How much surface current per unit length will be required to fully shield a superconductor from an applied magnetic field, parallel to the surface, of 100 gauss (10^{-2} tesla)?

6-13. If a sample is superdiamagnetic, it can be shown that the surface supercurrents increase the energy per unit volume of the sample by $\mu_0 H^2/2$. You apply a magnetic field of 100 gauss (10^{-2} tesla) along the length of a Type I superconducting rod 3 mm in diameter and 5 cm long. Assuming that this field is less than the critical field, calculate the strength of the surface supercurrent in amperes and the total energy in joules associated with this current.

Chapter 7

Elasticity, Springs, and Sonic Waves

7.1 STRESS AND STRAIN

In the earlier chapters, we considered the response of solids to applied electric and magnetic fields, including the alternating electric and magnetic fields present in electromagnetic waves. We now consider the elastic response of solids to mechanical forces—dimensional changes that on the atomic level correspond to departures of interatomic spacings from equilibrium spacings. Since interatomic forces are controlled by electrons, mechanical properties are also "electronic properties." As we'll discuss in detail in Chapter 10, equilibrium interatomic spacings—and the forces required to push atoms closer together or pull them further apart—are determined by a balance between attractive Coulomb forces (between electrons and nuclei) and repulsive Coulomb forces (between electrons and between nuclei).

At finite temperatures, atoms vibrate about their equilibrium lattice positions. (As we'll learn in Chapter 8, quantum mechanics requires that atomic vibrations do not completely cease even at $T = 0$.) In Chapter 1, we noted that lattice vibrations scatter free electrons and produce a decrease in collision time and electrical conductivity of metals with increasing temperature. Lattice vibrations appeared again in Chapter 3, when we considered the frequency dependence of ionic polarization, and in Chapter 6, where they played a role in the formation of superconducting Cooper pairs. The major goal of the present chapter is to consider lattice vibrations as another example of waves and as an important example of the effects of crystalline structure on wave propagation. We'll start by considering a continuous solid and later introduce the atomic structure.

To prepare for a consideration of waves, we will first consider uniform forces and uniform dimensional changes—uniform *stress* and *strain*. Suppose we have a solid rod of length L and uniform cross-sectional area A, and apply a tensile force $\mathbf{F}$ along its length. The rod will stretch a bit, with an increase in length ΔL proportional to the original length L. (It will also shrink a bit in the transverse directions, which we won't worry about here.) The fractional change in length $\varepsilon = \Delta L/L$ is called the *strain*, and *Hooke's law* of elasticity states that the strain will be proportional to the force per unit area $\sigma = F/A$, which is called the *stress*:

$$\varepsilon = \frac{\Delta L}{L} = \frac{\sigma}{E} \tag{7.1}$$

where E, the constant of proportionality between tensile stress and tensile strain, is known as *Young's modulus*. Strain is of course unitless. The SI units for stress and Young's modulus are newtons per square meter (N/m^2), or *pascals* (Pa).

Young's modulus for polycrystalline copper is about 10^{11} Pa, or 100 GPa. (The prefix G stands for *giga*, or 10^9—a billion.) Young's modulus for lead is nearly an order of magnitude lower, for diamond nearly an order of magnitude higher. Most ceramics and many metals have Young's moduli of 100 GPa or greater. Most polymers are much less resistant to mechanical stress, and have Young's moduli of less than 10 GPa, sometimes much less. The Young's moduli of rubbers can be as low as 10 MPa (10^7 Pa).

A solid that is stretched elastically under a tensile stress σ will, like a stretched spring, have stored elastic energy. By integrating force times distance, the stored elastic energy U per unit volume in the stretched rod is

$$U = \int \sigma d\varepsilon = \frac{E\varepsilon^2}{2} = \frac{\sigma^2}{2E} \tag{7.2}$$

If we had instead applied a compressive force along the length of the rod, it would of course have *decreased* slightly in length. The signs of stress and strain would be negative instead of positive, but Hooke's law would still be obeyed, and there would still be positive elastic energy stored in the rod. Springs are a common application of stored elastic energy.

Young's modulus and other elastic moduli (e.g., shear modulus, bulk modulus) describe the *stiffness* of solids—their resistance to *elastic* deformation. By definition, elastic deformation is *reversible*—the strain disappears and the sample returns to its original dimensions when the stress is removed. Except for a few materials such as rubbers, elastic strains in most solids are very limited, usually only a small fraction of a percent. At higher strains, most solids either deform plastically (producing permanent strain) or fracture.

Stiffness is one of the three "nesses" that define the mechanical properties of solids. The other two are *hardness*, resistance to plastic (permanent, irreversible) deformation, and *toughness*, resistance to fracture.

In crystalline materials, plastic deformation occurs primarily by the motion of linear structural defects called *dislocations*. Microstructural features like grain boundaries and second-phase particles can affect the mobility of dislocations, making hardness, like magnetic coercivity (Chapter 5) and superconducting critical current (Chapter 6), a very structure-sensitive property. Microstructural features can also affect the nucleation and propagation of cracks, so that toughness is also a structure-sensitive property.

We'll say little more about hardness and toughness in this text, but will occasionally refer to stiffness, which is the "ness" most directly related to interatomic bonding. It relates directly to the variation of interatomic bond energy with spacing. As we'll discuss in Chapters 10 and 14, interatomic bonds that are *stronger* tend also to be *shorter* and *stiffer*; this we call the *3S* rule.

Sample Problem 7.1

You hang a mass of 100 kg on a copper wire 50 cm long and 5 mm in diameter. By how much will the wire elongate? How much elastic energy is stored per unit volume in the wire? (Assume that the Young's modulus is 100 GPa and that the deformation remains elastic.)

Solution

On a mass of 100 kg, gravity exerts a force of 981 newtons. The wire area is 1.96×10^{-5} m^2, so $\sigma = 981/(1.96 \times 10^{-5}) = 50$ MPa.
From (7.1), $\varepsilon = \sigma/E = 50 \times 10^6/100 \times 10^9 = 5.0 \times 10^{-4}$,
so $\Delta L = \varepsilon L = (5 \times 10^{-4})(50 \times 10^{-2}) = 2.5 \times 10^{-4}$ m (0.25 mm)
From (7.2), $U = (100 \times 10^9)(5.0 \times 10^{-4})^2/2 = 12.5$ kJ/m^3.

7.2 HOOKE'D ON SONICS—ELASTIC WAVES

In the above case of uniform tensile stress and strain, if we consider one end of the rod ($x = 0$) as fixed and allow the other end to move, the *displacement* $u(x)$ of material from its unstressed position x varies linearly along the rod:

$$u(x) = \frac{du}{dx}x = \varepsilon x \tag{7.3}$$

and the change in length of the rod is the difference in displacement of the two ends:

$$\{u(L) - u(0)\} = \{\varepsilon L - 0\} = \Delta L \tag{7.4}$$

In this simple case, stress and strain were constant along the rod. We are interested, however, in a more general case where displacement, strain, and stress vary with position, that is, vary with x. In this case, we can generalize the relation implicit in (7.3), and the local strain $\varepsilon(x)$ will be the local value of the variation of displacement with distance

$$\varepsilon(x) = \frac{\partial u(x)}{\partial x} \tag{7.5}$$

and therefore, from Hooke's law, the local stress will be

$$\sigma(x) = E\frac{\partial u(x)}{\partial x} \tag{7.6}$$

We switched to partial derivatives in (7.5) and (7.6) because we want to consider a situation where displacement u also varies with time (we're interested in *waves* of displacement—waves of elastic stress and strain).

Consider an infinitesimal volume element of the rod lying between x and $x + dx$. The *net* force on that segment is

$$[\sigma(x + dx) - \sigma(x)]A = \frac{\partial \sigma}{\partial x} Adx \tag{7.7}$$

Using (7.6), the net force becomes

$$\frac{\partial \sigma}{\partial x} Adx = \frac{\partial^2 u}{\partial x^2} EAdx \tag{7.8}$$

The velocity of that volume element is $\partial u/\partial t$, and its *acceleration* is $\partial^2 u/\partial t^2$. If the mass density of our solid is ρ, the *mass* of the volume element is ρAdx, and *from* $F = ma$,

$$\frac{\partial^2 u}{\partial x^2} EAdx = \frac{\partial^2 u}{\partial t^2} \rho Adx \tag{7.9}$$

or

$$\frac{\partial^2 u}{\partial x^2} = \frac{\rho}{E}\frac{\partial^2 u}{\partial t^2} \tag{7.10}$$

This is familiar as a one-dimensional form of the *wave equation*, which we encountered in full three-dimensional form in equation (2.13). (The electromagnetic wave equation resulted from Maxwell's equations, the mechanical wave equation from Hooke's law and $F = ma$.) Solutions to (7.10) are *elastic waves* or *sonic waves* of displacement, and we can write a traveling wave in the form we used for electromagnetic waves in Chapter 2:

$$u(x, t) = u_0 e^{-i(\omega t - kx)} \tag{7.11}$$

(We should perhaps warn you that many texts use the symbol q for the wave number of elastic waves, and reserve k for the wave numbers of electromagnetic waves and electron waves, which we will meet in Chapter 8.) From (7.5), these waves of elastic displacement can also be viewed as waves of elastic strain:

$$\varepsilon(x, t) = \frac{\partial u(x, t)}{\partial x} = iku_0 e^{-i(\omega t - kx)} \tag{7.12}$$

where positive values of strain correspond to local tension, and negative values to local compression. (And, as usual, the i simply means that the strain is 90° out of phase with the displacement.)

The plane wave of displacements described by (7.11) will satisfy the wave equation (7.10) and travel down the rod with velocity

$$v_s = \frac{\omega}{k} = \left(\frac{E}{\rho}\right)^{1/2} \tag{7.13}$$

This is the velocity of mechanical vibrations, that is, the *velocity of sound*, in our solid. Of course, the frequency of the "sound" wave can be above or below those of audible sound (ultrasonic or infrasonic) just as the frequency of electromagnetic "light" waves can be above or below those of visible light. The physics of light and sound waves doesn't depend on the limited abilities of our eyes and ears.

Sample Problem 7.2

If copper has a Young's modulus of 100 GPa and a density of 8.9 g/cm^3, what is the velocity of sound in copper?

Solution

$v_s = (1 \times 10^{11}/8.9 \times 10^3)^{1/2} = 3.35 \times 10^3$ m/s (about an order of magnitude faster than in air)

The displacement u that we considered was in the direction of propagation x. This type of wave, where the vibrating quantity vibrates in the direction of propagation, is called a *longitudinal wave*. In the electromagnetic waves that we first considered in Chapter 2, the vibrating quantities were electric and magnetic fields that were perpendicular to the direction of motion; electromagnetic waves are *trans-*

verse waves. Sound waves in solids can also be transverse, with displacements perpendicular to the direction of propagation. For transverse elastic waves, Young's elastic modulus E is replaced by the shear elastic modulus G. In single crystals and other anisotropic solids, the various elastic moduli vary with direction, but we'll avoid that complication and assume our solid is isotropic.

7.3 VIBRATING ATOMS AND DISPERSION

Our treatment of sound waves in Section 7.2 assumed a continuous solid, and equation (7.13) gives no hint that the velocity of sound might vary with frequency or wavelength, that is, be dispersive. The continuous-solid approximation is valid only for sound waves with wavelengths long compared to interatomic spacings. For shorter wavelengths of our elastic waves, we can no longer ignore the fact that our solid actually consists of atoms coupled together by interatomic forces. We'll find that explicit consideration of atomic structure will lead to dispersion.

For simplicity, let's consider a simple one-dimensional lattice of atoms of mass m and an equilibrium interatomic spacing of a (Fig. 7.1). We assume that each atom is coupled to its two nearest neighbors by an elastic spring with spring constant K (not to be confused with our wave number k!). In a real solid, each atom will of course also be coupled to other atoms, but this simple model will be adequate to demonstrate dispersion. We again focus on the displacement u of the material, in this case of each atom, from its unstressed position. If the displacement of the nth atom is u_n, the force exerted on it by the $(n - 1)$th atom is thus $-K(u_n - u_{n-1})$. The nth atom will also experience a similar force from its bond to the $(n + 1)$th atom. Its motion will be determined by the *net* force and, as before, $F = ma$.

$$m \frac{d^2u_n}{dt^2} = -K(u_n - u_{n-1}) + K(u_{n+1} - u_n) = -K(2u_n - u_{n-1} - u_{n+1}) \quad \textbf{(7.14)}$$

At first glance, equation (7.14) looks different from the standard form of the wave equation that we saw in equation (7.10). However, that's only because our variable u_n is a discrete (rather than continuous) variable. Remembering that, the right side of (7.14) can be seen to be equivalent to a second derivative with respect to distance. We try an elastic wave solution of the form

$$u_n = u_0 e^{-i(\omega t - kna)} \quad \textbf{(7.15)}$$

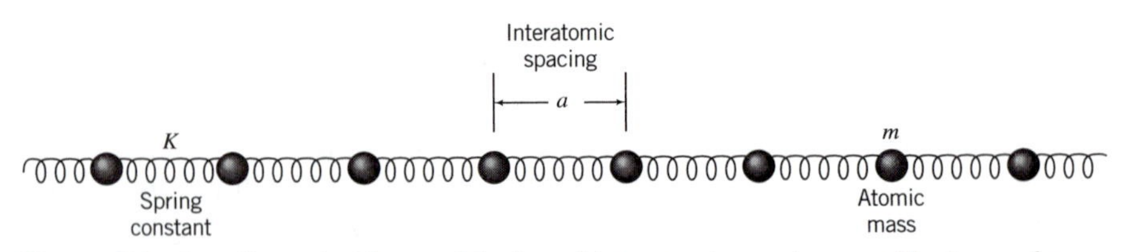

Figure 7.1. One-dimensional crystal lattice of interatomic spacing a, with atoms of mass m coupled to neighboring atoms by springs with spring constant K.

Note that na is the equilibrium position of the nth atom, from which the displacement u_n is defined, and thus it is the equivalent of x in equation (7.11). Substituting (7.15) into (7.14) and canceling the common factor $u_0 e^{-i\omega t} e^{ikna}$, we get

$$-m\omega^2 = -K[2 - e^{-ika} - e^{ika}] \tag{7.16}$$

From Euler's formula, $e^{-ika} + e^{ika} = 2\cos(ka)$. Since $1 - \cos(ka)$ is equivalent to $2\sin^2(ka/2)$, we find that (7.15) is a solution of (7.14) if

$$\omega = \left(\frac{4K}{m}\right)^{1/2} |\sin(ka/2)| \tag{7.17}$$

We consider only positive values of the sine (Fig. 7.2), since frequency is inherently positive. If the wave number k is so small that $ka \ll 1$ ($\lambda \gg 2\pi a$), $\sin(ka/2) \approx ka/2$ and

$$\omega = ka\left(\frac{K}{m}\right)^{1/2} \tag{7.18}$$

so that the velocity ω/k of this wave is constant at $a(K/m)^{1/2}$. This is equivalent to equation (7.13) for the sound velocity in a continuous solid if we consider m/a^3 as the equivalent of density ρ, and K/a as equivalent to a Young's modulus E for our one-dimensional lattice. (Since the units for the spring constant K are N/m, the units for K/a are N/m^2, or Pa.)

Sample Problem 7.3

At what wavelength will the velocity of sound given by (7.17) be 3% less than the long-wavelength limit given by (7.18)?

Solution

For small k (long wavelength), we can use the first two terms of the series expansion for sine: $\sin(ka/2) \approx (ka/2) - (ka/2)^3/6$.

The second term becomes 3% of the first when $ka = 0.85$, or when $\lambda = 7.4a$.

If we consider instead the full form of (7.17), we see (Fig. 7.2) that ω/k is no longer a constant. As discussed in Chapter 4 for electromagnetic waves, when the velocity of a wave varies with wave number or frequency, we call that *dispersion.* The dispersion of elastic waves in this one-dimensional crystal (the departure from a linear ω-k relation) becomes extreme as the wave number k approaches π/a. This is equivalent to the wavelength $\lambda = 2\pi/k$ approaching $2a$, that is, the Bragg diffraction condition, equation (4.13). At this wavelength, the frequency reaches a max-

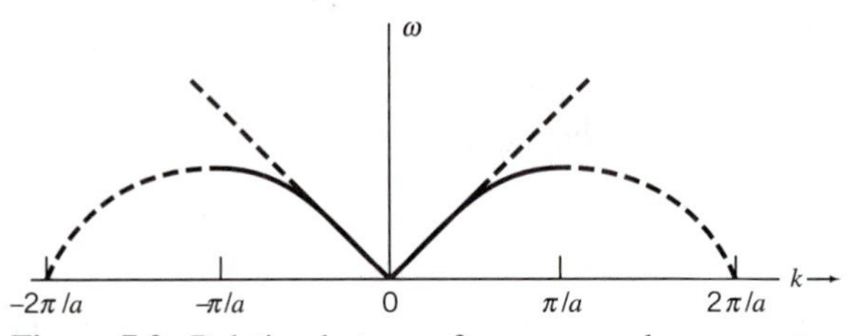

Figure 7.2. Relation between frequency and wave vector in one-dimensional crystal, according to equation (7.17). The nonlinear relation indicates *dispersion*, i.e., variation of wave velocity with frequency.

imum value ω_m of $(4K/m)^{1/2}$. (Using representative values for the spring constant K and the atomic mass m, ω_m is of the order of 10^{13} s^{-1}, a frequency in the infrared.)

7.4 GROUP VELOCITY AND BRILLOUIN ZONES

Whenever dispersion exists (whenever wave velocity varies with frequency and wavelength), we distinguish between two different velocities, the *phase velocity* v_p and the *group velocity* v_g, where

$$v_p = \frac{\omega}{k} \qquad \text{and} \qquad v_g = \frac{\partial \omega}{\partial k} \tag{7.19}$$

If we have a pure wave with a single frequency, it travels with the phase velocity v_p. However, if we have a wave *pulse*, the pulse (which can be Fourier-analyzed into a sum of many pure waves of different frequencies) travels at the group velocity v_g, with ω and k its average frequency and average wave number. Since in practice energy and momentum are transmitted with pulses rather than with pure waves, *it is the group velocity that is physically more significant.*

One way to grasp the physical significance of group velocity is to consider the superposition of two waves that differ only slightly in frequency and wave number

$$u_0 e^{-i(\omega t - kx)} + u_0 e^{-i(\omega' t - k' x)} \tag{7.20}$$

With the assumption that $\omega \approx \omega'$ and $k \approx k'$, a bit of mathematical manipulation will transform this into

$$2u_0 e^{-i(\omega t - kx)} \cos\left(\frac{\Delta\omega}{2} t - \frac{\Delta k}{2} x\right) \tag{7.21}$$

where $\Delta\omega = \omega' - \omega$ and $\Delta k = k' - k$. This is a wave much like one of the original waves with its amplitude "modulated" in space and time by the cosine term (Fig. 7.3). Considering the variation with time alone, this is the phenomenon known as

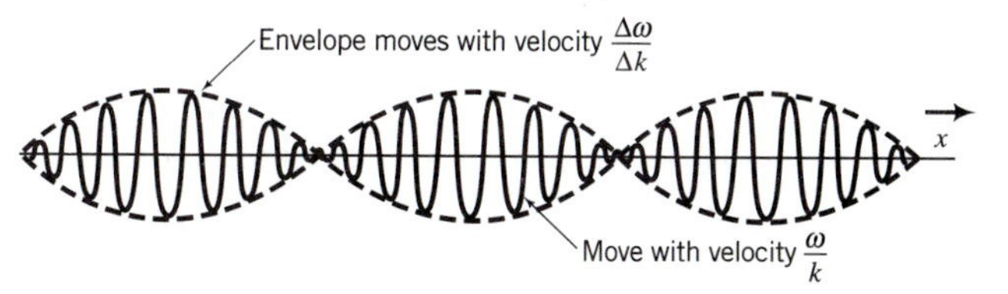

Figure 7.3. The superposition of two waves of nearly the same frequency and wave number, producing *beat* and an *envelope* (the dashed line) that moves with the group velocity $\Delta\omega/\Delta k$.

beat, such as you hear when you strike two of the low notes on a piano simultaneously. With a frequency of $\Delta\omega/2$, the two waves alternately add and subtract, that is, they interfere constructively and destructively. The "envelope" (the dashed lines in Fig. 7.3) provided by the cosine term can be seen to travel with the velocity $\Delta\omega/\Delta k$, the group velocity.

For waves in a nondispersive medium (e.g., electromagnetic waves in vacuum), ω and k remain proportional, and phase velocity and group velocity are identical. In our one-dimensional crystal, this remains true for elastic waves as long as wavelengths are very long compared to a, where (7.18) is a good approximation. However, as can be seen from Fig. 7.2, as k approaches π/a, the group velocity decreases. It is in this region that the discrete nature of the lattice interferes with the elastic waves, producing scattered waves that slow the propagating wave.

When $k = \pi/a$, or $\lambda = 2a$, the group velocity equals zero. Here the Bragg diffraction condition (4.13) is satisfied, and reflected waves are so strong that they become equal to the incident wave, and the two combined together become a *standing wave*. (As shown in Chapter 2, the addition of two equal waves traveling in opposite directions yields a standing wave.) A standing wave, of course, has zero group velocity.

What about waves with $k > \pi/a$? Consider specifically $k = 2\pi/a$, or $\lambda = a$. This corresponds to a wave in which all atomic displacements u_n are identical, and is therefore equivalent to a wave with infinite wavelength ($k = 0$). In fact, it can be seen from (7.15) that $u_n(k) = u_n(k - 2\pi/a)$, so that each wave in the range $\pi/a \leq k \leq 2\pi/a$ is equivalent to a wave in the range $-\pi/a \leq k \leq 0$. Similarly, other ranges of k can be similarly shifted, so that we can therefore *restrict our range of interest to* $-\pi/a \leq k \leq \pi/a$ *(Fig. 7.2)*. (Waves with positive k travel in the positive x-direction, and waves with negative k travel in the negative x-direction.)

Although we have considered a one-dimensional crystal for simplicity, similar considerations for a three-dimensional solid will lead to dispersion and diffraction of elastic waves in various directions, and a *limited region of three-dimensional "k-space" to which we can consider elastic waves confined.* We will later (Fig. 13.4b) find similar effects for electron waves traveling through crystals. These regions of k-space bounded by Bragg diffraction conditions are called *Brillouin zones*, named after Louis M. Brillouin, a French physicist who considered the interaction of waves with periodic structures.

Sample Problem 7.4

In the one-dimensional crystal where the velocity of elastic waves is defined by (7.17), what are the phase and group velocities of elastic waves at $k = 0$, $k = \pi/2a$, and $k = \pi/a$?

Solution

At $k = 0$, v_p and v_g are both equal to $a(K/m)^{1/2}$.
At $k = \pi/2a$, $v_p = (\omega/k) = 0.9a(K/m)^{1/2}$ and $v_g = \partial\omega/\partial k) = 0.71a(K/m)^{1/2}$
At $k = \pi/a$, $v_p = (\omega/k) = 0.64a(K/m)^{1/2}$ and $v_g = (\partial\omega/\partial k) = 0$

7.5 LATTICE HEAT CAPACITY AND PHONONS

Before leaving the topic of lattice vibrations, we can make a classical estimate of the *lattice* contribution to *heat capacity*. We describe the solid as a collection of vibrating atoms, each of which is bound to its lattice site by a restoring force, a classical harmonic oscillator. From Boltzmann statistics, the average energy of a one-dimensional oscillator is k_BT ($k_BT/2$ potential energy and $k_BT/2$ kinetic energy), and that of a three-dimensional oscillator is $3k_BT$. Since a mole contains N_A atoms, where N_A is Avogadro's number, the energy per mole associated with atomic vibrations is

$$E = 3N_Ak_BT = 3RT \qquad \textbf{(7.22)}$$

where R is the gas constant. Thus the lattice contribution to heat capacity per mole is

$$C_L = dE/dT = 3R \qquad \textbf{(7.23)}$$

This is in agreement with experiment at high temperatures but fails completely at low temperatures (Fig. 7.4), where heat capacity decreases rapidly as temperature decreases and approaches zero as temperature approaches zero. This discrepancy between classical theory and experiment was resolved in 1905, when Einstein applied the new principles of quantum mechanics to the problem of lattice vibrations. Quantized elastic lattice waves (sound waves) are called *phonons* (not to be confused with *photons*, which we meet in Chapter 8).

In this text, we will keep phonons in the background and devote most of our attention to electrons and photons. From the viewpoint of quantum mechanics, however, properties of solids that involve energy interchange between the lattice and electrons (such as electrical resistance, superconductivity, and some light-emitting and light-absorbing electron transitions) are properly described in terms of electron-phonon interactions.

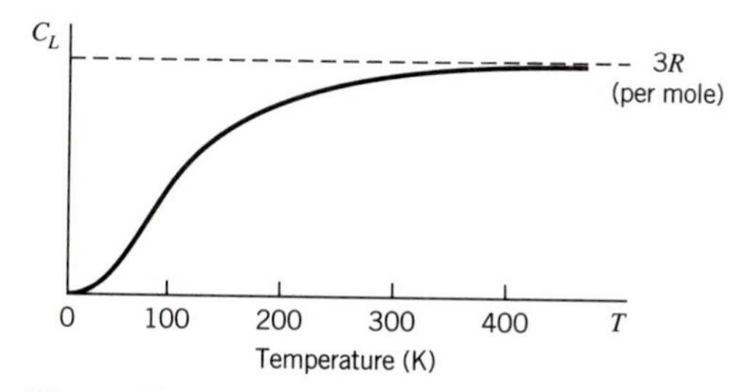

Figure 7.4. Temperature dependence of lattice specific heat. The fall-off at low temperatures results from the quantization of lattice vibrations (phonons).

SUMMARY

Hooke's law of elasticity and Newton's second law applied to longitudinal vibrations in a continuous solid lead to a wave equation, the solutions to which are elastic waves traveling with the sound velocity $(E/\rho)^{1/2}$, where E is Young's modulus and ρ is the mass density. Transverse elastic waves are also possible, with the shear modulus G replacing Young's modulus.

A one-dimensional model that specifically considers the atomic lattice leads to elastic wave solutions that are essentially equivalent to the continuous model at long wavelengths. However, at wavelengths that approach the interatomic spacing, elastic waves become highly dispersive. Frequency and wave number are no longer proportional, and phase velocity and group velocity differ substantially. The group velocity, $v_g = \partial\omega/\partial k$, which is physically more significant than the phase velocity $v_p = \omega/k$, approaches zero as the wave number k approaches π/a, the Bragg diffraction condition. Bragg diffraction conditions divide k-space into regions called *Brillouin zones.*

A classical model of the thermal energy associated with lattice vibrations leads to a lattice contribution to the heat capacity of solids of $3R$, where R is the gas constant. Although this agrees with experiment at high temperatures, it is far in excess of experimental values at low temperatures. This was another discrepancy of classical physics, one that quantum mechanics was later to resolve by the introduction of quantized lattice vibrations (phonons).

PROBLEMS

7-1. The Young's modulus of diamond is approximately 800 GPa (8×10^{11} Pa). If you are lucky enough to own a diamond cube 1 centimeter on a side, by how much will it compress if you apply a stress of 10 MPa? How many pounds of force must you apply to produce such a stress? How much elastic energy will be stored in the diamond?

7-2. Diamond has a Young's modulus of about 800 GPa and a density of 2.3 g/cm^3. Lead has a Young's modulus of about 15 GPa and a density of 11.4 g/cm^3. What is the velocity of sound in these two rather different materials? How do these velocities compare with the velocity of sound in air? What is the velocity of sound in vacuum?

7-3. The upper limit of human hearing is about 20 kHz. (a) Using equation (7.13) and values of modulus and density given in Problem 7-2, what is the wavelength of a sound wave with this frequency traveling through diamond? What if it was traveling through lead? (b) Equation (7.13) was derived using a continuum model that ignored the existence of atomic structure. Is this equation valid in these materials for this frequency? Would it be valid for supersonic waves of 1-MHz frequency? For 1 GHz? 1 THz (10^{12} Hz)?

7-4. Problem 1-7 dealt with a laminar composite that had an anisotropic electric conductivity. Similarly, a laminar composite with two phases having different elastic properties will have anisotropic elasticity. Suppose the two phases are tungsten carbide (E = 600 GPa) and aluminum (E = 69 GPa). If the two phases are of equal thickness, derive the Young's modulus of this composite parallel to and perpendicular to the lamellae. For each case, state clearly which quantities are the same in each phase and which are different. (In the electrical case, electric fields were the same in each phase when directed parallel to the lamellae, and current densities were the same in each phase when electric fields were directed perpendicular to the lamellae.)

7-5. At what value of ka will the phase velocity of elastic waves in the one-dimensional crystal considered in Section 7.2 differ by 5% from the velocity of elastic waves in the long-wavelength limit? (Hint: Use series expansion for sine.) What is the corresponding ratio of wavelength to interatomic spacing?

7-6. Over a wide range of frequencies, the dielectric constant of a particular polymer is found to be proportional to the inverse square root of frequency. (a) How does the phase velocity of electromagnetic waves vary with wavelength in this polymer? (b) What is the ratio between phase and group velocities of electromagnetic waves in this material?

7-7. Water waves are very dispersive, with the detailed relation between velocity and wavelength depending both on wavelength and on water depth. However, for long waves in deep water, frequency is proportional to the square root of wave number. (a) How do phase and group velocities vary with wavelength? (b) What is the ratio between phase and group velocities?

7-8. (a) From equation (7.17), derive a general formula for the ratio of phase velocity to group velocity for elastic waves in this model one-dimensional crystal. (b) At what ratio of wavelength to spacing (λ/a) will the group velocity be 5% less than the phase velocity? (Hint: See the Problem 7-5 hint.)

7-9. Figure 3.3 shows the frequency dependence of the dielectric constant of insulators based on the Debye relaxation model. (a) In the frequency range where dispersion occurs, does phase velocity increase or decrease with decreasing wavelength? (b) In this frequency range, will group velocity be greater or less than phase velocity?

7-10. The dielectric constant of water at frequencies of visible light results primarily from electronic polarization, and the relevant resonance frequencies are far enough into the ultraviolet that the maximum dielectric constant (corresponding to the maximum of displacement shown in Fig. 3.4b) occurs in the UV. (a) From that information, does the velocity of visible light in water increase or decrease with increasing wavelength? (b) Will the blue light in sunlight bend more or less than the red light when entering or leaving raindrops? (c) Is the group velocity of visible light in water greater or less than the phase velocity?

7-11. The velocity of sound in air at normal temperature and pressure is about 330 m/s, six orders of magnitude slower than the velocity of light. If you see a flash of lightning 5 seconds before you hear the thunder, the lightning struck about a mile away. Suppose the light and sound were traveling a mile through diamond instead of air. What would the time separation be then? Use diamond properties given in Problem 7-2.

7-12. You decide, in a wild moment, to jump off a tower 100 m high with a cord attached to your body that is 1 cm in diameter and 50 m long. You don't want to hit the ground, but you also don't want to experience any accelerations or decelerations greater than five times the acceleration of gravity. Assuming the cord remains elastic and follows Hooke's law, calculate the acceptable range of Young's modulus for the cord. Ignore the weight of the cord and the effect of changes of its cross-sectional area, and assume that equations (7.1) and (7.2) are valid to large strains. If you don't want to reveal your own mass, assume it to be 60 kg. (Hint: You may not want a Young's modulus as high as that of copper.)

Halftime Review

The Story so Far

We've considered the properties of materials so far with prequantum ideas—electrons as tiny particles with mass and charge, subject to Newtonian mechanics, and light as electromagnetic waves with oscillating electric and magnetic fields, subject to Maxwellian electromagnetics. We treated conductors simply as materials containing free electrons, and had some success in explaining electrical conductivity (as a product of charge carrier density and mobility) and the Hall effect (which allowed us to separate the two). The response of those free electrons to the AC electric fields of electromagnetic waves enabled us to explain skin depth (and the opacity of conductors to most electromagnetic radiation), and plasma frequency (the transparency of conductors to electromagnetic radiation of sufficiently high frequency).

Turning to nonconductors or dielectrics, we focused on the polarization produced by the displacement of bound electrons, the displacement of ions, and the rotation of permanent electric dipoles. This allowed us to develop at least a qualitative understanding of the dielectric constant and the related refraction of light (and the use of dielectrics in capacitors, lenses, and optical fibers), the frequency dependence of the dielectric constant (dispersion), and dielectric loss (the imaginary part of the dielectric constant).

We tackled the phenomenology of ferromagnetic materials in terms of the magnetization produced by unbalanced electron spins (coupled with those of neighboring atoms by exchange), and the phenomenology of superconducting materials in terms of the magnetization produced by diamagnetic supercurrents in Type I and Type II superconductors. We treated the technologically important structure-sensitive property of ferromagnets (coercivity) in terms of the interaction of magnetic domain walls with microstructure, and the technologically important structure-sensitive property of Type II superconductors (critical currents in the mixed state) in terms of the interaction of fluxoids with microstructure.

We explained lattice vibrations and the transmission of sound in solids in terms of elastic stress-strain relations, but noted dispersion of elastic waves when their wavelengths approach interatomic distances, an effect akin to Bragg diffraction of x-rays.

What we've done so far is not wrong, but it's not enough. There's a lot about the properties of materials that we have not yet been able to explain. Some of these problems we've noted explicitly, some we've not yet considered. The unanswered questions include:

- Why do some divalent metals and many semiconductors have Hall coefficients of the wrong sign? How can electrons carrying current sometimes act as if they have positive charge?
- Why does the electrical conductivity of nonmetals usually increase with increasing temperature and increase with the addition of impurities?
- Why is the contribution of free electrons to the heat capacity of metals much smaller than predicted by classical theory?
- Why is the lattice contribution to heat capacity at low temperatures much smaller than predicted by classical theory? (Actually, we jumped the gun here and already told you that the answer was related to quantized lattice vibrations, or phonons.)
- Why do measurements of cyclotron resonance suggest that free charge carriers in solids often have a different "effective mass" than that of totally free electrons?
- To get a bit more basic, why are some materials metals, some semiconductors, and some insulators? What controls their charge carrier density?
- Indium tin oxide goes from opaque to transparent when the light frequency exceeds the plasma frequency. But, as we'll learn later, it becomes opaque again in the ultraviolet. Window glass is transparent to visible light, but it also becomes opaque in the ultraviolet. Silicon is transparent in the infrared but opaque

in the visible. Why do insulators and semiconductors become opaque to electromagnetic waves of sufficiently high frequency?

- How do we explain the color of materials? Why is copper red? Why is grass green? Why is pure aluminum oxide colorless but aluminum oxide doped with chromium (ruby) red? Why does nitrogen turn diamonds yellow, while boron turns diamonds blue?
- Atoms and molecules are a lot simpler than solids, but classical physics can't explain their optical properties either (e.g., why are neon lights red?).
- In fact, classical physics can't even explain why atoms exist—how can an electron in the Coulomb field of a positively charged nucleus maintain a stable distance from the nucleus?

There are many other observations, a few of which we'll discuss in Chapter 8, that required the development of a more sophisticated view of the nature of electrons and light, the nature of matter and energy: quantum mechanics. In this modern view, light not only has a wavelike nature, it also has a particle-like nature. And electrons not only have a particle-like nature, they also have a wavelike nature. An elementary understanding of quantum mechanics will allow us to tackle most of the above questions, and to develop a more sound and more complete understanding of the properties of materials. In particular, it will be essential to our developing a basic understanding of *n*-type and *p*-type semiconductors and *p-n* junctions, and their role in electronic and optoelectronic devices, such as solar cells and LEDs (light-emitting diodes).

We'll find that much of the essence of quantum mechanics is embodied in a single constant: Planck's constant $h = 6.626 \times 10^{-34}$ Joule-seconds. Because h is pretty small in the units of energy and time that we deal with in our ordinary lives, we can usually deal with the mechanics of airplanes, baseballs, and such with classical physics. But because h is not zero, quantum mechanics plays a huge role in the properties of atoms and electrons.

For example, in Chapters 5 and 6, we were able to discuss the phenomenology of ferromagnetism and superconductivity with little reference to quantum mechanics. But the presence of quantum mechanics at the heart of magnetism and superconductivity was clear from the presence of Planck's constant in the formula for the Bohr magneton, equation (5.4), and the formula for the superconducting flux quantum, equation (6.5). And we'll find it appearing over and over again in formulas in the forthcoming chapters as we discuss various properties of atoms, molecules, and solids.

Although Planck's constant is tiny, it plays a very BIG role in the properties of engineering solids. For an amusing description of what life would be like if Planck's constant were much larger than it is in our universe, see George Gamow's enjoyable book, *Mr. Tompkins in Wonderland* (Cambridge University Press, 1993). (He also describes the effects of relativity in a world where the speed of light is less than a hundred miles per hour, instead of, as in our universe, 186,000 miles per second.)

You probably have already been exposed to some concepts of quantum mechanics elsewhere, perhaps in chemistry. And you will find a lot of the mathematics familiar.

But much of quantum mechanics is not consistent with the "common sense" we have developed in dealing with macroscopic objects. I often introduce quantum mechanics by referring to remarks made by the White Queen (a chess piece) to Alice in Lewis Carroll's *Through the Looking Glass*. When Alice says, "One *can't* believe impossible things," the White Queen replies: "I daresay you haven't had much practice. When I was your age, I always did it for half an hour a day. Why, sometimes I've believed as many as six impossible things before breakfast!"

If the White Queen could do it, you can do it! If you spend half an hour a day on the subject, whether it's before breakfast, during lunch, or after dinner, you'll soon be able to believe the strange results of quantum mechanics, and thereby understand more about the properties of engineering materials. Quantum mechanics is not really impossible; it just *seems* that way at first!

PART TWO

Quantum Mechanical Approach

Chapter 8

Light Particles, Electron Waves, Quantum Wells and Springs

8.1 THANK PLANCK

In 1900, one of the "hot problems" in physics was the frequency distribution of the electromagnetic radiation emitted from hot bodies. The problem was clearest when physicists considered an ideal radiator, one in which the radiation depended only on the temperature of the body and not on the specific material, the state of its surface, and so on. (These were pure physicists, not materials scientists.) The experimental approach was to measure radiation from a *cavity* inside a hot material by, for example, drilling a small hole in the wall of a hollow tungsten cylinder and measuring the radiation, called cavity or *blackbody* radiation, emitted through the hole when the tungsten was heated to high temperatures.

From the theorist's point of view, the problem was that of electromagnetic waves enclosed in a box, in thermal equilibrium with the walls of the box, which were at uniform temperature. (They assumed the hole was so small that it wouldn't disturb things much in the cavity.) As theorists do, they simplified the math by assuming the cavity was a cube of side a. They assumed the electromagnetic waves were standing waves with nodes (places where electric and magnetic fields were zero) at the surfaces of the cavity, so that in each of the three directions x, y, and z there was an integral number of the x, y, and z components of half-wavelengths across the cavity:

$$a = n_x(\lambda_x/2) = n_y(\lambda_y/2) = n_z(\lambda_z/2) \quad \textbf{(8.1)}$$

thus

$$\lambda^{-1} = (\lambda_x^{-2} + \lambda_y^{-2} + \lambda_z^{-2})^{1/2} = (n_x^2 + n_y^2 + n_z^2)^{1/2}/2a \quad \textbf{(8.2)}$$

and

$$\nu = \frac{c}{\lambda} = \frac{c}{2a}(n_x^2 + n_y^2 + n_z^2)^{1/2} = \frac{cn}{2a} \quad \textbf{(8.3)}$$

Each possible standing electromagnetic wave inside the cavity was called a *mode*, and each mode was represented by three integers: n_x, n_y, and n_z. This could be represented geometrically in *n-space* (Fig. 8.1), a space constructed with the three axes n_x, n_y, and n_z, in which each mode corresponds to a point in a cubic lattice of points as shown. As seen in the figure, we've considered only the space corresponding to positive values of the three integers.

In this n-space, a distance from the origin is given by $n = (n_x^2 + n_y^2 + n_z^2)^{1/2}$, and corresponds therefore to a single frequency given by (8.3). It is clear from Fig. 8.1 that the higher the frequency (the shorter the wavelength), the more modes the cavity can support. In other words, shorter wavelengths and higher frequencies correspond to a greater number of combinations of n_x, n_y, and n_z (a greater volume in n-space between ν and $\nu + d\nu$).

So far, so good. The problem for the physicists came in the next step. In thermal equilibrium inside the cavity, how would the total energy be distributed among the various possible modes of electromagnetic waves? Classically, each mode—each point in the cubic lattice in Fig. 8.1—should have an average energy of k_BT. So the amount of thermal energy associated with modes of frequencies between ν and $\nu + d\nu$ would be proportional to the volume in n-space corresponding to that frequency range. That volume is a spherical shell in n-space and therefore should increase as ν^2, a nice simple theoretical prediction. However, it was in serious disagreement with experiment (Fig. 8.2).

Although the energy distribution in the cavity (measured by the energy distri-

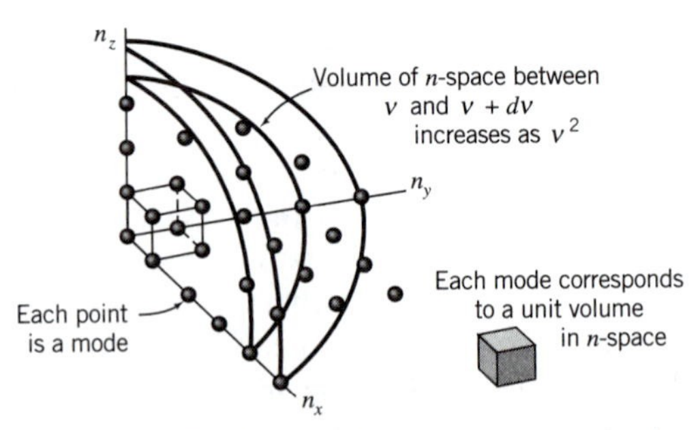

Figure 8.1. Each possible standing wave in the cavity is represented by a point in n-space. The number of available modes increases with frequency.

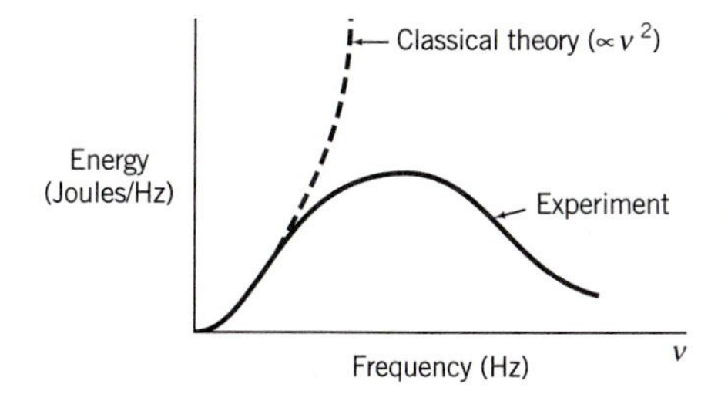

Figure 8.2. Frequency distribution of energy of cavity radiation. A huge discrepancy between classical theory and experimental results led Planck to hypothesize that radiation was quantized in energy units of $E = h\nu$.

bution of the radiation emitted from the little hole) varies quadratically at very low frequencies, it rapidly departs from the theoretical prediction, going through a maximum and then decreasing at high frequencies. Something is *very* wrong with the classical assumption about the distribution of energy among the various possible modes of electromagnetic radiation in the cavity. The problem is particularly acute at high frequencies, and was dramatically called the "ultraviolet catastrophe." For some reason, there is actually a lot less energy in the high-frequency, short-wavelength radiation than is predicted by classical theory.

Planck found that he could fit the experimental curve if he used a radical assumption: that the energy of the radiation in the cavity is quantized, specifically quantized into entities that depend linearly on frequency:

$$E = h\nu \tag{8.4}$$

The constant $h = 6.626 \times 10^{-34}$ J-s, which we will encounter over and over again in this and future chapters, is known as *Planck's constant.* Classically, at each frequency there would be a Boltzmann distribution over all possible values of energy, with an average energy per mode of k_BT. According to Planck, however, at each frequency the only possible energies were $h\nu$, $2h\nu$, $3h\nu$, $4h\nu$, and so on. This doesn't change things much at low frequencies, but it greatly decreases the average energy per mode at high frequencies, which becomes $h\nu/(e^{h\nu/k_BT} - 1)$. This average energy equals k_BT as frequency approaches zero, but decreases exponentially at high frequencies. This average energy per mode at each frequency, multiplied by the number of possible modes, which we recall varies as ν^2, gives an energy distribution in agreement with experiment. Planck himself remained doubtful of the proper interpretation of equation (8.4) for many years, but it was the first important step into quantum mechanics.

Sample Problem 8.1

For electromagnetic radiation in a cavity, compare the classical and quantum-mechanical average energy per mode for the following frequencies: $0.01\ k_BT/h$, k_BT/h, $4\ k_BT/h$, and $10\ k_BT/h$.

Solution

Classically, each of these modes is assumed to have an average energy of k_BT.

The average energy per mode according to Planck is given by $h\nu/(e^{h\nu/k_BT} - 1)$, which for the frequencies given equals k_BT, 0.58 k_BT, 0.075 k_BT, and 4.5×10^{-4} k_BT.

8.2 PHOTOELECTRONS AND PHOTONS

If you shine light on a metal surface under the right conditions, you can knock electrons out of the metal. We call such electrons *photoelectrons*. The phenomenon of photoelectricity has many applications, including automatic door openers, and, as we'll see later, it can be used to learn quite a bit about the energy distribution of electrons in atoms, molecules, and solids. But that's later. Right now our interest in the photoelectric effect derives from its historical importance in establishing the validity of Planck's equation (8.4), an achievement that gained Einstein the Nobel Prize.

Figure 8.3 shows how the maximum energy of photoelectrons ejected from a particular metal depends on the frequency of the light incident on the surface. Below a critical frequency ν_c, no photoelectrons are emitted, *no matter how intense the light.* Above that frequency, the maximum energy of the photoelectrons depends on the frequency of the light, *but not on the light intensity.* Neither of these observations can be explained by the classical wave theory of light, but both suggest that light can deliver energy to the electrons in the metal only in discrete amounts proportional to frequency. And the slope of the line in Fig. 8.3 turned out to be equal to the constant h that Planck had introduced to explain cavity radiation; that is, the maximum energy $E_{\max}$ of the photoelectrons was given by

$$E_{\max} = h(\nu - \nu_c) \tag{8.5}$$

The results were consistent with the strange and new idea that light consisted of particles of energy defined by equation (8.4), particles that came to be called *photons.* These particles had no mass and traveled, of course, at the speed of light. This didn't mean that light wasn't also an electromagnetic wave, since the wave nature of light was clearly demonstrated by diffraction experiments. Maxwellian theory wasn't wrong, it was just incomplete. Light had both wave-like *and* particle-like properties; it had *wave-particle duality.*

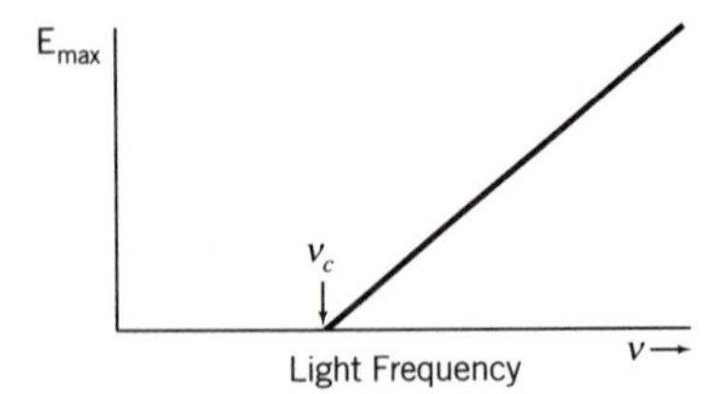

Figure 8.3. Maximum energy of photoelectrons as a function of light frequency.

Further experiments indicated that these massless entities also had *momentum*, requiring a broadening of that concept beyond the classical $p = mv$ momentum of particles with mass. Scattering of x-rays of a single frequency by electrons in a graphite target resulted in scattered x-rays with two different frequencies. The detailed results of this "Compton scattering" experiment could be interpreted in terms of a photon-electron collision in which both energy and momentum were conserved, if it was assumed that each photon had a momentum given by

$$p = h/\lambda \tag{8.6}$$

Equations (8.4) and (8.6) define the wave-particle duality of light, with the energy and momentum of the *particles* of light energy, the photons, dependent on frequency and wavelength, quantities familiar from the *wave* theory of light. It's often more convenient to express these relations in terms of $\hbar = h/2\pi = 1.05 \times 10^{-34}$ J-s (called "h-bar") and angular frequency ω and wave number k:

$$E = \hbar\omega \qquad p = \hbar k \tag{8.7}$$

Sample Problem 8.2

2 eV is the maximum kinetic energy of photoelectrons emitted from a metal when exposed to a beam of ultraviolet photons with a wavelength of 200 nm. What is the longest wavelength photon that can eject photoelectrons from this metal? What is the momentum of this photon?

Solution

Incident photon energy is $h\nu = hc/\lambda = 9.93 \times 10^{-19}$ J $= 6.21$ eV
From (8.5), $h\nu_c = h\nu - E_{\text{max}} = 4.21$ eV, which corresponds to $\lambda_c = 295$ nm
$p = h/\lambda_c = 2.25 \times 10^{-27}$ kg-m/s.

8.3 ELECTRON WAVES

French physicist Louis de Broglie then suggested that if light can have a particle-like nature, perhaps particles of matter, like electrons, could have a wave-like nature, with wavelength defined in terms of the ordinary momentum $p = mv$ of the electron and equation (8.6). The best way to demonstrate the wave-like nature of anything is by diffraction. And Davisson and Germer verified de Broglie's suggestion by directing a beam of electrons toward a single crystal of nickel and observing *electron diffraction*.

Like x-ray diffraction, electron diffraction in crystals follows the Bragg relation (4.13) and requires wavelengths of the order of interatomic spacings. Electrons accelerated through a voltage difference V to a kinetic energy of $E = Ve$ have a momentum of $(2mE)^{1/2}$ and therefore a wavelength of

$$\lambda_e = \frac{h}{p} = \frac{h}{(2mE)^{1/2}} = \frac{h}{(2mVe)^{1/2}} \tag{8.8}$$

An electron accelerated through a voltage of one volt reaches a kinetic energy of 1.6×10^{-19} Joules, or one *electron-volt*. As seen in Fig. 8.4, electrons with only a few electron-volts (eV) of energy have wavelengths of the order of interatomic spacings, allowing electron diffraction to be easily demonstrated.

Figure 8.4 also shows, on the same scale, the relation between wavelength and energy for photons. Whereas equation (8.8) shows that the wavelength of electrons varies as the inverse *square root* of energy, the wavelength of photons, from (8.4), varies instead inversely with energy:

$$\lambda_{ph} = \frac{c}{\nu} = \frac{hc}{E} \tag{8.9}$$

As is clear from Fig. 8.4, if the electron energy is well below 1 MeV (10^6 eV), electrons have a much smaller wavelength than photons of the same energy. The distinction between (8.8), appropriate for electrons (and protons and other particles with mass), and (8.9), appropriate for massless photons, is important, but can be a source of confusion. Although $p = h/\lambda$ is equally valid for electrons and photons, the relation between momentum and energy, and therefore the relation between wavelength and energy, is different for photons and most electrons, that is, nonrelativistic electrons.

We should warn you that equation (8.8) is valid only for electrons with $v \ll c$, and with E omitting the relativistic rest energy mc^2 (see Problem 8-8). As Fig. 8.4 shows, when electron kinetic energies exceed about 10^6 eV (1 MeV), that is, when electron velocities approach the photon velocity c, the wavelength-energy relation for electrons gradually transfers from (8.8) to (8.9). In this text we'll limit our attention to nonrelativistic ($v \ll c$) electrons, and let you learn about relativity else-

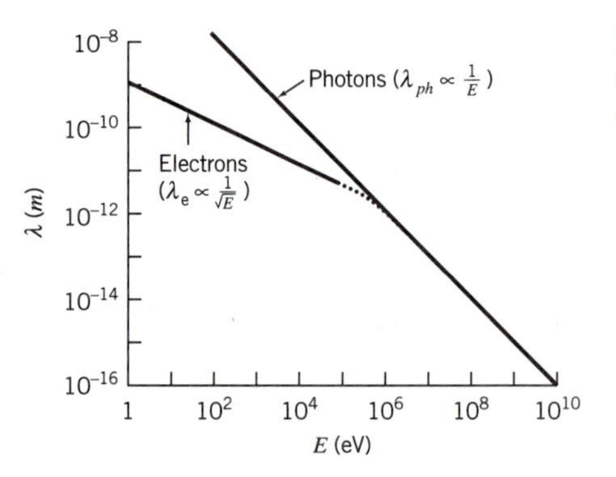

Figure 8.4. Wavelength vs. energy for electrons and photons. For nonrelativistic electrons ($E \ll 1$ MeV), $\lambda \propto E^{-1/2}$.

where. Most properties of engineering solids depend primarily on nonrelativistic electrons.

Okay, experimental evidence shows that both light and matter have wave-particle duality, sometimes displaying wave-like properties, sometimes displaying particle-like properties. But how do we make a connection between the two descriptions? For light, we noted earlier that the intensity and energy density were proportional to E_0^2, the square of the electric field amplitude; see equation (4.7). Since each photon has the same energy for a given frequency, we can look at E_0^2 as being proportional to the local density of photons—the square of the wave amplitude representing the local density of particles. For electrons, it was Schrödinger who invented a wave whose amplitude, squared, represented something close to a local density of electrons.

Sample Problem 8.3

An electron and a photon have the same energy, but the wavelength of the photon is a hundred times the wavelength of the electron. What is their common energy, and what are the two wavelengths?

Solution

$E = h^2/(2m\lambda_e^2) = hc/\lambda_{ph} = hc/100\lambda_e$

$\lambda_e = 50h/mc = 1.21 \times 10^{-10}$ m, $\lambda_{ph} = 1.21 \times 10^{-8}$ m,

$E = 1.64 \times 10^{-17}$ J $= 103$ eV

8.4 A NEW WAVE EQUATION

The wave equation for electromagnetic waves, equation (2.13), related second derivatives with respect to space (∇^2, the Laplacian) and time ($\partial^2/\partial t^2$) of electric field. The vibrating or waving quantity was electric field (and magnetic field). In the wave equation for lattice vibrations, equation (7.10), the vibrating or waving quantity was atomic displacement. Electron diffraction (now a standard characterization technique in studying crystal structures) shows clearly that electrons are waves. But what is "waving" in an electron wave?

The answer came from Erwin Schrödinger, and it was discovered during his 1925 Christmas vacation in the Swiss Alps, when he brought along an "old girlfriend" for inspiration. The mystery woman remains anonymous, but she must have been very inspiring because Schrödinger returned home with a beautiful wave equation for electrons. Three weeks later, he submitted his results for publication in a paper that earned him the Nobel Prize. According to Schrödinger, what is waving is the electron *wave function* Ψ, which satisfies the equation known as Schrödinger's equation:

$$-\frac{\hbar^2}{2m}\nabla^2\Psi + V\Psi = i\hbar\frac{\delta\Psi}{\delta t} \qquad \textbf{(8.10)}$$

where V *is the potential energy* of the electron, and m is its mass. Note that this equation differs from the wave equations we encountered for electromagnetic and sound waves, equations (2.13) and (7.10). Those involved a *second* derivative with respect to time, but this one involves a *first* derivative. Electron waves are very different from light waves and sound waves.

What is the physical meaning of Ψ? It is expressed in terms of the *square* of the wave function, Ψ^2, which represents the *probability distribution* of the electron. $\Psi^2(x, y, z, t)dxdydz$ gives the probability that the electron will be found at time t within an infinitesimal volume $dxdydz$ located at position (x, y, z). Rather than the firm classical picture of a particle that has a specific position at a specific time, quantum mechanics gives us a fuzzy picture of a wave-like, particle-like thing, and *we have only a statistical picture of its location at any given time*. It's a concept that takes some time to get used to, but it will gradually become clearer as we deal with more and more specific cases.

One reason that we stressed complex numbers in the earlier chapters is that Schrödinger's equation is expressed in complex terms, and wave functions are often complex. We remind you that a complex number $(a + bi)$ has a *complex conjugate* $(a - bi)$, and the product of $(a + bi)$ with its complex conjugate is $a^2 + b^2$. Whenever the wave function Ψ is complex, it has a complex conjugate Ψ^*. The probability distribution that we wrote simply as Ψ^2 earlier is really $\Psi^*\Psi$ when Ψ is complex.

It's convenient to separate the space and time variables in the wave function:

$$\Psi(x, y, z, t) = \psi(x, y, z)w(t) \tag{8.11}$$

Note that we use ψ to represent a time-independent (stationary) wave function, and Ψ to represent a time-dependent wave function.

Plugging this into (8.10) and dividing by ψw gives

$$-\frac{\hbar^2}{2m\psi}\nabla^2\psi + V = \frac{i\hbar}{w}\frac{\partial w}{\partial t} \tag{8.12}$$

The left side varies with position, and the right side varies with time. The only way the two can be equal is if both sides are separately equal to the same constant. Using some foresight, we'll call that constant E. The right side then yields a differential equation for $w(t)$:

$$Ew = i\hbar\frac{\partial w}{\partial t} \tag{8.13}$$

a solution to which is

$$w(t) = \exp\left(\frac{-iEt}{\hbar}\right) \tag{8.14}$$

Remembering from (8.7) that energy $E = \hbar\omega$, if we are to identify the constant that we conveniently called E with energy, w becomes

$$w(t) = e^{-i\omega t} \tag{8.15}$$

That's the form of time-dependence that we used in earlier chapters for electromagnetic waves and elastic waves. As you can probably guess, that's no coincidence. The left side of equation (8.12), multiplied by ψ, becomes

$$-\frac{\hbar^2}{2m}\nabla^2\psi + V\psi = E\psi \tag{8.16}$$

This is the *time-independent Schrödinger's equation.* The wave functions that are solutions to this time-independent equation correspond to "stationary states." Their square, ψ^2, will describe the spatial dependence of the probability density of electrons. The wave functions will tell us the probable locations of the electrons, and the E will tell us their total (kinetic plus potential) energy. In a sense, most of the rest of the text will be finding or approximating solutions to (8.16) for different potential energy functions $V(x, y, z)$.

If we identify the E on the right side with the total energy of the electron, and the V with potential energy, that suggests that we should somehow relate $(-\hbar^2/2m)\nabla^2$ with kinetic energy. To connect that with our classical view that kinetic energy is $p^2/2m$, we have the rather bewildering relationship that momentum p is related to $-i\hbar\nabla$. More about that later, but it's important at least to note now that the gradient of the wave function, $\nabla\psi$, is related to momentum and kinetic energy. *Electron wave functions that vary rapidly with distance correspond to greater kinetic energy than wave functions that vary slowly with distance.*

8.5 FREE ELECTRON WAVES AND TUNNELING

To acquaint ourselves with various solutions of equation (8.16), we might as well start with the simplest possible potential energy function, $V = 0$. That's for a totally free electron traveling in the absence of any potential energy or forces. And since we want to ease into the mysterious world of quantum mechanics as gently as possible, in this chapter we'll limit ourselves to wave functions that vary only in one dimension, z. In this case, if $V = 0$, equation (8.16) becomes simply

$$-\frac{\hbar^2}{2m}\frac{d^2\psi}{dz^2} = E\psi \tag{8.17}$$

That's not all that difficult to solve. Suppose we try $\psi = e^{ikz}$. That's a solution of (8.17) if

$$E = \frac{\hbar^2k^2}{2m} \qquad k = (2mE)^{1/2}\hbar^{-1} \tag{8.18}$$

Hmmm. Since $V = 0$, the energy is all kinetic energy, $p^2/2m$. We noted before a connection between momentum p and $-i\hbar\nabla$. For this particular wave function, $-i\hbar\nabla\psi = -i\hbar(d\psi/dz) = \hbar k\psi$. This all hangs together with $p = \hbar k$, the version of the de Broglie relation expressed in equation (8.7). It's known that Schrödinger took de Broglie's thesis with him on that 1925 vacation, and he surely wrote his equation (8.16) to be consistent with de Broglie's relation in this simplest possible case, the totally free electron.

Of course, $\psi = e^{-ikz}$ is also a solution of (8.17). Including the time-dependent $w(t)$, a general solution of Schrödinger's one-dimensional time-dependent equation (8.10) for the case $V = 0$ is

$$\Psi(z, t) = Ae^{-i(\omega t - kz)} + Be^{-i(\omega t + kz)} \quad \textbf{(8.19)}$$

These mathematical forms should be familiar by now. They correspond to plane electron waves traveling in the positive-z direction (the A term) and in the negative-z direction (the B term). Each wave has a definite energy E and a definite momentum $\hbar k = (2mE)^{1/2}$. But *where* is the electron located in each case? Since the complex conjugate of $e^{-i\theta}$ is $e^{i\theta}$, and the product of these two is unity, the probability density $\Psi^*\Psi$ for each wave is a constant, independent of z! This means that there is an equal probability of the electron being *anywhere* along the z-axis! This is our first glimpse of what became known as *Heisenberg's uncertainty principle*, which tells us that the product of the uncertainty in momentum Δp and the uncertainty in position Δz can be no smaller than Planck's constant:

$$\Delta p \Delta z \geq h \quad \textbf{(8.20)}$$

With a definite k and therefore a definite momentum $p = \hbar k$, there is no uncertainty in momentum ($\Delta p = 0$), which results in an infinite uncertainty in position ($\Delta z = \infty$). We know the energy and momentum of the electron precisely, but we have no idea where it is. More about uncertainty later. (Some texts present the uncertainty principle as $\Delta p \Delta z \geq \hbar$, and others use other multiples of h, corresponding to slightly different mathematical definitions of uncertainty. Consider the precise value of $\Delta p \Delta z$ slightly uncertain.)

Now let's tackle a problem one step more complicated. We start with an electron traveling in the positive z-direction, with $V = 0$, which meets a potential-energy barrier of $V > 0$ (Fig. 8.5). Equation (8.19) remains a general solution for the region on the left where $V = 0$, the first term corresponding to the incident electron wave,

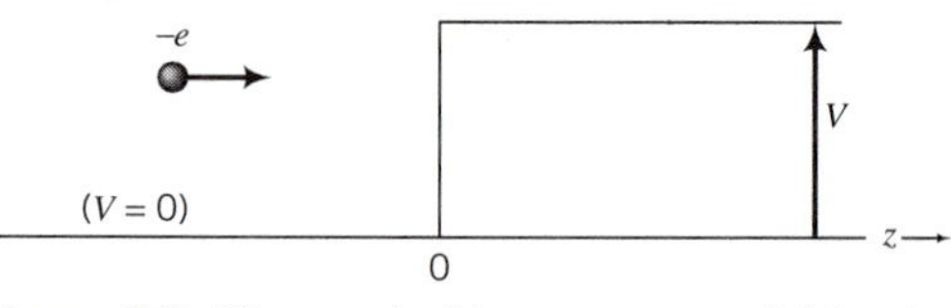

Figure 8.5. Electron incident on a potential barrier of height V.

the second term to a possible reflected electron wave. The solution of Schrödinger's time-independent equation (8.16) for the region on the right, where V is nonzero (but constant), is $\psi = Ce^{ik'z}$, where

$$k' = \{2m(E - V)\}^{1/2}\hbar^{-1} \quad \textbf{(8.21)}$$

This is a transmitted electron wave traveling to the right (we ignore the solution traveling to the left, which has no physical significance in this problem). The wave number k' is decreased below the wave number k of the incident wave, corresponding to a decrease in the kinetic energy of the electron from E to $(E - V)$. This decrease in kinetic energy is, of course, just what a classical treatment would predict for an electron moving from a potential energy of 0 to a potential energy $V > 0$.

Sample Problem 8.4

A 6 eV electron traveling as in Fig. 8.5 encounters a step potential barrier of 2 eV. What is the wavelength and wave number of the electron to the left and to the right of the barrier?

Solution

To the left of the barrier, the electron kinetic energy is 6 eV,
so from (8.8), $\lambda = h(2mE)^{-1/2} = 5.01 \times 10^{-10}$ m and $k = 2\pi/\lambda = 1.25 \times 10^{10}$ m^{-1}
To the right of the barrier, the kinetic energy $(E - V) = 4$ eV,
so from (8.21), $\lambda = 6.13 \times 10^{-10}$ m, $k' = 1.02 \times 10^{10}$ m^{-1}.

However, the quantum-mechanical solution to this simple barrier problem does have some surprises in it. Classically, if a beam of many electrons were incident on this potential barrier with the same $E > V$, they would *all* enter the region at the right (with reduced kinetic energy). If instead $E < V$, the electrons would not have enough energy to overcome the potential barrier, and they would *all* be reflected. Quantum mechanics will instead tell us that in the case $E > V$, there is a finite probability that an electron will be reflected. (Of course, back in Chapter 4, it didn't bother us too much that a certain fraction of the intensity of a light wave, given by equation (4.8), was reflected from the surface of a dielectric. Now that we know that light consists of photons, we can look at equation (4.8) as representing the probability that a particular photon, in a collection of identical photons, will be reflected.)

More surprising, in the case $E < V$, quantum mechanics tells us that there is a finite probability that an electron will enter the region on the right. In this region, since the electron's total energy is less than its potential energy, it would have (thinking classically) negative kinetic energy! At this point, I ask you to remember the White Queen's comments to Alice or—even better—to try to avoid thinking too classically.

To solve quantitatively the quantum-mechanical problem of the potential barrier, we have to determine the relation between the coefficients A, B, and C from *boundary conditions at the interface*. (As with any problem in differential equations, boundary conditions will select which among the various mathematical solutions to Schrödinger's equation are the solutions for a given physical problem.) We define the interface as $z = 0$. From the boundary condition that ψ must be continuous across the interface, we know that $A + B = C$. If we apply the additional boundary condition that $d\psi/dz$ (proportional to momentum) must also be continuous, we find that $k(A - B) = k'C$. From these two relationships, and from (8.18) and (8.21),

$$\frac{B}{A} = \frac{k - k'}{k + k'} = \frac{\sqrt{E} - \sqrt{(E - V)}}{\sqrt{E} + \sqrt{(E - V)}}$$
$$\frac{C}{A} = \frac{2k}{k + k'} = \frac{2\sqrt{E}}{\sqrt{E} + \sqrt{(E - V)}} \qquad \textbf{(8.22)}$$

For $V = 0$ (no barrier), $B = 0$ and $C = A$. But for $V > 0$, $B > 0$, that is, a reflected wave exists even though, for $E > V$, no electrons would be reflected classically. The statistical nature of quantum mechanics becomes apparent. If a beam of identical electrons is incident on a potential barrier, some will be reflected and some transmitted.

Since the probability density is proportional to the square of the wave function, the electron "reflectivity" of this barrier is proportional to $(B/A)^2$, and it is very similar mathematically to equation (4.8) for the reflectivity of electromagnetic waves, with the indices of refraction n and 1 replaced by the wave numbers k and k'.

Sample Problem 8.5

What is the reflectivity of the barrier in Sample Problem 8.4? How much of the incident electron wave is transmitted past the barrier?

Solution

From (8.22), $B/A = (\sqrt{6} - \sqrt{4})/(\sqrt{6} + \sqrt{4}) = 0.10$, $(B/A)^2 = 0.01$

Also from (8.22), $C/A = (2\sqrt{6})/(\sqrt{6} + \sqrt{4}) = 1.10$. However, this does not mean that the probability of transmission is greater than one. Just as the intensity of electromagnetic waves equals the energy density times the wave velocity—equation (4.7)—the electron flux equals the probability density ψ^2 times the electron velocity, which is proportional to k. The velocity decreases in passing the barrier by the factor $k'/k = (\sqrt{4}/\sqrt{6})$. Thus the fraction of transmitted electron wave is $(C/A)^2(\sqrt{4}/\sqrt{6}) = 0.99$.

The strangeness of quantum mechanics becomes especially apparent for the case $E < V$, where, classically, all electrons would be totally reflected; no electrons would penetrate the barrier. But quantum mechanics tells us that there's a finite probability that some electrons *will* penetrate the barrier. From (8.21), k' is imaginary beyond the barrier, that is, $k' = \pm i|k'|$. Hence the wave function to the right of the barrier will be

$$\psi = Ce^{-i\omega t}e^{-|k'|z} \tag{8.23}$$

We have chosen the sign of k' that yields an exponentially *decaying* wave function, as an exponentially increasing wave function would not satisfy the implicit boundary condition of $\psi \rightarrow 0$ for $z \rightarrow \infty$. So in the region to the right, where the electron classically should never enter, there is a finite probability (proportional, of course, to ψ^2) of the electron's existence. If the barrier region where $V > E$ is thin enough, there's a reasonable probability that an electron can get through to the other side. This quantum-mechanical phenomenon of electrons penetrating thin barriers from which they are classically forbidden is called *tunneling*. It is important to many physical problems and is the basis of practical electron devices known as *tunnel diodes* (discussed in Chapter 16).

Again there is a partial mathematical analogy to one of our earlier results for electromagnetic waves, with the exponential damping of the electron wave function in (8.23) analogous to the exponential damping of the electric field within the skin depth of a conductor, equation (2.26). The associated physics, however, is of course quite different. And the mathematical analogy is not complete, since the k here is pure imaginary, whereas for the electromagnetic wave penetrating a conductor the k was complex, having both real and imaginary parts (representing electric fields both cyclic and decaying).

8.6 WAVE PACKETS AND UNCERTAINTY

That totally free electron wave we started with in the last section certainly embraced wholeheartedly the wave-like nature of the electron. But what about its particle-like nature? There are experiments in which electrons seem to be behaving like particles, giving us at least some notion of where they are. The location of that free electron wave with a precise k was totally uncertain. Is there some way to return some particle-like nature to the electron while retaining its wave-like nature?

You've probably seen audiences at baseball or football games producing the "wave," a propagating region of fans briefly rising out of their seats with arms raised. Despite its name, this wave has a particle-like nature, located in a range of space while moving around the stadium. Mathematically, such a *wave packet* (a wave-like entity located in a region of space) can be constructed by a Fourier series, that is, by summing a large number of pure sine waves, or, in complex form, a large number of e^{ikz} waves of different k.

To locate a wave packet *precisely* in space ($\Delta z = 0$)—mathematically speaking, to create a *delta-function*—requires an infinite range of Fourier components, that is, it requires $\Delta k = \infty$ (corresponding to $\Delta p = \infty$). However, if the wave packet, like the wave in the stadium, has a finite width Δz, you can get away with a finite range of k values, a finite Δk. The exact relationship between Δz and Δk depends on your exact definition of each. However, using one common definition, you get $\Delta z \Delta k \approx 2\pi$. From this and $p = \hbar k$, you get the uncertainty relationship (8.20).

Other definitions can yield products of uncertainties differing from h by modest factors, but the main point remains: if you want to create a localized, particle-like entity from a collection of waves, you'll need more and more waves in your Fourier series (greater uncertainty in k and momentum) as you try to decrease the uncertainty in position. Electrons don't like to be pinned down; they need their space. The more you try to pin them down, that is, to localize them, the more "nervous" (higher momentum, higher kinetic energy) they get.

With what velocity will this wave packet travel? From our discussion in Chapter 7, the appropriate velocity of the wave packet is not the phase velocity ω/k (appropriate for individual waves), but the *group velocity* $\partial\omega/\partial k$. If our wave packet is built from free-electron waves, from equations (8.7) and (8.18), $\omega = E/\hbar = \hbar k^2/2m$. So the group velocity will be $\hbar k/m$, or p/m, which is consistent with the classical relation between velocity and momentum of a particle.

But note that our electron wave is *very* different from an electromagnetic wave! An electromagnetic wave traveling in vacuum is nondispersive; its velocity is independent of wavelength. But when traveling in vacuum, the velocity of an electron is proportional to k, that is, it is inversely proportional to wavelength. *An electron wave exhibits strong dispersion, even in vacuum.* Although the group velocity is consistent with the classical electron velocity, the phase velocity equals $p/2m$, *half* the group velocity. In Chapter 13, we'll consider the more complicated case of electron waves traveling through a crystal lattice, where the electron wave is also dispersive, but where the electron's group velocity can have a very different dependence on k.

8.7 BOXED-IN ELECTRON WAVES: ENERGY LEVELS

There are many potential energy functions $V(z)$ that will tend to confine an electron, but let's start with a very simple one. Let's say $V = 0$ inside a one-dimensional energy well or "box" between $z = 0$ and $z = L$ (Fig. 8.6) and $V = \infty$ outside. With infinite potential energy outside the box (the box has "rigid walls"), the electron is totally confined to the box, that is, $\psi = 0$ outside the box. Appropriate boundary conditions are thus $\psi = 0$ at $z = 0$ and $z = L$. Solutions to Schrödinger's time-

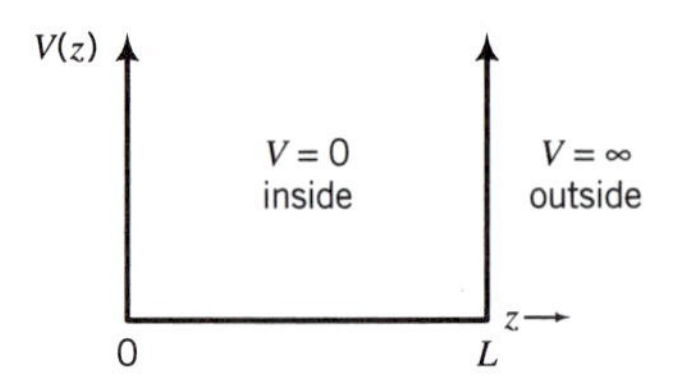

Figure 8.6. Potential energy well ("box") for an electron.

independent wave equation (8.16) inside the box are therefore (Fig. 8.7) sine waves, with an integral number of half-wavelengths inside the box (much like the modes of vibration of a guitar string or like the electromagnetic waves in the cavity considered earlier in this chapter). Since $L = n\lambda/2$, $k = 2\pi/\lambda = n\pi/L$, and

$$\psi_n = A_n \sin(n\pi z/L) \qquad (n = 1, 2, 3, \ldots) \tag{8.24}$$

Each of these *standing-wave* solutions ψ_n corresponds to a particular energy E_n given by

$$E_n = \frac{\hbar^2 k^2}{2m} = \frac{n^2 h^2}{8mL^2} \qquad (n = 1, 2, 3, \ldots) \tag{8.25}$$

Our solution for the totally free electron, and for the free electron encountering an energy barrier, had no restrictions on E. However, the imposition of the boundary conditions at the edges of the box restricted the choice of k values and thus restricted the energy E (which is all kinetic energy in this case, since $V = 0$ in the box). So the electron inside the box has only a *limited set of allowed energy levels* defined by (8.25). The lowest possible energy is $h^2/8mL^2$ (for $n = 1$). This solution is called the *ground state*. All the other solutions, which correspond to higher energies, are called *excited states*. Each allowed energy level is identified by a single integer n called a *quantum number*.

We have confined the electron inside a box of width L, so the uncertainty in position is $\Delta z = L$. In the ground state, we can view the standing wave as consisting of two equal traveling waves moving in opposite directions. Their momentum $p = \hbar k = \pm h/2L$ (the different signs corresponding to the different directions), so the range of uncertainty $\Delta p = h/L$, and therefore $\Delta p \Delta z = h$, which satisfies the uncertainty relation (8.20). The more we confine the electron (the smaller we make L), the more nervous the electron gets (the higher its kinetic energy).

Figure 8.7 shows the first few wave functions corresponding to allowed energy levels in the box. For $n = 1, 3, 5$, etc., the functions are *symmetric* across the center of the box, while for $n = 2, 4, 6$, etc., they are *antisymmetric* (equal in magnitude but opposite in sign). Of course, the probability distributions are given by ψ^2, and

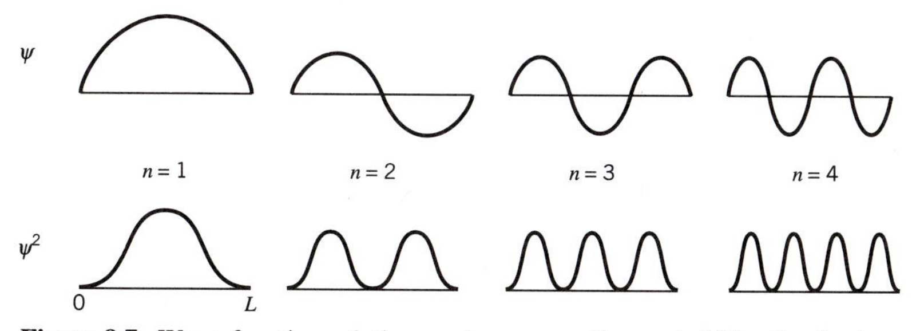

Figure 8.7. Wave-function solutions and corresponding probability distributions for an electron in a box with rigid walls ($V = \infty$ outside).

therefore are symmetric for all states (Fig. 8.7). This is a result of the symmetry of our potential function $V(z)$. (If we had placed our origin at the center of the box, as some texts do, our symmetric solutions would have been cosine functions and only the antisymmetric solutions would have been sine functions. One origin's sine is another origin's cosine.)

This problem has shown us a very important consequence of the wave-like nature of the electron. When you confine an electron to a limited region of space, it can have only certain values of energy, identified by quantum numbers. All other energies—energies between the allowed levels—are forbidden to it. Combined with our knowledge that light consists of photons with $E = h\nu$, the existence of quantized energy levels for electrons is the basis of the quantum-mechanical interpretation of the interaction of light and matter—the optical properties of atoms, molecules, and solids. *Electrons can make transitions between allowed energy levels by emitting or absorbing photons of the appropriate energy.*

In this particular case of an "electron in a box," the allowed energy levels are described by (8.25). The electron can make transitions between these levels only by absorbing or emitting photons with energies that correspond to the difference in energy ΔE between the final and initial energy levels. The emission and absorption spectra of the electron in the box are thus limited to specific frequencies:

$$\nu_{if} = \frac{\Delta E}{h} = \frac{h}{8mL^2}(n_i^2 - n_f^2) \qquad \textbf{(8.26)}$$

If an electron falls from an initial level n_i to a lower final level n_f, it *emits* a photon whose frequency is described by (8.26). The same equation will hold for the *absorption* spectra of the photon, although here $n_i < n_f$ (the electron uses the photon's energy to jump to a higher allowed energy level), so you'd have to reverse the terms in (8.26) to keep the frequency positive. Energy is conserved in each case.

Sample Problem 8.6

What wavelength photons are required to excite electrons from the ground state to the first excited state in a one-dimensional energy well 7 nm wide? In eV, what are the energies of the two electron states?

Solution

$\nu = h(2^2 - 1^2)/8mL^2 = 5.57 \times 10^{12}$ Hz $= c/\lambda$, $\lambda = 5.39 \times 10^{-5}$ m

$E_1 = h^2/8mL^2 = 1.23 \times 10^{-21}$ J $= 7.68 \times 10^{-3}$ eV

$E_2 = 4E_1 = 3.07 \times 10^{-2}$ eV, $E_2 - E_1 = 2.30 \times 10^{-2}$ eV $= h\nu$

There is a third possibility: A photon of the appropriate energy can stimulate the emission of a second photon. This *stimulated emission* is the basis of the laser (Light

Amplification by Stimulated Emission of Radiation). The stimulated photon is in phase (light is still also a wave!) with the stimulating photon, the two in-phase photons can stimulate more, and so on. More about lasers in Chapter 16.

If there's only one electron in the box, it will normally prefer to be in its lowest-energy state, the ground state. For the electron to have a nonzero probability of occupying one or more of the excited states, you'll have to bombard it with enough photons of appropriate energies or heat it up to a high enough temperature (we'll discuss the effect of temperature on the occupation of states in Chapter 12). Another way to produce occupation of the higher states is to add more electrons to the box. The rules of quantum mechanics (the *Pauli principle*) allow you to put only two electrons in each level, the two electrons paired with opposite spins. So if you put three electrons in the box, at least one of them will have to occupy one of the excited energy levels.

By assuming an infinite potential energy outside the box, we got an infinite number of allowed energy levels in the box. A more realistic box is one that goes instead to a finite potential energy V outside the box, still keeping $V = 0$ inside (Fig. 8.8). Provided we keep $E < V$, we'll probably still get some allowed energy levels in the box (that will depend on both V and L), but the answer won't be as easy as (8.25) because now the wave functions can extend outside the box, exponentially damping as in (8.23). (Instead of rigid walls, our box now has soft, or "leaky," walls.)

Instead of the simple boundary condition that $\psi = 0$ at the edges of the box, our boundary conditions for this energy well of finite depth will be those we used before in the barrier problem, that is, continuity of both ψ and $d\psi/dz$ across the walls. We'll get equations with sines or cosines in them that we'll have to solve numerically or graphically. (See Problem 8-15.) The net effect, however, will be an effective increase in the wavelengths of wave functions (compared to those in the box with rigid walls) and a corresponding decrease in kinetic energy. And, of course, with a finite V there will be only a finite number of allowed localized energy levels in the box, since an electron with $E > V$ won't be confined to the box (although its wave function will be modified in the vicinity of the box).

As abstract as this "electron in a box" problem may seem, it pertains to a real configuration used in semiconductor technology called a *quantum well*. The chemical composition of a semiconductor produced layer by layer by a deposition process can be locally modified to produce a layer in which electrons have a somewhat lower energy—in effect, a one-dimensional "box," or energy well, of finite depth, like the

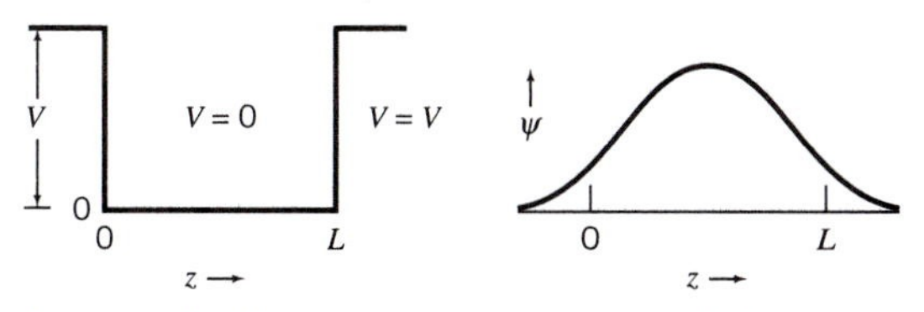

Figure 8.8. Electron potential energy well (box) with finite walls and corresponding ground-state wave function that "leaks" out of the box.

one discussed in the preceding paragraph. As we have seen, this electron confinement produces a series of allowed energy levels (only a few levels if the energy depth of the well is not great), which can be used to control local optical and electronic properties. Quantum wells have been used as the basis of several technological devices, including semiconductor lasers and resonant tunnel diodes (both discussed in Chapter 16). Semiconductor technologists have also studied regions in which electron waves are confined in two dimensions, called *quantum wires*, and in which electron waves are confined in three dimensions, called *quantum dots*. In Chapter 9, we'll consider electron energy wells in two and three dimensions.

One more simple matter before we let that poor electron out of the box. In (8.24), our ψ_n wave-function solutions for the box with rigid walls contained an as-yet unspecified constant A_n called a *normalization constant.* We can determine that constant if we recognize that the total probability that the electron will be somewhere in the box is one. Thus

$$\int_0^L \psi^2 dz = 1 = A_n^2 \int_0^L \sin^2(n\pi z/L)dz = \frac{A_n^2}{2} L \tag{8.27}$$

so $A_n = \sqrt{(2/L)}$. Any wave-function solution to Schrödinger's equation can be *normalized* (i.e., the normalization constant determined) by integrating the probability density ψ^2 over the entire appropriate volume. In this case, it was sufficient to integrate over z from 0 to L. In some problems, it may be necessary to integrate over all space.

8.8 THE HARMONIC OSCILLATOR—A QUANTIZED SPRING

Confining an electron or any other matter wave with other forms of potential energy will also tend to produce a series of wave functions and corresponding allowed energy levels. Only a few specific forms of V, however, will yield wave functions and energy levels of simple mathematical form. Most problems will have to be solved numerically, a procedure that is becoming easier and easier in these days of powerful computers. However, one form of potential that produces a simply described set of energy levels and wave functions that are well-defined (even if not as simple as those in the box with rigid walls) is that for an important problem we have already discussed classically in Chapters 3 and 7—the *harmonic oscillator.*

Suppose we have a mass m confined to the region surrounding $z = 0$ by a restoring force of $-Kz$. (If our mass m is an electron, this is the electronic polarization problem we considered in Chapter 3. But we could also consider our mass m to be something heavier—perhaps a complete atom or ion. Electrons aren't the only bits of matter that must satisfy Schrödinger's equation.) The potential energy V for this situation is $Kz^2/2$, and therefore the time-independent Schrödinger's equation to be solved is

$$-\frac{\hbar^2}{2m}\frac{d^2\psi}{dz^2} + \frac{Kz^2\psi}{2} = E\psi \tag{8.28}$$

Substituting $\psi_1 = A \exp(-z^2/2a^2)$ into this equation shows that ψ_1 is a solution if $a^2 = \hbar/(mK)^{1/2}$, and the corresponding energy is $E_1 = \hbar\omega_c/2$, where $\omega_c = (K/m)^{1/2}$, the resonant frequency of a classical harmonic oscillator, which we encountered earlier as equation (3.16). (If you have ever studied statistics, this particular exponential function may be familiar to you as representing the normal or Gaussian distribution, with the "standard deviation" equal to a.) ψ_1 is the ground-state (lowest energy) wave function for the harmonic oscillator, and the probability distribution and ground-state energy corresponding to this wave function are

$$\psi_1^2 = A^2 \exp(-z^2/a^2) = A^2 \exp\{-z^2(mK)^{1/2}/\hbar\} \tag{8.29}$$

$$E_1 = \frac{\hbar\omega_c}{2} = \frac{\hbar}{2}\left(\frac{K}{m}\right)^{1/2} \tag{8.30}$$

As with our electron in a box, we find that the lowest possible energy in a quantum-mechanical system is not zero, as it would be classically. Even at $T = 0$, there will be vibration and a *zero-point energy*, another embodiment of the uncertainty principle. You could lower potential energy further by lowering the average displacement z, but that would raise the kinetic energy; the zero-point energy represents the minimum in *total* energy that is compatible with the uncertainty principle.

We can calculate the maximum displacement z_m for the *classical* harmonic oscillator by setting this zero-point energy E_1 equal to $Kz_m^2/2$, since at maximum displacement of a classical oscillator, the energy is all potential energy. We find that $z_m = a$. Yet equation (8.29) tells us that a quantum-mechanical harmonic oscillator has a nonzero probability of having displacements greater than a, positions at which, thinking classically, the kinetic energy would be negative (since the potential energy V is greater than the total energy E). This is equivalent to our earlier observations of "tunneling" into barriers where $E < V$, and of wave functions "leaking" out of quantum wells of finite depth.

There are, of course, many other solutions to (8.28), Schrödinger's equation for the harmonic oscillator. They are all of the form $A \exp(-z^2/2a^2)$ multiplied by specific polynomials $f(z)$. The next two, for example, are $\psi_2 = Az \exp(-z^2/2a^2)$ and $\psi_3 = A(z^2 - a^2/2) \exp(-z^2/2a^2)$. Plots of the first six wave function solutions are shown in Fig. 8.9.

The corresponding energies are given by

$$E_n = \left(n - \frac{1}{2}\right)\hbar\omega_c = \left(n - \frac{1}{2}\right)\hbar\left(\frac{K}{m}\right)^{12} \qquad (n = 1, 2, 3, \ldots) \tag{8.31}$$

(Many texts, for historical reasons, label the ground-state level of the harmonic oscillator as $n = 0$, which changes the $(n - 1/2)$ in this formula to $(n + 1/2)$. We have chosen instead to be consistent with the notation used in the previous section and in later sections, and have labeled the ground-state level as $n = 1$.)

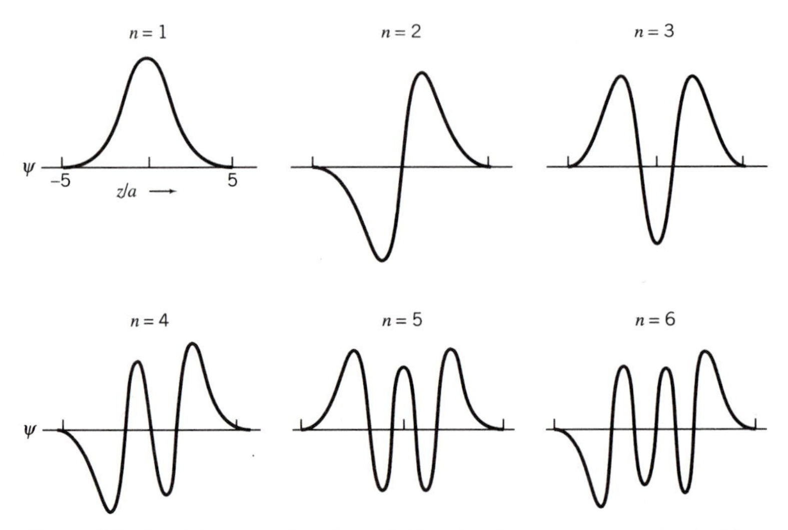

Figure 8.9. First six wave-function solutions for the quantum-mechanical harmonic oscillator. (Adapted from L. Pauling and E. B. Wilson, Jr., *Introduction to Quantum Mechanics*, McGraw-Hill, New York (1935).)

Sample Problem 8.7

In a one-dimensional harmonic oscillator with a classical resonant frequency ω_c of $10^{15}\ s^{-1}$, what energy photon is required to produce a transition from the ground state to the first excited state? In eV, what are the energies of the two electron states?

Solution

$E_2 - E_1 = \hbar\omega_c = (1.05 \times 10^{-34})(10^{15}) = 1.05 \times 10^{-19}\text{ J} = 0.656\text{ eV}$
$E_1 = \hbar\omega_c/2 = 5.25 \times 10^{-20}\text{ J} = 0.328\text{ eV}$
$E_2 = 3\hbar\omega_c/2 = 0.984\text{ eV}$

As can be seen from Fig. 8.9, the wave functions for the harmonic oscillator, although very different mathematically from those for the electron in a box, are similar in that: (a) they are either *symmetric* or *antisymmetric* (again because the potential energy function is symmetric), and (b) in addition to vanishing at their extremes, they have *(n − 1) nodes* (positions where the wave function changes sign). Since we associate $\nabla\psi$ with momentum and kinetic energy, it is clear from the form of the wave functions that increasing quantum number n corresponds to increasing kinetic energy. *The more wiggles in the wave function, the higher the kinetic energy.*

For the electron in the box, the increase in energy from level to level was entirely kinetic energy, since we took $V = 0$ in the box. For the harmonic oscillator, increasing quantum number n also corresponds to an increase in potential energy (more stretching of the spring). How would we calculate that increase? The potential energy is $Kz^2/2$, so the average potential energy in any given energy level can be calculated by calculating the average value of z^2. Since the probability of any displacement z is proportional to $\psi_n^2(z)$, the average value of z^2 can be calculated by summing $z^2\psi_n^2(z)$, which will give the desired *weighted average*. If the wave functions are normalized, the average value of z^2 is given by

$$(z^2)_{\text{avg}} = \int_{-\infty}^{+\infty} z^2\psi_n^2(z)dz \quad \textbf{(8.32)}$$

Using the normalized ground-state wave function, this will yield $(z^2)_{\text{avg}} = a^2/2$ for the ground state of the harmonic oscillator, corresponding to an average potential energy of $\hbar\omega_c/4$, half of the total zero-point energy. Thus the zero-point energy is *half potential energy, half kinetic energy*. The same is true for all the other energy levels of the harmonic oscillator.

In quantum mechanics, an average value calculated as in (8.32) is called an *expectation value*. The example given is a simple one for three reasons: first, because the wave functions of the harmonic oscillator are real (otherwise we must replace ψ^2 by $\psi^*\psi$); second, because we have assumed the wave functions were normalized (otherwise, we must divide by $\int\psi^*\psi dV$); and third, because the quantity we were averaging was one that did not "operate" on wave functions. In contrast, a quantity like momentum, represented by the quantum-mechanical operator $-i\hbar\nabla$, has to be placed *between* ψ^* and ψ, and $\psi^*\nabla\psi$ will appear in the integrand. In the general case, the formula for the expectation value of some quantity S is

$$\langle S\rangle = \frac{\int\psi^*S\psi dV}{\int\psi^*\psi dV} \quad \textbf{(8.33)}$$

Figure 8.10 shows the probability distribution ψ^2 corresponding to the $n = 11$ energy level of the harmonic oscillator. Also shown, as a dashed line, is the equivalent of the "probability distribution" of a *classical* harmonic oscillator of the same energy. (Imagine making a movie of a vibrating classical oscillator and counting the number of frames that the object appears in each position. This number will be inversely proportional to the velocity at each position, and it is the inverse of the classical velocity that is plotted.) The similarity of shape is evident, and suggests that as $n \rightarrow \infty$, the quantum-mechanical probability distribution will approach the classical. That's comforting.

The energy levels of the quantum-mechanical harmonic oscillator given by equation (8.31) are *equally spaced*, each differing from the level above and below by the same quantity: $\hbar\omega_c$. Any emission or absorption of photons associated with transitions between energy levels will therefore all be at the same frequency or integral multiples thereof. Measurements of these frequencies will determine ω_c and, presuming that m is known, allow experimental determination of the spring constant K. If our oscillator is not simply an electron but represents, for example, the oscillating separation between the two atoms in a diatomic molecule, optical measurements can determine the separation between the "vibrational energy levels" of the molecule

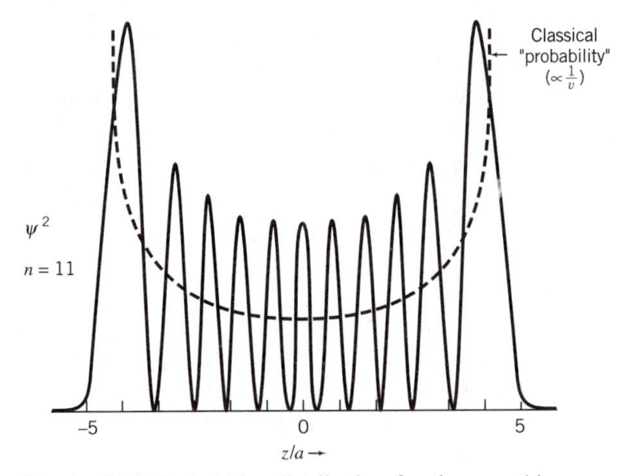

Figure 8.10. Probability distribution for the $n = 11$ wavefunction solution for the quantum-mechanical oscillator compared to the classical limit. (Adapted from L. Pauling and E. B. Wilson, Jr., *Introduction to Quantum Mechanics*, McGraw-Hill, New York (1935).)

and thereby determine the spring constant K of the interatomic bond (see Chapter 10). If we further extend the quantized harmonic oscillator to the problem of lattice vibrations in solids, we are led to the concept of *phonons*, the quanta of elastic waves mentioned in Chapter 7 in connection with low-temperature lattice heat capacity.

SUMMARY

The frequency distribution of cavity radiation and the photoelectric and Compton effects were among the experiments that led to the realization that electromagnetic radiation was quantized into particles we call *photons*, whose energy and momentum are given by $E = h\nu$ and $p = h/\lambda$ (or the equivalent relations $E = \hbar\omega$ and $p = \hbar k$). Since light had both a wave-like and a particle-like nature, de Broglie suggested that matter might also have *wave-particle duality* described by the same equations, a suggestion soon confirmed by the demonstration of electron diffraction.

Although the energy-frequency and momentum-wavelength relations for electrons and photons are identical, the relations between energy and wavelength are quite different because electrons have mass and photons do not. Photon wavelengths are inversely proportional to energy, but the wavelengths of nonrelativistic electrons ($E \ll 1$ MeV) are instead inversely proportional to $\sqrt{E}$.

Schrödinger developed a wave equation for matter that in time-dependent form is

$$-\frac{\hbar^2}{2m}\nabla^2\Psi + V\Psi = i\hbar\frac{\delta\Psi}{\delta t}$$

and in time-independent form is

$$-\frac{\hbar^2}{2m}\nabla^2\psi + V\psi = E\psi$$

where $\Psi(x, y, z, t) = \psi(x, y, z)\exp(-i\omega t)$. The first term represents kinetic energy $p^2/2m$, identifying the momentum p with the operator $-i\hbar\nabla$. The square of the wave function ψ represents the probability distribution of the electron (or other particle of matter). We considered solutions of Schrödinger's time-independent wave equation for several potential energy functions $V(x, y, z)$.

The kinetic energy of a totally free electron ($V = 0$) is $\hbar^2k^2/2m$. If the wave number k of an electron wave is precisely known, the position of the electron is totally unknown, an example of the uncertainty principle $\Delta p\Delta z \geq h$. The electron can instead be localized in a "wave packet" of width Δz if it is formed by a Fourier series of waves with a range of wave numbers Δk and a range of momenta $\Delta p \approx h/\Delta z$. The group velocity $\partial\omega/\partial k$ of a free electron in vacuum corresponds to the classical electron velocity p/m. Electron waves exhibit extreme dispersion, even in vacuum.

A free electron wave of energy E encountering a potential energy barrier of height V will have a nonzero probability of reflection even if $E > V$, and it will have a nonzero probability of transmission ("tunneling") even if $E < V$. The relative strengths of incident, reflected, and transmitted waves can be calculated using the boundary conditions that both the wave function and its derivative should be continuous across the interface.

When the potential energy function $V(x, y, z)$ confines the electron or other matter particle to a limited region of space, only a limited number of energies are available to the particle. Schrödinger's equation plus boundary conditions lead to a series of wave functions ψ_n and a series of corresponding discrete energy levels E_n, identified by a quantum number n. The two examples considered were an electron in an energy well of $V = 0$ (electron in a box) and the harmonic oscillator.

An energy well with rigid walls ($V = \infty$ outside) yields wave functions that are simple sines (standing waves), and energy levels given by $n^2h^2/8mL^2$. Even the lowest-energy state, the "ground state" ($n = 1$), has some kinetic energy, another effect of the uncertainty principle. The energy difference ΔE between two electron energy levels defines the energy of the photons emitted or absorbed when electrons make transitions from one level to another. Normalization constants can be determined by requiring that the total probability, that is, the integral of ψ^2 over all pertinent space, be equal to one. Wave functions in an energy well with nonrigid walls (V finite outside the box) will extend outside the box and have lower energy than corresponding wave functions in a box with rigid walls.

The quantum-mechanical harmonic oscillator has equally spaced energy levels given by

$$\left(n - \frac{1}{2}\right)\hbar\omega_c = \left(n - \frac{1}{2}\right)\hbar\left(\frac{K}{m}\right)^{1/2}$$

Like the wave functions in the box, those of the oscillator are either symmetric or antisymmetric, and the nth wave function has $(n - 1)$ nodes where the wave function changes sign. With increasing quantum number n, both kinetic energy (related to $\nabla\psi$) and potential energy (related to average displacement) of the oscillator increase. *Expectation values*, weighted averages of various quantities while in a particular quantum state, can be calculated by integration.

Many of the results of quantum mechanics are nonintuitive and seem inconsistent with our experience gained working with objects of macroscopic size. But hang in there—the more you work with quantum mechanics, the less strange it becomes. And remember the White Queen!

PROBLEMS

8-1. An electron and a photon traveling in vacuum have the same wavelength, but the energy of the photon is fifty times the energy of the electron. (a) For each, what is the energy, wavelength, momentum, velocity, mass, and charge? (b) For the photon, what is the frequency in Hz? What type of light is this? (c) For the electron, how many volts are required to accelerate it from rest to this energy?

8-2. We perform the photoelectric experiment on a sample of tantalum, for which the barrier to electron escape from the surface (work function) is 4.2 eV. We use light of 150 nm wavelength with an intensity of 3 W/m^2. (a) What is the wavelength of the fastest photoelectrons emitted? (b) How many photons per cm^2 strike the tantalum surface per second? (c) What is the maximum wavelength of electromagnetic radiation capable of producing photoelectrons from tantalum?

8-3. For electromagnetic waves in a cavity at $T = 300$ K, calculate the average energy per mode, both classically and according to Planck, for the following energies: 0.01 eV, 0.1 eV, 1 eV, 10 eV.

8-4. Calculate the velocity, momentum, and wavelength of the following objects: (a) a photon of 50 eV energy, (b) an electron accelerated through 50 volts, (c) an α-particle accelerated through 50 volts, (d) a neutral uranium atom with a kinetic energy of 50 eV, (e) a uranium nucleus stripped of all its electrons and accelerated through 50 volts, (f) a 1-gram object after falling through a distance of 1 m.

8-5. We want to use a giant space sail, one km^2 in area, to move a kilogram of cargo from Earth to Pluto (5.8 billion km). What flux of photons incident on the sail will we need to get to Pluto within ten years? Assume a photon wavelength near the midpoint of the visible spectrum, and for simplicity, assume a constant flux of photons. (Obviously, the last is an unreasonable assumption, since the distance from the sun, our presumed source of photons, is steadily increasing, but it does simplify the $F = ma$ part of the problem.)

8-6. Write a mathematical expression in complex form, including time-dependence, for the wave function of an electron with 10 eV total energy traveling in the positive x-direction (a) in a region with $V = 0$, (b) in a region with $V = 5$ eV, (c) in a region with $V = 15$ eV.

8-7. An electron is trapped in a one-dimensional energy well of width L with rigid walls. (a) For $L = 1$ nm, calculate the wavelengths of the three lowest-energy photons capable of exciting electrons from the ground state. (b) At what value of L will the energy gap between the two lowest energy levels equal k_BT at $T = 300$ K? (c) For $L = 1$ cm, what would be the quantum number of the energy level k_BT above the ground state at $T = 300$ K?

8-8. In relativity theory, the "rest energy" of an electron is given by $E_0 = m_0c^2$, where $m_0 = 9.11 \times 10^{-31}$ kg is the "rest mass" of the electron. As electron velocity v increases, the relativistic mass increases according to the formula

$$m = m_0\left(1 - \frac{v^2}{c^2}\right)^{-1/2}$$

momentum increases as

$$p = m_0v\left(1 - \frac{v^2}{c^2}\right)^{-1/2}$$

and the total relativistic energy (rest energy plus kinetic energy) increases as

$$E_{\text{tot}} = E_0\left(1 - \frac{v^2}{c^2}\right)^{-1/2}$$

From those relations, it follows that $E_{\text{tot}}^2 = p^2c^2 + E_0^2$. (a) An electron accelerated through a voltage V acquires a kinetic energy eV. Whereas equation (8.8) gives the relation between acceleration voltage V and wavelength λ for a nonrelativistic electron, derive the corresponding formula for a relativistic electron. (Note: The formula must be equivalent to (8.8) for low kinetic energies.) (b) If you operate an electron microscope with a voltage of 30 kV, what error in electron wavelength will result if we use the classical expression instead of the (correct) relativistic one? What will the error be for an acceleration voltage of 300 kV? (c) What is the limiting expression for wavelength for $eV \gg E_0$? Compare this to the wavelength-energy relation for photons.

8-9. A 10 eV electron is incident upon a step potential barrier of 12 eV. By how much will the wave function and the probability density of the electron decrease (a) after penetrating 0.1 nm into the barrier? (b) after penetrating 1 nm?

8-10. Nine electrons are injected into a one-dimensional energy well with a width of 2 nm. (a) What will be the quantum numbers and energies of each of the nine electrons? (b) What is the lowest-energy photon that can now be absorbed by electrons in this energy well?

8-11. Write the wave function, including the normalization constant, for the second-lowest energy level in a 1-D energy well of width L and rigid walls. With the origin at one edge of the well (as in Fig. 8.6), what is the probability density at $z = L/8$? at $z = L/4$? at $z = 3L/8$? What is the probability of the electron being more than $L/4$ from the center of the energy well?

8-12. (a) Demonstrate that the ground-state wave function for the one-dimensional harmonic oscillator satisfies the appropriate Schrödinger's equation. (b) Do the same for the first excited state.

8-13. In the text, the edges of the one-dimensional energy well were defined by $z = 0$ and $z = L$, and the resulting wave functions for a well with rigid walls, given by equation (8.24), were all sine functions. It is often convenient instead to set the origin for z at the center of the well. (a) With this choice of origin, what are the wave functions for the well? (b) With this choice of origin, derive the expectation values of z and z^2 for the ground state and first two excited states.

8-14. Electrons are incident upon a 5 eV step energy barrier. Calculate the reflectivity of the barrier (the ratio of the probability density of the reflected electron wave to that of the incident electron wave) for 10 eV and for 20 eV electrons. How does this result differ from what would be expected classically?

8-15. Calculate the possible energy levels of an electron in a one-dimensional energy well 1 nm in width and 10 eV in depth. (Because the well is finite in depth, the wave function is nonzero outside the box, the simple solution for the well with rigid walls no longer applies, and you must use the boundary conditions described in the text. For simplicity, consider only symmetric wave functions. You will end up with an equation including a trigonometric function, which can be solved graphically or numerically. You should end up with only three energy levels, although there are also two allowed energy levels associated with antisymmetric wave functions.) Compare the resulting energy levels with the corresponding energy levels of a well of the same width with rigid walls.

8-16. Equation (8.20) gives one form of Heisenberg's uncertainty principle. Another way to formulate this principle is in terms of energy and time: $\Delta E \Delta t \geq h$. A laser beam is normally monochromatic but becomes less so when operated in pulses of shorter and shorter times. At what pulse length Δt will the frequency range of a green laser cover the full range of the visible spectrum, making the laser "white"?

8-17. For the simple harmonic oscillator, (a) calculate the positions of the nodes and the positions of local maximum probability density for the first two excited states. (b) Calculate the maximum displacements of a classical harmonic oscillator with the same total energies as these two quantized states. (c) For each of these two excited states, calculate the ratio of the probability density at the positions calculated in (b) to the maximum probability density.

8-18. A light wave with an electric field amplitude E_0 of 10^{-3} V/m and a wavelength of 500 nm traveling in vacuum is in normal incidence on a surface. How many photons strike the surface per second?

8-19. Assume that from your optical measurements, the vibrational energy levels of the diatomic nitrogen molecule have been determined to be about 0.2 eV apart. Since both atoms are oscillating about the molecule's center of mass, the "effective mass" of this quantized harmonic oscillator should be taken as one-half the atomic mass. What is the spring constant K of this molecule?

Chapter 9

The Periodic Table, Atomic Spectra, and Neon Lights

9.1 DEGENERACY AND SYMMETRY

We've used Schrödinger's wave equation to tackle some simple one-dimensional problems: a free electron, a free electron encountering a barrier, an electron in a box (a "quantum well"), and a harmonic oscillator. The last two, with potential energy functions $V(z)$ that confined the matter wave, led to quantized energy levels and a series of wave functions characterized by a quantum number n. *Wave mechanics produced quantum mechanics.*

In this chapter, we're interested in a problem that's inherently three-dimensional: electron wave functions and energy levels in atoms. But before tackling that, a brief digression while we consider something we'll often encounter in three dimensions—*degeneracy*.

As we've noted earlier, words often have a different meaning in physics than in common usage. If a physicist says that you and your friend are degenerate, he probably only means that the two of you have the *same energy*, because that's how the word is used in quantum mechanics.

Let's return to the electron-in-a-box problem, but now make it a *two*-dimensional box, a "quantum wire." We'll assume it's a square L on a side, with our origin at one corner of the square. As before, we assume $V = 0$ inside the box ($0 < y < L$,

$0 < z < L$) and $V = \infty$ outside. Our boundary conditions now are $\psi = 0$ at $y = 0$, $y = L$, $z = 0$, and $z = L$. In analogy to our solutions (8.24) for the one-dimensional box, our wave-function solutions of Schrödinger's equation for the two-dimensional box are

$$\psi = A \sin(n_y \pi y/L) \sin(n_z \pi z/L) \tag{9.1}$$

where n_y and n_z are positive integers. We now have *two* quantum numbers (plus, of course, the spin quantum number $s = \pm 1/2$). The corresponding energy levels are

$$E(n_y, n_z) = \frac{\hbar^2 k^2}{2m} = \frac{(n_y^2 + n_z^2)h^2}{8mL^2} \qquad (2 - D) \tag{9.2}$$

Since energy is proportional to $(n_y^2 + n_z^2)$, an electron with $n_y = 1$ and $n_z = 2$ has the same energy as one with $n_y = 2$ and $n_z = 1$. These two energy levels are said to be *degenerate*. Different wave-function solutions to Schrödinger's equation (different quantum numbers) that correspond to the same energy E are defined as degenerate.

Similarly, if we consider a *three*-dimensional box (a "quantum dot") in the shape of a cube L on a side, the wave-function solutions will be a product of *three* sine functions, and we will have *three* quantum numbers. Energy is proportional to $(n_x^2 + n_y^2 + n_z^2)$, and levels with quantum numbers (n_x, n_y, n_z) equal to (1, 2, 3), (2, 1, 3), (1, 3, 2), (3, 1, 2), (2, 3, 1), and (3, 2, 1) are degenerate. All six energy levels have the same energy. Most energy levels in the three-dimensional cube are sixfold degenerate, although levels like (1, 1, 2) are just threefold degenerate and levels like (2, 2, 2) are nondegenerate.

Degeneracy of energy levels is closely *related to symmetry*. Those six energy levels had the same energy because we assumed our box was a cube, with the same length L in the x, y, and z directions. If we instead allowed our box to have different lengths in the different directions, this degeneracy would be removed. Decreasing symmetry tends to remove degeneracies. For example, molecules have lower symmetry than single atoms, and we shall find in Chapter 10 that some degeneracies of atomic energy levels are removed when atoms combine to form molecules.

Since each energy level can accommodate two electrons with opposite spin, and these two electron "states" in the same level have the same energy in the absence of magnetic field, each level also has a twofold *spin degeneracy*. In the absence of a magnetic field, the two electrons of opposite spin in the same energy level will have the same energy. An applied magnetic field would favor one spin orientation over the other, lowering the energy of one orientation and raising the energy of the other, thereby removing the spin degeneracy. But we will generally consider problems with no applied magnetic field, so that the number of available electron *states* will always be twice the number of available energy *levels*.

Sample Problem 9.1

What wavelength photons are required to excite electrons from the ground state to the first excited state in a three-dimensional energy well of dimensions 7 nm × 7 nm × 6 nm? In joules, what are the energies of the two electron states? Is either of these states degenerate?

Solution

$E = (h^2/8m)[(n_x/L_x)^2 + (n_y/L_y)^2 + (n_z/L_z)^2]$

$E_{111} = 4.13 \times 10^{-21}$ J, $E_{211} = E_{121} = 7.81 \times 10^{-21}$ J ($E_{112} = 9.14 \times 10^{-21}$ J)

$E_{211} - E_{111} = 3.68 \times 10^{-21}$ J $= hc/\lambda$, so $\lambda = 5.40 \times 10^{-5}$ m

The ground state is nondegenerate, but the first excited state is twofold degenerate (because $L_x = L_y$). It would have been threefold degenerate if all three dimensions had been equal.

9.2 SCHRÖDINGER DOES HYDROGEN

Now we consider the simplest atom of all—hydrogen. The demonstration that Schrödinger's equation led directly to the quantum numbers n, l, and m and energy levels that explained the electronic structure and optical properties of the hydrogen atom was what first convinced physicists that Schrödinger's wave equation accurately represented the nature of matter (and won him the Nobel Prize).

The potential energy function for the electron in the hydrogen atom is pretty simple, and represents the Coulomb attraction between the electron and the positively charged nucleus—which is just a proton in the dominant isotope of hydrogen. With r the distance between the electron and the proton, the potential energy is

$$V(r) = -\frac{e^2}{4\pi\varepsilon_0 r} \tag{9.3}$$

So Schrödinger's time-independent equation (8.16) for the hydrogen atom is

$$-\frac{\hbar^2}{2m}\nabla^2\psi - \frac{e^2}{4\pi\varepsilon_0 r}\psi = E\psi \tag{9.4}$$

The appropriate coordinates for the hydrogen atom (Fig. 9.1) are spherical coordinates (r, θ, ϕ), where θ is the polar angle (latitude, measured from the pole) and ϕ is the azimuthal angle (longitude). The Laplacian ∇^2 was simply

$$\left(\frac{\partial^2}{\partial x^2} + \frac{\partial^2}{\partial y^2} + \frac{\partial^2}{\partial z^2}\right)$$

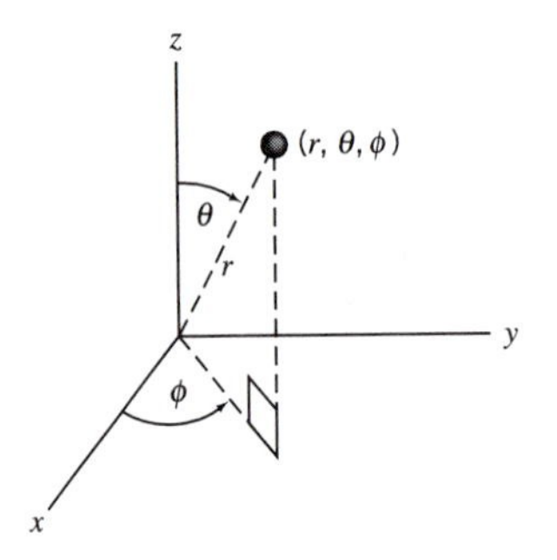

Figure 9.1. Spherical coordinates.

in Cartesian coordinates, but it is more complex in spherical coordinates:

$$\nabla^2 = \frac{1}{r^2}\frac{\partial}{\partial r}\left(r^2\frac{\partial}{\partial r}\right) + \frac{1}{r^2 \sin\theta}\frac{\partial}{\partial\theta}\left(\sin\theta\frac{\partial}{\partial\theta}\right) + \frac{1}{r^2\sin^2\theta}\frac{\partial^2}{\partial\phi^2} \quad \textbf{(9.5)}$$

Boundary conditions for the hydrogen atom are simply that the wave function should vanish as $r \to \infty$ and should be single-valued, so that $\psi(r, \theta, \phi) = \psi(r, \theta, \phi + 2\pi)$. Those interested in all the hairy details of the mathematics can consult any introductory text on quantum mechanics. We'll just outline the approach and the results. Equation (9.4) is solved using *separation of variables*, taking the wave function to be a product of separate functions, each dependent on only one of the variables:

$$\psi(r, \theta, \phi) = R(r)\Theta(\theta)\Phi(\phi) \quad \textbf{(9.6)}$$

As in Chapter 8, where we separated the space and time variables, this leads to separate differential equations for each of the separate functions. The simplest is for $\Phi(\phi)$:

$$\frac{d^2\Phi}{d\phi^2} = -m^2\Phi \quad \textbf{(9.7)}$$

the solutions to which are

$$\Phi = Ae^{im\phi} + Be^{-im\phi} \quad \textbf{(9.8)}$$

Since $\Phi(\phi)$ must equal $\Phi(\phi + 2\pi)$, $m = 0, \pm 1, \pm 2, \ldots$ The quantum number m is commonly called the *magnetic quantum number*. The electron energy in a free hydrogen atom does not vary with m (levels with different m are degenerate) unless a magnetic field is applied.

The differential equation for $\Theta(\theta)$ involves both m and another quantum number l, called the *angular momentum quantum number*, which must meet the condition $l \geq |m|$. The solutions to the equation for $\Theta(\theta)$ are called *associated Legendre functions* and are written as $P_m^l(\cos\theta)$. The products of the two angularly dependent functions, $\Theta(\theta)\Phi(\phi)$, are called *spherical harmonics*, and they appear in many problems of spherical symmetry. More about them later.

The differential equation for $R(r)$ involves m, l, and another quantum number n, called the *principal quantum number*, which must meet the condition $n \geq (l + 1)$. The solutions $R(r)$ are the products of polynomials *and the exponential function* $exp(-r/na_0)$. The quantity a_0 is known as the *Bohr radius*, and is given by

$$a_0 = \frac{4\pi\varepsilon_0\hbar^2}{me^2} = 0.0528 \text{ nm} \qquad \textbf{(9.9)}$$

The form of the damped exponential shows that wave functions will extend to larger radii (and therefore have higher, i.e., less negative, average potential energies) for larger values of n.

Solving equation (9.4) gives us not only the electron wave functions for hydrogen, which we describe in the next section, but also the electron *energies* corresponding to each wave function. The electron energy is found to depend only on the principal quantum number n:

$$E_n = -\frac{me^4}{8\varepsilon_0^2h^2n^2} = -\frac{13.6 \text{ eV}}{n^2} \qquad (n = 1, 2, 3, ..) \qquad \textbf{(9.10)}$$

Thus the energy levels are -13.6 eV, -3.4 eV, -1.51 eV, -0.85 eV, and so on, getting closer and closer together with increasing n. Energy values are negative because the potential energy function $V(z)$ is negative, with zero energy corresponding to an electron infinitely separated from the proton.

In contrast, in the one-dimensional box treated in Chapter 8, E_n is positive and proportional to n^2, and the spacing between energy levels increases with increasing energy. In the harmonic oscillator, energy levels remain evenly spaced. The differences, of course, derive from the different shapes of the potential energy functions, as shown schematically in Fig. 9.2.

Equation (9.10) had previously been obtained by Bohr with a semiclassical model that assumed an orbiting electron with quantized angular momentum, and had been shown to be consistent with the optical emission spectra of excited hydrogen atoms.

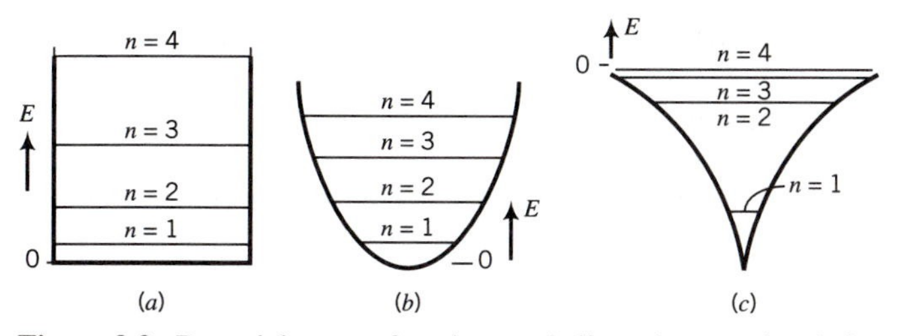

Figure 9.2. Potential energy functions and allowed energy levels for electrons in (a) one-dimensional energy well, (b) harmonic oscillator, (c) Coulomb energy well (hydrogen atom).

From (9.10), an electron transition from an initial energy level with n_i to a final level with n_f will emit a photon with energy

$$h\nu = (13.6 \text{ eV})\left(\frac{1}{n_f^2} - \frac{1}{n_i^2}\right) \tag{9.11}$$

Electrons falling into the ground state ($n_f = 1$) from various outer shells ($n_i > 1$) produce the *Lyman series* of emission lines for the hydrogen atom. Electrons falling into the second shell ($n_f = 2$) yield the *Balmer series*, and those falling into the third shell ($n_f = 3$) yield the *Paschen series*. The Paschen series of lines is in the infrared, and the Lyman series is in the ultraviolet, but several lines of the Balmer series are in the visible. (The visible range includes photons with energies from roughly 1.8 eV to 3.1 eV.)

Sample Problem 9.2
What wavelength photons are required to excite electrons from the ground state to the first excited state in hydrogen?

Solution
$h\nu = (13.6 \text{ eV})(3/4) = 1.63 \times 10^{-18} \text{ J} = hc/\lambda$, $\lambda = 1.22 \times 10^{-7}$ m

9.3 ATOMIC ORBITALS

Although the Schrödinger solution for the hydrogen atom gives the same energy levels as the simple Bohr model of the atom, the wave functions tell us that the position of the electron is not as simple as Bohr assumed. Consider the wave function for the 1s ground state ($n = 1$, $l = 0$, $m = 0$):

$$\psi_{1s}(r) = A_1 e^{-r/a_0} \tag{9.12}$$

The square of this wave function represents the probability that the electron will be located at a particular point in space. To get the probability that the electron will be at any given radius r, that is, the *radial probability function*, we must multiply the square of the wave function (probability *per unit volume*) by the volume of the spherical shell that corresponds to the radius r, $4\pi r^2 dr$. We know that the electron must be somewhere, so we can determine the normalization constant A_1 by setting the integral of the radial probability function from $r = 0$ to $r = \infty$ equal to one.

$$\int_0^\infty \{\psi_{1s}(r)\}^2 4\pi r^2 dr = \int_0^\infty A_1^2 e^{-2r/a_0} 4\pi r^2 dr = 1 \tag{9.13}$$

Carrying out this integral yields $A_1 = (\pi a_0^3)^{-1/2}$.

Figure 9.3 shows a plot of the integrand as a function of r. Although the wave function defined in (9.12) decreases monotonically with increasing r, the radial probability function $\psi^2 4\pi r^2$ goes through a maximum. The maximum of the radial probability function is reached at $r = a_0$, the radius at which Bohr pictured the electron orbiting the nucleus. However, the electron has a finite probability of being closer to, or further from, the nucleus. And it should not be pictured as orbiting, since the angular momentum is zero ($l = 0$).

Sample Problem 9.3

Show that the maximum in the radial probability function for the 1*s* electron state in hydrogen occurs at $r = a_0$.

Solution

$$\frac{d}{dr}(\psi^2 4\pi r^2) = \frac{d}{dr}(4r^2 e^{-2r/a_0}/a_0^3) = (4/a_0^3)(2r - 2r^2/a_0)e^{-2r/a_0}$$

$$(2r - 2r^2/a_0) = 0 \text{ at } r = a_0$$

The wave functions for the 2*s* ($n = 2, l = 0$) and 3*s* ($n = 3, l = 0$) wave functions are

$$\psi_{2s}(r) = A(2 - r/a_0)e^{-r/2a_0} \quad \textbf{(9.14)}$$

$$\psi_{3s}(r) = A\{27 - 18(r/a_0) + 2(r^2/a_0^2)\}e^{-r/3a_0} \quad \textbf{(9.15)}$$

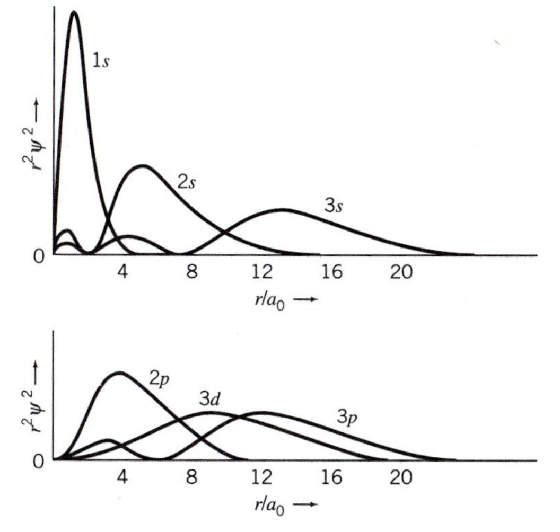

Figure 9.3. Radial probability functions for (a) 1*s*, 2*s*, and 3*s* and (b) 2*p*, 3*p*, and 3*d* electrons of hydrogen.

The normalization factors A will be different for each wave function, and can be determined in the same manner we determined A_1. The polynomials in front of the exponentials are known to mathematicians as Laguerre polynomials (*c'est Laguerre*).

Sample Problem 9.4

Write the integral used to determine the normalization factor for the 2*s* wave function of hydrogen.

Solution

$$\int_0^\infty \{\psi_{2s}(r)\}^2 4\pi r^2 dr = \int_0^\infty A_{2s}^2(2 - r/a_0)^2 e^{-r/a_0} 4\pi r^2 dr = 1$$

The radial probability functions for these wave functions are also plotted in Fig. 9.3. They show n maxima and $(n - 1)$ radial nodes (radii where $\psi = 0$, not counting the node as $r \rightarrow \infty$). The shifting of the probability function to larger radii with increasing n shows that the *potential* energy of the electron increases with increasing n.

Calculation of the expectation value of potential energy in the 1*s* state involves integration of $V(r)$ times the radial probability function. This integration yields -27.2 eV. Since the *total* energy of this state is -13.6 eV, the average kinetic energy must be $+13.6$ eV. It is clear from equation (9.3) that the electron could significantly lower its potential energy by shifting to smaller radii. *However, that decreased uncertainty in position would, via the uncertainty principle, lead to increased momentum and kinetic energy.* The ground state of -27.2 eV $+ 13.6$ eV $= -13.6$ eV, and the corresponding radial probability function in Fig. 9.3, represent a minimum in the *total* energy E.

Fig. 9.3 also shows the radial probability functions for the 2*p* ($n = 2, l = 1$), 3*p* ($n = 3, l = 1$) and 3*d* ($n = 3, l = 2$) wave functions. (Electrons and their wave functions are traditionally designated with their principal quantum number n, followed by *s*, *p*, *d*, or *f*, for $l = 0, 1, 2,$ or 3, respectively.) The *p* functions have one less radial node, and the *d* functions two less radial nodes, than their corresponding *s* functions. However, these angularly dependent functions make up for that with nodes in the spherical harmonics $\Theta(\theta)\Phi(\phi)$. For the magnetic quantum number $m = 0$, the 2*p* wave function is

$$\psi_{2p,m=0}(r) = A(r/a_0)e^{-r/2a_0} \cos\theta \tag{9.16}$$

This function has a node in the equatorial plane where $\theta = \pi/2$. If we identify the pole with the z-axis, we call this the p_z wave function (Fig. 9.4). The other two 2*p* wave functions, with $m = \pm 1$, are

$$\psi_{2p,m=\pm 1}(r) = A(r/a_0)e^{-r/2a_0} \sin\theta \exp(\pm i\phi) \tag{9.17}$$

The sum and difference of these two solutions can be formed if we wish them to represent p_x and p_y specifically (Fig. 9.4):

$$\psi_{2p_x}(r) = A(r/a_0)e^{-r/2a_0} \sin\theta \cos\phi \tag{9.18}$$

$$\psi_{2p_y}(r) = A(r/a_0)e^{-r/2a_0} \sin\theta \sin\phi \tag{9.19}$$

The angular dependence of d functions (Fig. 9.4) includes an additional node. The simplest to represent mathematically is the $3d$ wave function for $m = 0$:

$$\psi_{3d,m=0}(r) = A(r/a_0)^2 e^{-r/3a_0}(3\cos^2\theta - 1) \tag{9.20}$$

The nodes for this function are two cones with $\theta = \cos^{-1}(\pm 1/\sqrt{3})$.

As with the box and the harmonic oscillator, confining a matter wave in the hydrogen atom leads to quantized energy levels and a series of corresponding wave functions that develop more nodes with increasing energy. The three-dimensionality of the problem leads to three quantum numbers (plus the spin quantum number). Although the s, p, d, and f functions all have very different shapes, *the energy, as given in (9.10), depends only on the principal quantum number n.* Thus, in addition to the twofold spin degeneracy of each solution, there is additional degeneracy in the hydrogen atom: the $2s$ and $2p$ levels are degenerate, the $3s$, $3p$, and $3d$ levels are degenerate, and so on. This degeneracy is related to the spherical symmetry of the potential energy function (9.3). We shall see later that some of this degeneracy is removed in multielectron atoms and in molecules.

Although quantum mechanics and the solutions of Schrödinger's equation tell us that Bohr's model of particle-like electrons traveling in orbits around the nucleus was naïve, we retain the name *orbital* for electron wave functions like those in

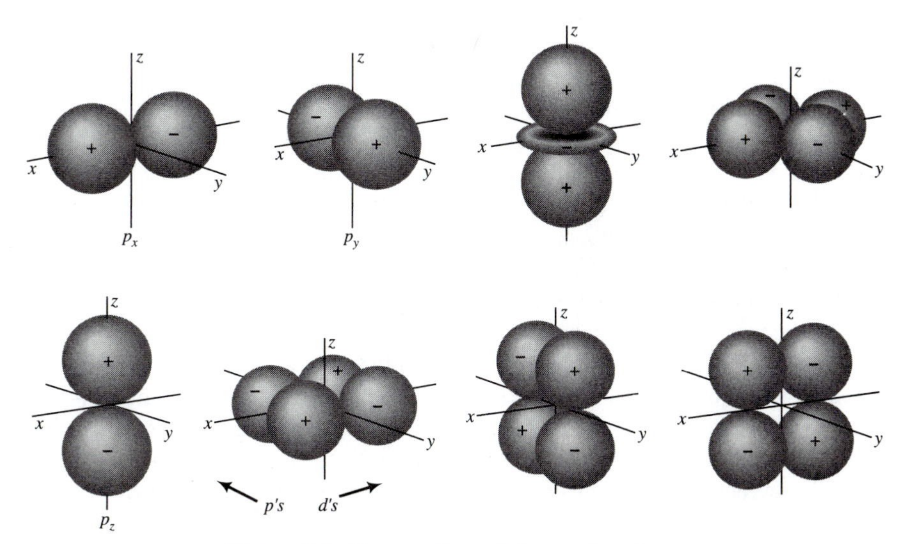

Figure 9.4. Shapes of p and d wave functions.

equations (9.12), (9.14), (9.15), and (9.16). Since these wave functions specifically describe electron distributions in an atom, we call them *atomic orbitals*, abbreviated as AOs. We will later encounter wave functions that describe electron distributions in molecules (*molecular orbitals* or MOs) and in crystals (*crystal orbitals* or COs).

9.4 BEYOND HYDROGEN—SCREENING AND THE PERIODIC TABLE

Schrödinger's equation gave us well-defined mathematical forms for the AOs of the hydrogen atom. That's not true for the other hundred-odd elements of the periodic table (some odder than others), unless we consider them in the form of *one-electron ions* (He^{+}, Li^{+2}, Be^{+3}, etc.). In this case, for a nucleus with atomic number Z and charge +Ze, the potential energy for the one electron is simply

$$V(r) = -\frac{Ze^2}{4\pi\varepsilon_0 r} \tag{9.21}$$

For these one-electron ions, all the wave functions are simply the hydrogen AOs with the Bohr radius a_0 replaced by a_0/Z (and the normalization factors increased by $Z^{3/2}$). The increased charge on the nucleus pulls electrons closer to the nucleus than in the hydrogen atom. And the energy levels are given by

$$E_n = -\frac{Z^2 me^4}{8\varepsilon_0^2 h^2 n^2} = -\frac{Z^2(13.6 \text{ eV})}{n^2} \tag{9.22}$$

The binding energies are seen to increase by the factor Z^2, with one Z coming directly from equation (9.21) and the other from the decrease in radius by a factor Z.

However, the neutral atoms are another story, since they have more than one electron, and electron-electron repulsion must be taken into account. Consider an atom of sodium (Z = 11), which has two 1*s* electrons, two 2*s* electrons, six 2*p* electrons, and one 3*s* electron. If we were to use formula (9.22) blindly, we would predict energies of the 1*s*, 2*s*, 2*p*, and 3*s* electrons of sodium to be −1650, −412, −412, and −183 eV, respectively. It is possible to measure each of these binding energies, and the experimental results are −1060, −71, −38, and −5.1 eV. Obviously the interelectron forces have a big effect, especially on the electrons in the outer shells!

The biggest discrepancy between equation (9.22) and experiment was for the 3*s* valence electron, which has a binding energy of only 5.1 eV, while equation (9.22) predicted 183 eV. We can understand this qualitatively in terms of the concept of *screening*. We know that, most of the time, the 3*s* electron is further from the nucleus than the 10 electrons in the inner shells. In effect, the 10 negatively charged core electrons "screen" the 3*s* electron from the full +11e charge of the nucleus. If they *completely* screened the 3*s* electron, it would see only a net charge of +11e − 10e = 1e, and (9.22) would predict a binding energy of only 13.6/9 = 1.5 eV. But the 3*s* electron is not *always* farther from the nucleus than all the other electrons, since the wave functions represent only probability distributions. The 3*s* electron is not *always* completely screened. If we set the experimental values of binding energy equal to equation (9.22) with an *effective nuclear charge* eZ_{eff}, the experimental value of 5.1 eV indicates that the 3*s* electron of sodium sees, on the average, an effective nuclear charge of 1.8e.

Sample Problem 9.5

From the data given above, what effective nuclear charge is seen by the 1*s*, 2*s*, and 2*p* electrons of sodium?

Solution

$E_{1s} = -1060 \text{ eV} = -Z_{\text{eff}}^2(13.6 \text{ eV})/1$, so $eZ_{\text{eff}} = 8.83\text{e}$
$E_{2s} = -71 \text{ eV} = -Z_{\text{eff}}^2(13.6 \text{ eV})/4$, so $eZ_{\text{eff}} = 4.57\text{e}$
$E_{2p} = -38 \text{ eV} = -Z_{\text{eff}}^2(13.6 \text{ eV})/4$, so $eZ_{\text{eff}} = 3.34\text{e}$
Z_{eff} decreases, that is, screening increases, with increasing n and, for a given n, with increasing l.

The discrepancy between experimental binding energies and (9.22) is least for the 1*s* electrons, because, being on average closer to the nucleus, they are less screened than the other electrons. The two 1*s* electrons do, however, screen each other significantly.

For the electrons with $n = 2$, screening has not only produced a weaker binding energy than predicted by equation (9.22), but it has also *removed the degeneracy between 2s and 2p electrons*. The energy of a 2*s* electron in Na is −71 eV and that of a 2*p* electron is −38 eV, whereas in the hydrogen atom the electron energy depended only on n. This removal of degeneracy is a result of the different radial dependences of the wave functions. The 2*s* electrons have a nonzero probability of being near the origin whereas the 2*p* electron does not, as can be seen by comparing (9.14) with (9.16). On the average, the 2*s* electrons are therefore *less screened* than the 2*p* electrons.

Similarly, *differential screening* removes the *s*-*p*-*d*-*f* degeneracies in other shells, leading to the relative energies that, along with the Pauli principle, are used in introductory chemistry to explain the organization of the periodic table. For example, if equation (9.22) were valid, 4*s* electrons would have energies considerably higher than 3*d* electrons. However, differential screening raises the energy of 3*d* electrons more than the energy of 4*s* electrons, so that, in K ($Z = 19$) and Ca ($Z = 20$), 4*s* levels are filled before 3*d* levels. Degeneracies associated with different values of the magnetic quantum number m, and with the spin quantum number s, remain (in the absence of an applied magnetic field).

9.5 THE SCF APPROXIMATION

Although the concept of screening and an effective nuclear charge Z_{eff} can *qualitatively* explain the binding energies of electrons in multielectron atoms, it is also possible to calculate their binding energies quantitatively and accurately using Schrödinger's equation. However, these calculations require the use of an approximation method known as the *self-consistent field* (SCF).

Consider helium ($Z = 2$). Each electron will not only be attracted to the nucleus, but will also be repelled by the other electron. If r_1 is the distance between electron 1 and the nucleus, and r_{12} is the distance between the two electrons (Fig. 9.5), the total potential energy function for electron 1 is

$$V_1 = -\frac{e^2}{4\pi\varepsilon_0}\left(\frac{2}{r_1} - \frac{1}{r_{12}}\right) \tag{9.23}$$

A similar equation describes the potential energy function for electron 2. Schrödinger's equation for even this simple atom cannot be solved exactly, since the potential energy function for electron 1 depends on knowing where electron 2 is, and vice versa. In equation (9.23), the second term can be replaced by an averaged position distribution for electron 2, that is, the second term in the bracket replaced by the integral

$$\int \frac{\psi_2^2 dV}{r_{12}}$$

where ψ_2 is the wave function for electron 2 and dV is a differential volume element. (This term accounts for the Coulomb repulsion between the electrons. In the more advanced Hartree-Fock treatment, an additional "exchange" term must be added.)

In the case of helium, electron 2 and electron 1 are both 1*s* electrons and, except for spin, they have the same wave function, that is, $\psi_1 = \psi_2$. We can't solve for these wave functions until we know the potential energy function V, but we don't know the potential energy function V until we know the wave function. It's a catch-22.

The SCF method starts with a good guess for the wave function, calculates the resulting potential energy function V, and then solves Schrödinger's equation with that V. The solution is another wave function, which allows us to calculate another V, and then we solve Schrödinger's equation with that new V. This gives us a third

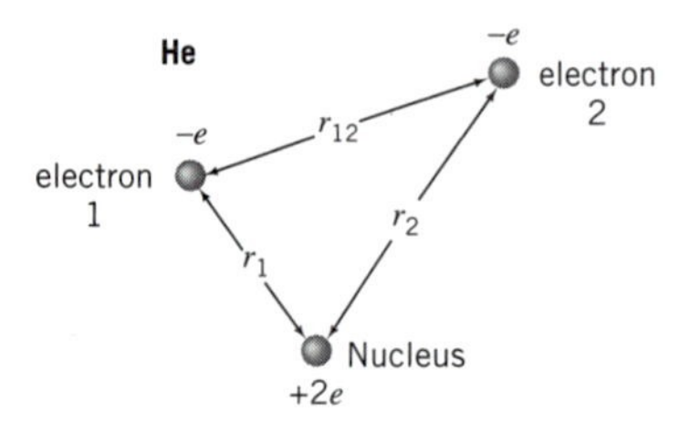

Figure 9.5. Helium atom contains two electrons, which repel each other.

wave function, from which we can calculate a third V, from which. . . . This process is reiterated until self-consistency is obtained, that is, until the wave-function solution to Schrödinger's equation for a particular V yields, to a good approximation, the same V.

With the help of a powerful computer, the same SCF process can be used for atoms with *many* electrons, in which the V for each electron depends on the wave functions of all the others. A good guess is made for the wave functions, V is calculated, Schrödinger's equation is solved to produce a new set of wave functions, from which a new V is calculated, and so on. The angular-dependent parts of the resulting wave functions are identical to those of the one-electron systems, yielding orbitals with familiar s, p, d, and f symmetries. The SCF *radial* wave functions are qualitatively similar to the one-electron functions but quantitatively different.

Examples of radial probability functions ($\propto r^2\psi^2$) resulting from SCF calculations are shown in Fig. 9.6a for the 1s, 2s, 2p, 3s, and 3p electrons of argon ($Z = 18$). The *total* radial probability function for argon is shown in Fig. 9.6b, indicating that to a fair approximation one can still consider the electrons in each shell (each value of the principal quantum number n) to be located primarily in a restricted range of r. The figure also shows the function $Z(r)$ that represents the effective nuclear charge seen at each radius and shows directly the effects of *screening*. Consistent with our qualitative arguments given earlier, the effective nuclear charge Z_{eff} approaches Z as $r \to 0$ and approaches one as $r \to \infty$.

Similar results have been obtained for every atom in the periodic table. Along each row of the periodic table, atoms get smaller with increasing Z because of in-

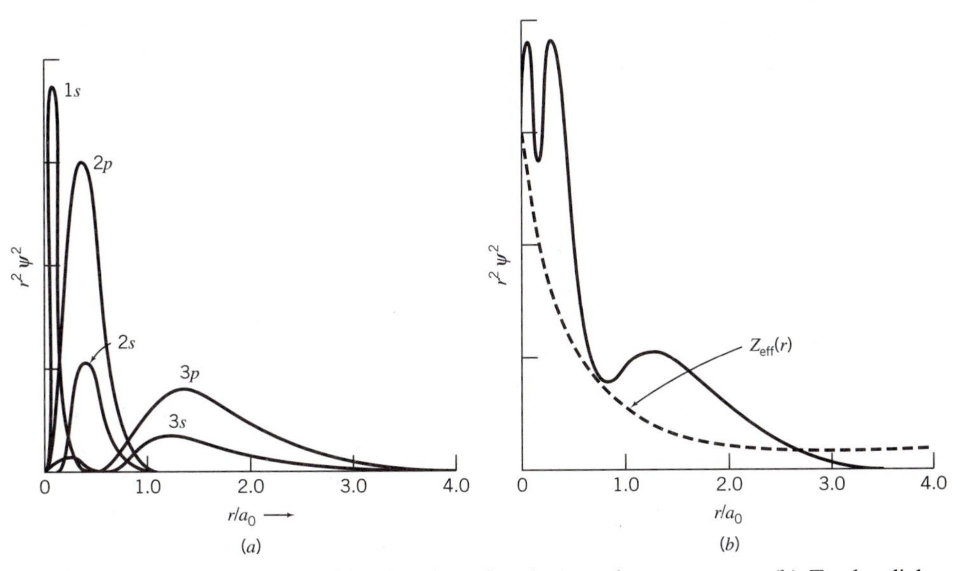

Figure 9.6. (a) Radial probability functions for electrons in argon atom. (b) Total radial probability function and $Z_{\text{eff}}(r)$ for argon atom. (Adapted from R. M. Eisberg, *Quantum Physics of Atoms, Molecules, Solids, Nuclei, and Particles*, John Wiley & Sons, New York, 1985.)

creased nuclear charge, but atoms get larger when the increase in Z leads to the occupation of additional shells. Thus the largest neutral atoms are at the lower left of the periodic table, the smallest at the upper right. Perhaps more important than the radial electron density distributions resulting from SCF calculations are the calculated *energy levels*. Electron energy levels are known experimentally from optical spectra and other measurements, and agreement between SCF calculations and experiment is good.

Just as hydrogen has its Lyman, Balmer, and Paschen series of emission lines, each atom of the periodic table has characteristic emission and absorption lines that are dependent on the differences between electron energy levels. These lines serve as "fingerprints" that enable scientists to identify atomic species, even in faraway stars. (Helium was actually detected from solar spectra before it was discovered on earth, and its name is derived from *helios*, Greek for the sun.) When strong emission lines are in the visible, electron transitions in excited atoms can be used to produce illumination. In neon lights, many electrons in excited neon atoms make energy transitions that emit red photons. Similarly, excited neon atoms emit red photons in helium-neon lasers, which you may have encountered at supermarket checkouts. Other atoms of the periodic table emit different colors (and many nonvisible photons), but neon atoms, like many of us, turn red when excited.

Although hydrogen is the only atom of the periodic table for which neat mathematical expressions can be written for the electron wave functions and their energy levels, qualitative features such as quantum numbers and the general shapes of s, p, d, and f AOs are preserved in multielectron atoms. *Iterative SCF calculations allow us to make accurate quantitative determinations of the energy levels and corresponding wave functions of all the electrons of any element.* These AOs and their energies can be used as a basis for quantitative considerations of electronic structure and bonding in molecules and solids.

SUMMARY

Sometimes wave functions with different quantum numbers have the same energy. These wave functions, and their corresponding energy levels, are said to be *degenerate*. Degeneracy is closely related to symmetry, with greater symmetry generally leading to greater degeneracy.

Schrödinger's equation for the hydrogen atom involves a simple Coulomb potential energy function ($\propto -1/r$). The equation is solved in spherical coordinates using separation of variables, yielding separate solvable differential equations for each of the variables. Solutions to the equations in the two angular variables θ and ϕ are a series of spherical harmonics, while solutions to the radial differential equation are polynomials multiplied by the damped exponential $\exp(-r/na_0)$. The mathematics produces three quantum numbers—the principal quantum number n, the angular momentum quantum number l, and the magnetic quantum number m—but energy depends only on n and varies as $-1/n^2$. Thus levels with the same principal quantum number n, but different values of l and/or m, are degenerate in the hydrogen atom.

Schrödinger's equation gave exact mathematical forms of the wave functions, or *atomic orbitals* (AOs), of the hydrogen atom and the corresponding energy levels. The solutions can also be extended to one-electron ions of other elements, like He^+ or Li^{+2}. However, for multielectron atoms or ions, the Coulomb repulsion between electrons complicates the problem. Electrons partially screen each other from the full nuclear charge. The degeneracy of energy levels with the same principal quantum number n but different angular momentum quantum number l (e.g., $2s$ and $2p$ levels) that existed in the hydrogen atom is removed by the effects of screening in multielectron atoms (e.g., s electrons are less screened than p electrons of the same shell, and therefore have lower energy). Differential screening leads to $4s$ electron levels being occupied before $3d$ levels, and so on, and explains the atomic sequence in the periodic table.

The wave functions and energy levels of multielectron atoms and ions can be calculated with the self-consistent field (SCF) method, an iterative approach capable of determining approximate wave functions for all electrons of any atom. Energy levels calculated by SCF are in excellent agreement with experiment.

PROBLEMS

9-1. What wavelength photons are required to excite electrons from the ground state to the first excited state in (a) a one-dimensional energy well 6 nm wide? (b) a two-dimensional energy well 6 nm on a side? (c) a two-dimensional energy well 6 nm by 5 nm? (d) a three-dimensional energy well 6 nm on a side? (e) In each of the four cases, what is the degree of degeneracy of the excited state?

9-2. For the $1s$ electron in hydrogen, what is the probability that the electron will be located no more than two Bohr radii from the nucleus?

9-3. Using the $1s$ wave function for hydrogen, calculate the expectation value of potential energy of this electron. From this result and total $1s$ energy, what is the average kinetic energy of this electron?

9-4. Calculate the locations of the nodal surfaces of the $2s$ and $3s$ electrons in (a) H, (b) He^{+1}, and (c) Li^{+2}.

9-5. Calculate the average (expectation value) of the radius of the $1s$ electron of hydrogen.

9-6. From data given in Section 9.4, calculate the effective nuclear charge eZ_{eff} for the $1s$, $2s$, $2p$, and $3s$ electrons of sodium. In each case, what percentage of the full nuclear charge is "screened" by other electrons?

9-7. (a) The first ionization energies of Li, K, and Cs are 5.39, 4.34, and 3.89 eV, respectively. Compare these values with what the first ionization energies would be if there were no screening and equation (9.22) applied. (b) In each case, what is the effective nuclear charge eZ_{eff} seen by the valence electron, and what percentage of the full nuclear charge is screened by the inner electrons?

9-8. Briefly explain how $4s$ electrons can have lower energy than $3d$ electrons despite having a higher principal quantum number n.

9-9. The most probable radius of $1s$ electrons among the elements varies approximately as $1/Z$, and the binding energy of $1s$ electrons increases approximately as Z^2. Estimate the most probable radius and binding energy of $1s$ electrons in uranium ($Z = 92$). Assuming that the kinetic energy of the $1s$ electron, like that of hydrogen, is equal to its binding energy, what is the approximate velocity of this electron?

9-10. The potential energy of an electron in a hydrogen atom continually decreases with decreasing r. However, as r decreases, the uncertainty principle leads to an increasing uncertainty in momentum and a corresponding increase in kinetic energy. Assuming that the electron momentum can be approximated by h/r, at what radius is the total energy a minimum? How does this compare to the Bohr radius?

9-11. Calculate the expectation value of the potential energy of a $2s$ electron in hydrogen. From this and the total energy given by (9.10), what is the kinetic energy?

9-12. What is the most probable radius of $1s$ electrons in He^+ and Li^{+2}?

9-13. The neutral Ne atom and the Na^+ ion are *isoelectronic* (each contains 10 electrons). How will the radial extent and the energies of the electrons in Na^+ compare with those of the corresponding electrons in Ne?

9-14. From (9.4) and (9.5), find the differential equation that $R(r)$ for s atomic orbitals of hydrogen must satisfy, and show that (9.12) satisfies this equation.

Chapter 10

The Game Is Bonds, Interatomic Bonds

10.1 THE SIMPLEST MOLECULE

With Schrödinger's equation and the self-consistent field (SCF) method, approximate wave functions and accurate energy levels have been determined for any electron of any isolated atom or ion. Our ultimate goal, however, is to understand electron wave functions and energy levels in solids, which may contain 10^{23} atoms or more. One approach to solids is to start with the atomic orbitals (AOs) and energy levels and to see how they are modified by bonding with other atoms. We'll use this approach, specifically the *molecular orbital* method popular with chemists, in this and the next chapter. This is related to what physicists call the *tight binding* method.

We started our consideration of atoms with the simplest atom of all—hydrogen, which has only one proton and one electron. We start our consideration of interatomic bonding with the simplest molecule of all—the hydrogen molecular ion H_2^+. This molecule has two separate nuclei (protons) separated by a distance R, and one electron, separated by distances r_1 and r_2 from the two protons (Fig. 10.1).

Two protons separated by a distance R repel each other with a Coulomb force of $e^2/4\pi\varepsilon_0 R^2$. If we put an electron exactly at the midpoint between the two protons, it will attract each proton toward it with a force of $e^2/\pi\varepsilon_0 R^2$, which is four times larger than the repulsive force between the protons because the distance between the electron and each proton is only $R/2$. Properly placed electrons produce attractive

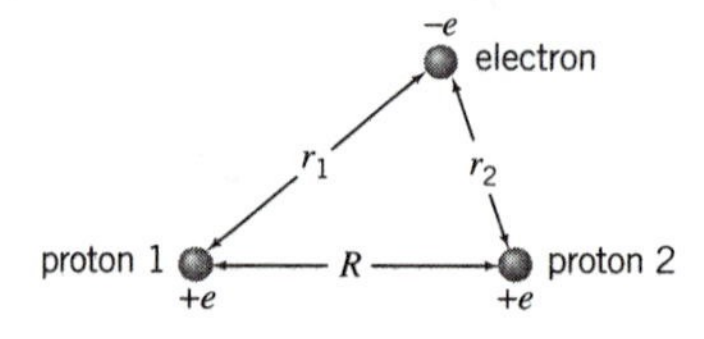

Figure 10.1. The hydrogen molecular ion H_2^+.

Coulomb forces that can overcome repulsive Coulomb forces. *This is the basis of chemical bonding.*

Classically, of course, the two protons would converge on the electron. But just as quantum mechanics and the uncertainty principle kept the electron of the hydrogen atom from converging on the nucleus, here they will keep H_2^+ from collapsing to a point. This chapter will focus on an approximation method used to solve Schrödinger's equation for this and other interatomic bonds.

Like the helium atom (pictured in Fig. 9.5), H_2^+ is a "three-body" problem. Whereas He has two electrons and one nucleus, H_2^+ has one electron and two nuclei. This is actually a simpler problem. Because this one electron is much lighter than the protons, it can move much more rapidly, and it can adjust its wave function rapidly in response to any changes in the internuclear separation R. Thus we can use the *Born-Oppenheimer approximation*:

1. Assume the protons have a fixed separation R and solve Schrödinger's equation for the electron, with the potential energy function for the electron expressed by

$$V = -\frac{e^2}{4\pi\varepsilon_0}\left(\frac{1}{r_1} + \frac{1}{r_2}\right) \tag{10.1}$$

   This will give us electron wave functions and $E_{el}(R)$, the electron energy levels as a function of the internuclear distance.

2. To get the total energy E_t of the molecule, add the electron energy $E_{el}(R)$ to the energy term representing the Coulomb repulsion between the two protons:

$$E_t(R) = E_{el}(R) + \frac{e^2}{4\pi\varepsilon_0 R} \tag{10.2}$$

Using the fact that the protons are heavy and clumsy compared to the nimble electron, the Born-Oppenheimer approximation allows us to turn the three-body problem into solving the Schrödinger equation for the one electron, with the electron's potential energy expressed in equation (10.1), and then adding the proton-proton repulsion term separately. It turns out that for this simple molecule, Schrödinger's equation can actually be solved exactly using elliptical coordinates. However, this isn't true for other molecules or solids, and it will be more instructive to attack this problem with an approximation method of very general applicability—the method of *linear combination of atomic orbitals* (LCAO).

10.2 MOS FROM LCAO

The LCAO method is a variant of the *variation method* used to find approximate solutions to differential equations like Schrödinger's wave equation. The general variation method involves taking trial wave functions that include some adjustable parameters, calculating the expectation value of energy for that trial wave function, and then minimizing the energy with respect to variations in the adjustable parameters. The better your trial wave function, and the more parameters you include, the closer you can get to the actual solutions.

It will be convenient here to introduce a bit of mathematical shorthand. We write the time-independent Schrödinger's equation (8.16) simply as

$$\mathcal{H}\psi = E\psi \qquad \text{where} \qquad \mathcal{H} = -\frac{\hbar^2}{2m}\nabla^2 + V \tag{10.3}$$

The operator $\mathcal{H}$ is known as the *Hamiltonian operator.* (The name derives from a method of analysis in classical mechanics developed in the 19th century by the Irish mathematician and astronomer William Hamilton.) The first term represents kinetic energy and the second term represents potential energy. Using equation (8.33), the expectation value of the sum of kinetic and potential energies of the electron is

$$\langle E \rangle = \frac{\int \psi^* \mathcal{H} \psi dV}{\int \psi^* \psi dV} \tag{10.4}$$

It is this quantity that you minimize with respect to various adjustable variables to obtain approximate solutions via the variation method.

For our hydrogen molecular ion H_2^+, our Hamiltonian is defined by the potential energy function in (10.1). When the electron is near proton 1, its behavior is presumably dominated by that proton, and its lowest-energy wave function there would be similar to the 1s wave function centered on proton 1. On the other hand, when the electron is near proton 2, its wave function would be expected to be similar to the 1s wave function centered on proton 2. Recalling our expression (9.12) for the 1s wave function, and the normalization constant we determined with (9.13), these two normalized AOs are

$$\phi_1 = \frac{E^{-r_1/a_0}}{(\pi a_0^3)^{1/2}} \qquad \text{and} \qquad \phi_2 = \frac{e^{-r_2/a_0}}{(\pi a_0^3)^{1/2}} \tag{10.5}$$

This suggests a trial wave function for the H_2^+ molecule:

$$\psi = c_1\phi_1 + c_2\phi_2 \tag{10.6}$$

This *molecular orbital* (MO, the electron wave function for a molecule) is a linear combination of atomic orbitals (LCAO). When the electron is near either proton, ψ will be dominated by one of the AOs. However, at positions near the midplane between the two protons, both AOs will make significant contributions.

The MO in (10.6) is then plugged into (10.4) to calculate the expectation value of the energy. In a more general case, this energy would then be minimized with respect to c_1 and c_2. However, in this particular molecule, symmetry across the midplane can be used to conclude that $c_1 = \pm c_2$. Simplifying the results a bit, the positive sign ($c_1 = c_2$) corresponds to an energy given by

$$E_{el}(R) = H_{11} + H_{12}(R) \qquad \textbf{(10.7)}$$

while the negative sign ($c_1 = -c_2$) corresponds to a *higher* energy

$$E_{el}(R) = H_{11} - H_{12}(R) \qquad \textbf{(10.8)}$$

The energy (10.8) is higher than (10.7) because H_{12} is negative. This term is called the *bond integral* and is defined by

$$H_{12} = \int \phi_1 \mathcal{H} \phi_2 dV = \int \phi_2 \mathcal{H} \phi_1 dV \qquad \textbf{(10.9)}$$

These two integrals are equal by symmetry, and they are negative because the potential energy (10.1) is negative. The term H_{11} ($= H_{22}$) represents a corresponding integral called the *Coulomb integral*, which for present purposes we can consider a constant (for a particular Hamiltonian $\mathcal{H}$). A more complete treatment also includes the overlap integral $S_{12} = \int \phi_1 \phi_2 dV$, which we can ignore for current purposes.

For widely separated protons ($R \to \infty$), the two AOs are widely separated (since one is centered on proton 1, the other on proton 2). Thus at every point in space either ϕ_1 or ϕ_2 is nearly zero, and therefore the bond integral H_{12} is nearly zero. Both (10.7) and (10.8) therefore approach

$$E_{el}(R \to \infty) = H_{11} \qquad \textbf{(10.10)}$$

This is the energy of the reference state where there is no overlap, no bonding. The electron is bound either to one proton or to the other, and we have two degenerate states, each corresponding to one neutral hydrogen atom plus one hydrogen ion (a lone proton). Instead of the molecular ion H_2^+, we have $H+H^+$ (or $H^+ + H$). If we now imagine the protons gradually approaching each other, that is, R gradually decreasing, there will eventually be some overlap of the two AOs, positions between

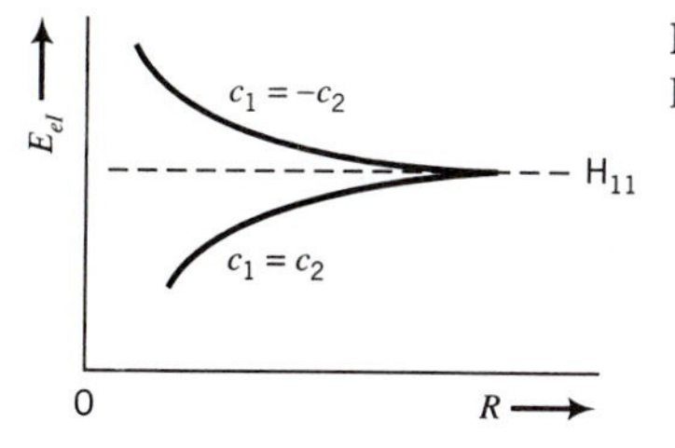

Figure 10.2. Electron energies E_{el} of the two MOs for H_2^+ as a function of internuclear separation R.

the two protons where both ϕ_1 and ϕ_2 are nonzero. Therefore H_{12} will no longer be zero. As R decreases, (10.7) will gradually decrease to energies below H_{11}, and (10.8) will gradually *increase* to energies *above* H_{11} (Fig. 10.2).

It is primarily the bond integral H_{12} defined in equation (10.9) that produces the driving force for the formation of molecules and solids—for forming interatomic bonds. Much of this chapter and the next will be focused on bond integrals and the factors that determine them. Examination of the integral in (10.9) shows that H_{12} depends directly on the spatial overlap of AOs from adjoining atoms.

We have completed what we described earlier as step 1 of the Born-Oppenheimer approximation. We have calculated $E_{el}(R)$, and have gotten two separate curves, one corresponding to $c_1 = c_2$ and the other corresponding to $c_1 = -c_2$. Next is step 2, represented above by equation (10.2). To get the *total* energy of the molecule corresponding to each of these two electron states, we add the term $e^2/4\pi\varepsilon_0 R$ corresponding to the repulsive energy between the two protons. As seen in Fig. 10.3, the lower-energy electron state yields a total energy function $E_t(R)$ that has a minimum, while $E_t(R)$ for the higher-energy state instead rises monotonically with decreasing R.

The energy minimum in the lower curve indicates a stable molecule, and determines the equilibrium bond energy E_b and equilibrium internuclear spacing R_0 for the molecule. This lower curve corresponds to the molecular orbital with $c_1 = +c_2$, which is called the *bonding MO*. The upper curve corresponds to the molecular orbital with $c_1 = -c_2$, which is called the *antibonding MO*. With an electron in this higher energy state, the molecule is unstable.

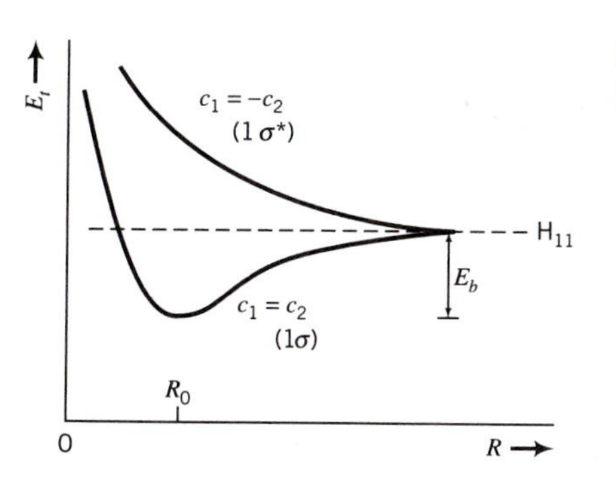

Figure 10.3. Total energy of the two MOs of H_2^+ as a function of internuclear separation R.

Sample Problem 10.1

The equilibrium internuclear separation R_0 for the hydrogen molecular ion is 0.108 nm. What is the attractive force produced by the electron at this spacing?

Solution

At equilibrium, there is no net force, that is, the attractive force produced by the electron just balances the repulsive Coulomb force between the two protons, which is $e^2/4\pi\varepsilon_0 R_0^2 = 1.97 \times 10^{-8}$ newtons.

These MOs were built from 1*s* AOs. Because the 1*s* AOs have spherical symmetry, the MOs built from them have cylindrical symmetry, that is, circular symmetry around the internuclear axis. Such MOs are called *sigma orbitals*. The bonding MO we label 1σ, and the antibonding MO we label $1\sigma^*$. Figure 10.4a shows the variation of ψ with distance along the molecular axis, and Fig. 10.4b shows a schematic 3-D representation of the shape of these two MOs. The bonding MO is symmetric across the molecular midplane, while the antibonding MO is antisymmetric.

The region *between* the two protons is where the attractive force between the electron and the protons overcomes the repulsive force between the protons (Fig. 10.5). An electron in the bonding MO has substantial probability of being there, but an electron in the antibonding MO does not. In fact, for the antibonding MO the midplane of the molecule is a nodal plane (a plane where the electron wave function and hence the probability density are zero). An electron in the antibonding MO instead spends more time in locations where the attractive force between the electron and the protons actually tends to pull the protons apart.

The MOs are the electron wave functions for the molecule. As with AOs, the *square* of the wave function represents the probability distribution. Thus the probability distribution of *both* MOs is symmetric. At any given time, the electron in the bonding MO could be near one proton or near the other. But averaged over time, since the wave function is symmetric, the electron is *equally shared* by the two protons. This bond is a *pure covalent* bond.

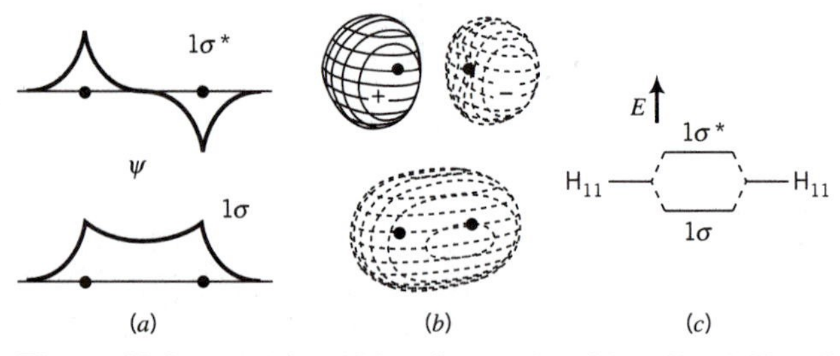

Figure 10.4. (a) Value of bonding and anti-bonding MOs of H_2^+ along molecular axis, (b) Schematic 3-D shape of 1σ and $1\sigma^*$ MOs, (c) Energy levels of MOs.

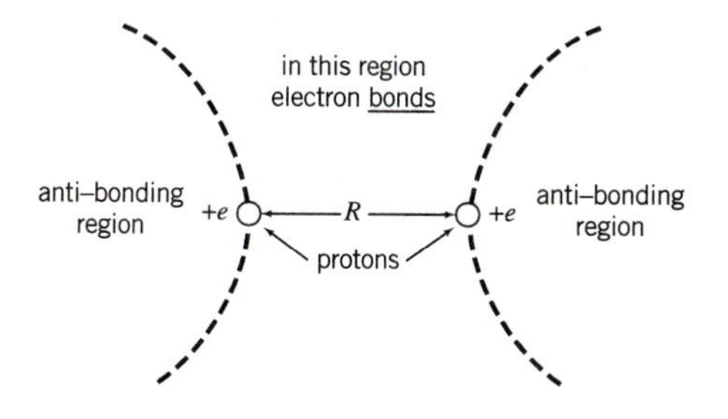

Figure 10.5. Showing the electron locations where it contributes to pulling the two protons together and where it contributes to pulling them apart.

Figure 10.4c shows the two energy levels corresponding to the two MOs for the equilibrium spacing R_0. The two degenerate energy levels H_{11}, corresponding to separated protons with the electron bound to one or the other of the protons, have been replaced by two MO energy levels, one of lower energy and one of higher energy. This is our first example of the rule of conservation of energy levels, usually expressed as *conservation of states*. Two AO levels, one from each atom, are replaced by two MO levels.

Many students have trouble at first with the concept of antibonding MOs and antibonding energy levels. It is clear from Fig. 10.3 that if the electron were in the antibonding MO, the molecule would not be stable. But electrons have no mandate to produce stable molecules. They are required only to satisfy Schrödinger's wave equation, and antibonding MOs satisfy that equation as well as bonding MOs do. If, as in this case, we have only one electron, it will certainly prefer to be in the bonding MO. But it prefers to be there because it has lower energy, *not* because it feels any requirement to provide bonding. Fortunately for us, the state that corresponds to lower energy for the electron corresponds to interatomic bonding. That allows atoms to form molecules, to form solids, and to form us.

The simple calculation we have described yields a bond energy and internuclear spacing for H_2^+ that is in rough agreement with experiment. A slightly more sophisticated LCAO calculation, using additional parameters, yields essential agreement with the experimental values of $E_b = 2.7$ eV and $R_0 = 0.108$ nm.

10.3 THE NEUTRAL HYDROGEN MOLECULE

If we add a second electron to our two-proton system, we have the neutral H_2 molecule. We must then account for the repulsion between the two electrons with the SCF method discussed in Chapter 9, but we end up with similar results. In the ground state (lowest-energy state of the molecule), the two electrons, with opposite spins, will both occupy the bonding MO. With *two* electrons providing the bonding, the equilibrium internuclear spacing decreases to 0.076 nm and the bond energy increases to 4.7 eV. Figure 10.6 shows schematically the $E_t(R)$ curves for both H_2^+ and H_2 with electrons in the bonding MO. (For the neutral H_2 molecule, the reference state at $R \rightarrow \infty$ is two neutral hydrogen atoms (H+H), whereas it was $H+H^+$ for the H_2^+ molecular ion.)

Although the purely covalent bond in H_2 involves *two* bonding electrons, chemists

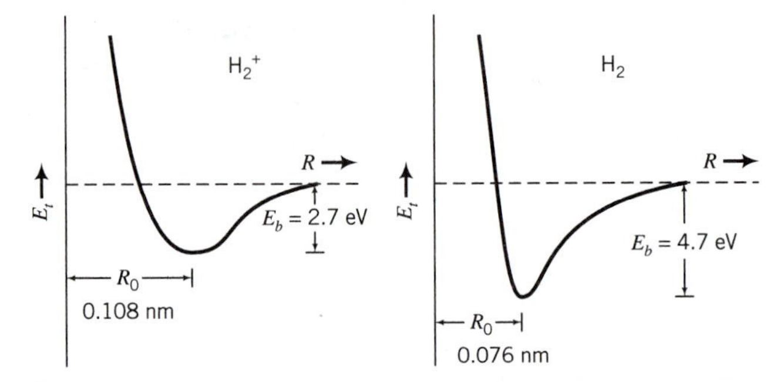

Figure 10.6. Energy vs. separation curves for H_2^+ and H_2. The single bond (two bonding electrons) of H_2 is *shorter*, *stiffer*, and *stronger* than the half-bond (only one electron) of H_2^+.

call it a *single bond*. In this terminology, the bond in H_2^+ is only a half-bond, since it involves only *one* bonding electron.

These curves also yield another result that can be compared with experiment, namely the *stiffness* of the interatomic bond. Near the minimum of the $E_t(R)$ curve, the curve can be approximated by a parabola:

$$E_t(R) - E_t(R_0) \approx \frac{K}{2}(R - R_0)^2 \qquad \textbf{(10.11)}$$

Near the equilibrium spacing R_0, a departure from this spacing will produce a restoring force of $F = -dE_t/dR = -K(R - R_0)$, and we can regard the interatomic bond as a "spring" with a *spring constant* K given by the curvature of $E_t(R)$ at the equilibrium spacing:

$$K = \left(\frac{\partial^2 E_t}{\partial R^2}\right)_{R=R_0} \qquad \textbf{(10.12)}$$

K determines the force required to pull the two atoms further apart or to push them closer together, and is therefore the equivalent of a Young's modulus (see Chapter 7) for the molecule. As seen in Fig. 10.6, the H_2 bond is stiffer than the H_2^+ bond, since the $E_t(R)$ curve has more curvature at $R = R_0$. The two electrons in H_2 make a bond that is shorter, stronger, and stiffer than the bond produced by the one electron in H_2^+. This is an example of the 3S rule of interatomic bonds: Bonds that are *s*horter also tend to be *s*tronger and *s*tiffer.

The parabolic energy of (10.11) is reminiscent of the quantum-mechanical harmonic oscillator that we considered in Chapter 8, which led to a series of energy levels with a spacing of $\hbar\omega_c = \hbar(K/m)^{1/2}$. Molecules not only have quantized electron energy levels, but their vibrations—variations in R—are also quantized. (As noted in Chapter 7, the vibrational states in solids, called *phonons*, are also quantized.) Since the energy difference between different vibrational states is proportional to

$(K/m)^{1/2}$, spectroscopic techniques can be used to measure K, the stiffness of the interatomic bond.

Thus the $E_t(R)$ curve for a diatomic molecule, calculated with the LCAO and SCF methods, yields three quantities that can be compared with experiment—the bond energy E_b, the equilibrium spacing (or bond length) R_0, and the bond stiffness (spring constant) K. Although a diatomic molecule is a long way from the solids we are ultimately interested in, *these three molecular quantities are analogous to three important properties of a solid*—its *cohesive energy* (the difference between the energy of a solid and the energy of its isolated atoms), its *equilibrium lattice spacing*, and its *bulk elastic modulus*. In fact, with modern computers and advanced quantum-mechanical calculations, we can now compute curves like Fig. 10.6 for simple solids and predict properties that are in agreement with experiment. Thus in a sense, the $E_t(R)$ curve for a single interatomic bond holds the essence of the thermodynamic stability, crystal structure, and elastic properties of solids. All these properties result, like E_b, R_0, and K, from a balance between attractive Coulomb forces produced by bonding electrons and repulsive Coulomb forces between ions (in H_2^+, between the two protons).

Sample Problem 10.2

The energy of a diatomic molecule is given by $E_t(R) = \mathrm{A}R^{-1} - \mathrm{B}R^{-1/2}$. In terms of the coefficients A and B, derive the equilibrium spacing R_0, the bond energy E_b, and the spring constant K.

Solution

$dE_t/dR = -\mathrm{A}R^{-2} + \mathrm{B}R^{-3/2}/2 = 0$ at $R = R_0$. So $\mathrm{B}R_0^{-3/2} = 2\mathrm{A}R_0^{-2}$, $R_0 = (2\mathrm{A}/\mathrm{B})^2$

$E_b = E_t(\infty) - E_t(R_0) = -E_t(R_0) = -\mathrm{A}R_0^{-1} + \mathrm{B}R_0^{-1/2} = -(\mathrm{B}^2/4\mathrm{A}) + (\mathrm{B}^2/2\mathrm{A}) = \mathrm{B}^2/4\mathrm{A}$

$K = (d^2E_t/dR^2)$ (at $R = R_0$) $= 2\mathrm{A}R_0^{-3} - 3\mathrm{B}R_0^{-5/2} = \mathrm{B}^6/128\mathrm{A}^5$

(Note that increasing the attractive coefficient B or decreasing the repulsive coefficient A makes the bond *s*horter, *s*tronger, and *s*tiffer.)

10.4 BEYOND HYDROGEN—COVALENT DIATOMIC MOLECULES

Just as we designate the ground-state electronic structure of atoms by their occupied AOs (H = $1s$, He = $1s^2$, Li = $1s^22s$, etc.), we designate the ground-state electronic structure of molecules by their occupied MOs ($H_2^+ = 1\sigma$, $H_2 = 1\sigma^2$). Going beyond hydrogen, we next consider other simple homonuclear diatomic molecules and molecular ions.

The *helium molecular ion* He_2^+ similarly will have 1σ (bonding) and $1\sigma^*$ (antibonding) MOs built from $1s$ AOs. However, this molecule has *three* electrons, and, by the Pauli exclusion principle, only two can occupy the bonding MO. Thus one electron must go into the $1\sigma^*$ MO, and the ground state of He_2^+ is $1\sigma^2 1\sigma^*$. (Electrons *do* occupy antibonding states when they have to!) With two bonding electrons and one antibonding electron, He_2^+ can be viewed as having a *net* bond corresponding to only *one* electron. It is therefore not surprising that the bond energy and bond length of He_2^+ are very similar to those of H_2^+. We will soon encounter important neutral molecules in which antibonding states are occupied in the ground state. As in He_2^+, the net bonding will depend on the number of bonding electrons minus the number of antibonding electrons.

The neutral He_2 molecule would have a $1\sigma^2 1\sigma^{*2}$ electronic configuration, and with two bonding electrons offset by two anti-bonding electrons, this molecule would be expected to be unstable. It is.

Moving along to *lithium* ($Z = 3$), the $2s$ AOs will form 2σ and $2\sigma^*$ MOs in the Li_2 diatomic molecule. The ground state of the neutral molecule will be $1\sigma^2 1\sigma^{*2} 2\sigma^2$. The internuclear spacing will be determined primarily by the $2s$ electrons, which extend much further than the $1s$ electrons. Thus the $1s$ AOs will have little overlap, and the corresponding bond integral H_{12} and the resulting splitting between the 1σ and $1\sigma^*$ energy levels will be small. The four inner electrons (two bonding, two antibonding) will contribute nothing to the bonding, but will partially *screen* the 2σ bonding electrons from the $+3e$ nuclear charges. Since the 2σ bonding electrons are much further from the Li nuclei than the 1σ bonding electrons are from the H nuclei, the bond energy of Li_2 is considerably lower than that of H_2. It is an important general rule that *bond strengths of analogous molecules decrease as you go down columns of the periodic table*; thus the bond strength of Na_2 is weaker than Li_2, K_2 weaker than Na_2, and so on.

A diatomic *beryllium* ($Z = 4$) molecule would have a $1\sigma^2 1\sigma^{*2} 2\sigma^2 2\sigma^{*2}$ electronic configuration. With an equal number of bonding and anti-bonding electrons, the Be_2 molecule, like He_2, is unstable (well, nearly so—a weak dipolar interaction does produce a bond, but an extremely weak one). So far, so dull. But things get more interesting once we get past beryllium, and begin to encounter p AOs.

Atoms heavier than Be have $2p$ electrons in the ground state. Since p AOs have different symmetries (Fig. 9.4) than spherically symmetric s AOs, MOs built from p and s AOs can have different symmetries. Figure 10.7 shows schematically the shape of bonding and antibonding MOs formed by $2p$-$2p$ overlap. If we define the molecular axis as the z-axis, MOs formed by $2p_z$-$2p_z$ overlap, like those formed from s-s overlap, have circular symmetry around the molecular axis. They are called, you'll recall, *sigma orbitals*.

However, MOs formed by $2p_x$-$2p_x$ overlap or $2p_y$-$2p_y$ overlap have a *different symmetry* (Fig. 10.7). They change sign on rotation about the molecular axis by half a turn, and have a *nodal plane parallel to the z-axis* separating positive and negative lobes. They are called *pi orbitals*. No MOs are produced by p_x-p_y, p_y-p_z, or p_x-p_z overlap, since, by symmetry, their net overlap integrals are zero. MOs that change sign by rotation of 90° about the molecular axis, formed from d-d overlap, are called *delta orbitals* (AOs are labeled s, p, and d, and MOs are labeled with the corresponding Greek letters—σ, π, and δ.) However, we limit our considerations to s and p AOs, which can only form σ and π MOs.

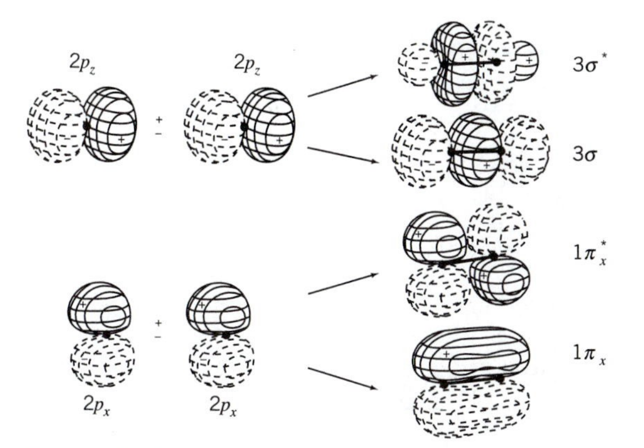

Figure 10.7. Schematic shapes of MOs built from p AOs.

As with sigma orbitals, pi orbitals are labeled in order of increasing energy (1π, 2π, etc.), and asterisks are used to indicate antibonding orbitals ($1\pi^*$, $2\pi^*$, etc.). The MOs produced by $2p_z$-$2p_z$ are 3σ and $3\sigma^*$. The MOs formed by $2p_x$-$2p_x$ and $2p_y$-$2p_y$ are $1\pi_x$, $1\pi_y$, $1\pi_x^*$ and $1\pi_y^*$. Since the x and y axes are equivalent, the energy levels corresponding to $1\pi_x$ and to $1\pi_y$ are equal, as are the energy levels of $1\pi_x^*$ and $1\pi_y^*$. These pi levels are *doubly degenerate* (Fig. 10.8).

We see again the *conservation of states*. Each atom had three $2p$ AOs. The *six* $2p$ AOs of a diatomic molecule build *six* MOs: 3σ, $3\sigma^*$, $1\pi_x$, $1\pi_y$, $1\pi_x^*$ and $1\pi_y^*$. (If you count in addition the MOs resulting from the $1s$ and $2s$ AOs, a total of five AOs per atom—ten in all—created ten MOs.) We can also see the reduction of degeneracy that results from a reduction in symmetry of the potential energy V. The individual atoms had spherical symmetry, and the $2p$ AOs were *threefold* degenerate—the x, y, and z axes were equivalent. The diatomic molecule has reduced symmetry—cylindrical symmetry. This removes some of the degeneracy, but the equivalence of the x and y axes retains *twofold* degeneracy.

Back to our march through the early part of the periodic table (don't worry—we'll stop at fluorine). Table 10.1 lists the homonuclear diatomic molecules from B_2 to F_2 (plus two molecular ions, N_2^+ and O_2^+). Also shown are the electron configurations (omitting the inner electrons $1\sigma^2 1\sigma^{*2} 2\sigma^2 2\sigma^{*2}$), the net numbers of bonding electrons (bonding minus antibonding) in these MOs, and the strengths and lengths of the interatomic bonds.

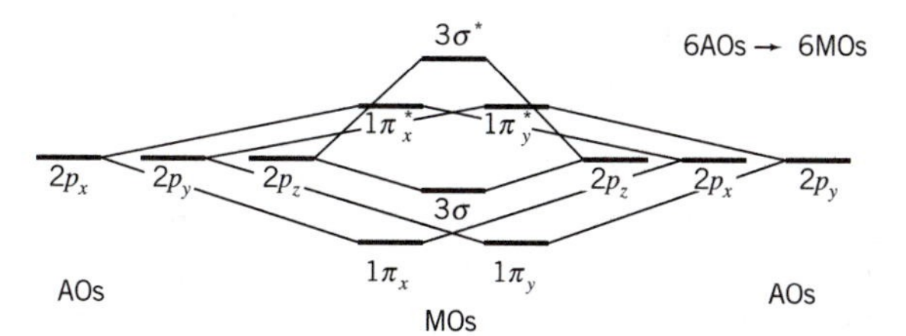

Figure 10.8. Energy levels of MOs built from $2p$ AOs.

Table 10.1. Homonuclear Diatomic Molecules from B_2 to F_2

Molecule	Electron Configuration	Net Bonding e's	Strength (eV)	Length (nm)
B_2	$1\pi_x^1 1\pi_y^1$	$2 - 0 = 2$	3.0	0.159
C_2	$1\pi_x^2 1\pi_y^2$	$4 - 0 = 4$	6.1	0.124
N_2^+	$1\pi_x^2 1\pi_y^2 3\sigma^1$	$5 - 0 = 5$	8.7	0.112
N_2	$1\pi_x^2 1\pi_y^2 3\sigma^2$	$6 - 0 = 6$	9.8	0.109
O_2^+	$1\pi_x^2 1\pi_y^2 3\sigma^2 1\pi_x^{*1}$	$6 - 1 = 5$	6.7	0.112
O_2	$1\pi_x^2 1\pi_y^2 3\sigma^2 1\pi_x^{*1} 1\pi_y^{*1}$	$6 - 2 = 4$	5.1	0.121
F_2	$1\pi_x^2 1\pi_y^2 3\sigma^2 1\pi_x^{*2} 1\pi_y^{*2}$	$6 - 4 = 2$	1.6	0.144

The two outer electrons in B_2 provide a *single bond.* It turns out that the two degenerate 1π levels have lower energy than the 3σ level in this molecule, and, *by Hund's rule*, one electron occupies $1\pi_x$ and the other occupies $1\pi_y$—*with parallel spins*. (By occupying different orbitals, the two electrons can stay, on average, farther apart than if they both occupied the same orbital.) Thus, although the B_2 molecule has an even number of electrons, it has a *net spin*, so it is *paramagnetic*. In contrast, in C_2 the 1π levels are now completely filled, each occupied with a pair of electrons of opposite spins. With no net spin, the molecule is *diamagnetic*. This is a *double bond* that is both stronger and shorter than the single bond in B_2. The bond stiffness also goes up. Compared to B_2, C_2 is *stronger*, *shorter*, and *stiffer* (the 3S rule).

The diatomic *nitrogen* ($Z = 7$) molecule, N_2, is the most common molecule in the air you breathe. There are *six* bonding electrons—a *triple bond* (one sigma bond and two pi bonds). Spins are balanced, so it's diamagnetic. Its bond strength of 9.8 eV makes the N_2 triple bond the strongest bond of all homonuclear diatomic molecules. The bond is also very short and very stiff.

You can get along without food for weeks, and without water for days, but there's one thing you can't get along without for more than a few minutes. It's the next diatomic molecule in this series—*oxygen* ($Z = 8$). Two oxygen atoms have *eight p* electrons, but the corresponding bonding MOs can only handle six—they're already filled. So two electrons go into *antibonding* states—one in $1\pi_x^*$ and one in $1\pi_y^*$ (Hund's rule again). With six electrons in bonding MOs and two in antibonding MOs, O_2 has a *net* of *four* ($6 - 2$) bonding electrons, giving it a *double bond.* Compared to N_2, O_2 is weaker and wider. Thanks to Hund's rule, it has a net electron spin and is therefore *paramagnetic*. It was long known that O_2 was paramagnetic, but it was not until molecular-orbital theory was developed that this was understood.

Further evidence for the occupation of anti-bonding MOs in O_2 comes from comparison with the molecular ion O_2^+. Despite having one less electron than the neutral molecule, O_2^+ has the stronger (and shorter and stiffer) bond. In contrast, neutral N_2 is stronger than the molecular ion N_2^+, because the six outer electrons in N_2 all occupy bonding states. Both O_2^+ and N_2^+ are of course paramagnetic, because each contains an odd number of electrons (and therefore *must* have a net spin).

Finally, F_2 has six bonding and four anti-bonding electrons, and therefore has a

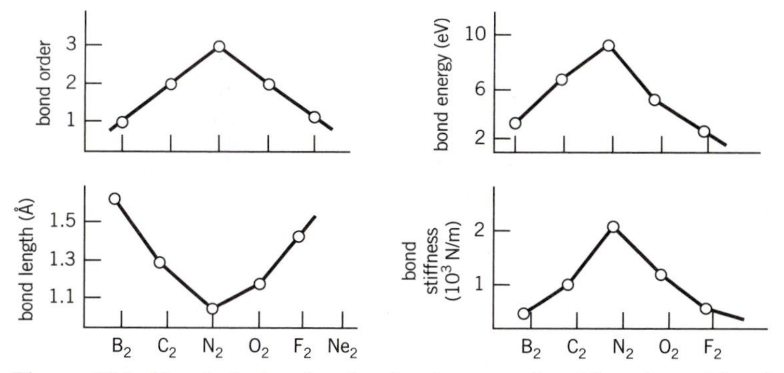

Figure 10.9. Trends in bond order, bond energy, bond length, and bond stiffness in first-row homonuclear diatomic molecules. Bonds get *stronger*, *shorter*, and *stiffer* as the bonding MOs are filled, and the trend is reversed as the antibonding MOs are filled. (Adapted from G. C. Pimentel and R. D. Spratley, *Chemical Bonding Clarified Through Quantum Mechanics*, Holden-Day, San Francisco, 1969.)

net of *two* bonding electrons—a *single bond*. It is weaker and wider than O_2, and is diamagnetic (no net spin). Figure 10.9 shows the change in bond order, bond energy, bond length, and bond stiffness of homonuclear diatomic molecules across this row of the periodic table. The interatomic bonds become *s*tronger, *s*horter, and *s*tiffer as more and more bonding states are occupied, but the trend is reversed when the additional electrons occupy antibonding states.

A similar pattern occurs for homonuclear diatomic molecules bound by *p* electrons of other shells; for example, P_2 is the strongest of such molecules bound by 3*p* electrons. However, as noted earlier, the strengths of analogous molecules decrease as you descend columns of the periodic table because of increasing atomic size, so P_2 is not as strongly bound as N_2.

Sample Problem 10.3

The diatomic molecules P_2, S_2, and Cl_2 are bound by electrons in MOs formed from 3*p* AOs. Of these three molecules, which has the strongest bond, which has the longest bond, and which is paramagnetic?

Solution

Like N_2, P_2 has a triple bond and is the strongest. Like O_2, S_2 is paramagnetic because it has unpaired spins in two degenerate antibonding MOs. Like F_2, Cl_2 has only a single bond (six bonding electrons and four antibonding electrons in the MOs formed from 3*p* AOs, with a net of only two bonding electrons) and has the weakest and longest bond.

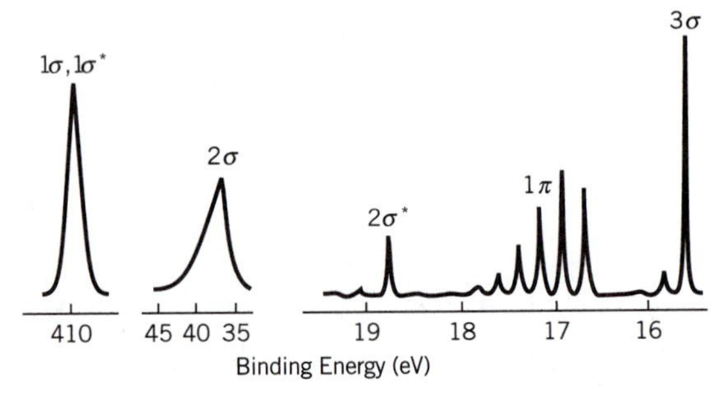

Figure 10.10. Photoelectron spectrum of N_2. (Adapted from J. N. Murrell, S. F. A. Kettle, and J. M. Tedder, *The Chemical Bond*, John Wiley & Sons, New York, 1985.)

Atomic spectra, including absorption and emission of photons and photoelectron spectra, provide direct evidence of electron energy levels in atoms that can be compared with quantum-mechanical SCF calculations. Similarly, *molecular spectra* have been widely studied and provide direct evidence of MO energy levels. Figure 10.10, for example, shows the experimental photoelectron spectrum of N_2. Plotted along the abscissa are the binding energies of electrons in different molecular energy levels. (Electron binding energies are calculated by subtracting the kinetic energy of the emitted photoelectrons from the energy of the photons that knocked them out. Photoelectron spectroscopy is discussed in more detail in Chapter 12.)

Because the $1s$ electrons from the two N atoms have little overlap, there is no measurable separation between the 1σ and $1\sigma^*$ levels. All four electrons in these lowest-energy states are bound to the molecule by about 408 eV, essentially the same energy with which the $1s$ electrons are bound to isolated N atoms. In contrast, the $2s$ electrons have substantial overlap, and the 2σ and $2\sigma^*$ levels can be seen to be widely split. The two electrons in 2σ have a binding energy of about 37 eV, while the two in $2\sigma^*$ are bound by less than 19 eV. The aforementioned eight electrons, however, contribute no net bonding. The triple bond of N_2 comes from four electrons in 1π and two in 3σ. The former have a binding energy of about 17 eV, the latter of about 15.5 eV. LCAO-SCF calculations for the N_2 molecule yield energy levels for the various MOs in good agreement with experiment. The several lines associated with the 1π MOs come from different vibrational states of the molecule. The energy separation of these lines can be used to estimate K, the spring constant, as discussed earlier.

10.5 UNEQUAL SHARING—POLAR BONDS

The diatomic molecules considered so far have been *homo*nuclear molecules, and we built our MOs from *matching* AOs: $1s + 1s \rightarrow 1\sigma + 1\sigma^*$, $2s + 2s \rightarrow 2\sigma + 2\sigma^*$, $2p_z + 2p_z \rightarrow 3\sigma + 3\sigma^*$, $2p_x + 2p_x \rightarrow 1\pi_x + 1\pi_x^*$, $2p_y + 2p_y \rightarrow 1\pi_y + 1\pi_y^*$. All MOs were either symmetric or antisymmetric across the midplane of the molecule, so that the electron probability distributions were all symmetric. The mathematics of the LCAO calculations were simplified by the symmetry. Turning to *heteronuclear* diatomic molecules, the mathematics will be a bit more complex, and we must consider other pairings of AOs, guided by two rules based on our knowledge that bonding is determined primarily by the bond integral H_{12}:

1. The AOs ϕ_1 and ϕ_2 must have the *same symmetry* with respect to the molecular axis. (If ϕ_1 is an s AO, then ϕ_2 cannot be p_x or p_y, since the positive and negative lobes of the p AOs will cancel each other in the overlap with the s AO, and the overlap and bond integrals will vanish. Similarly, p_x-p_y, p_x-p_z, and p_y-p_z pairings are ruled out. However, s-p_z is allowable.)
2. The AOs should have *substantial overlap.*

We'll delay discussion of the second rule until Chapter 11, but will consider the first rule as applied to three heteronuclear molecules: HF, LiH, and LiF. We will generally expect the interatomic bonding to involve the highest energy (least bound) electron of each atom.

The only AO available in H for bonding is $1s$, which has an energy of -13.6 eV. In F, the electron energies are: $1s$, -721 eV; $2s$, -39 eV; $2p$, -18.6 eV. We choose a $2p$ electron of F to pair with the $1s$ of H. From the symmetry rule above, we choose $2p_z$. The six $2s$, $2p_x$, and $2p_y$ electrons of F will be *nonbonding* orbitals, traditionally called *lone pairs* by chemists.

Calculation of the bonding produced by the $1s$(H)-$2p_z$(F) overlap in HF will follow the same mathematical procedure outlined earlier for H_2^+. We construct an MO, as we did in (10.6), by LCAO: $\psi = c_1\phi_1 + c_2\phi_2$. We plug that into the equation for the expectation value of energy, equation (10.4), where the Hamiltonian now includes the Coulomb attraction to the hydrogen nucleus and the more complex Coulomb attraction to the fluorine ion (a nuclear charge of +9e, screened by the two $1s$ electrons and partially screened by the six $2s$, $2p_x$, and $2p_y$ electrons of F). We then minimize the expectation value to get values of c_1 and c_2 and corresponding MO energy levels.

Although the mathematics is straightforward, it is much messier than for the homonuclear molecule, for which the two Coulomb integrals H_{11} and H_{22} were equal. Now the two AOs, ϕ_1 and ϕ_2, are different, so the two Coulomb integrals are now also different. As with H_2^+, the Coulomb integral is a reference state of energy for $R \rightarrow \infty$. However, here H_{11} corresponds to the electron on one atom (say the $1s$ level in H, or -13.6 eV), and H_{22} to the electron on the other atom (the $2p$ level in F, or -18.6 eV). As shown schematically in Fig. 10.11, in a heteronuclear molecule we get an energy level for the bonding MO that is lower than H_{22}, and an energy level for the antibonding MO that is higher than H_{11}.

Instead of having $c_1 = \pm c_2$, as we had for the homonuclear molecule, we find that $c_1 < c_2$ for the bonding MO (weighting the AO of lower energy, in this case the $2p$ of F, more heavily), but $c_1 > c_2$ for the antibonding MO (weighting the AO of higher energy, in this case the $1s$ of H, more heavily). This asymmetry is shown schematically in Fig. 10.11. This means that the two electrons occupying the bonding

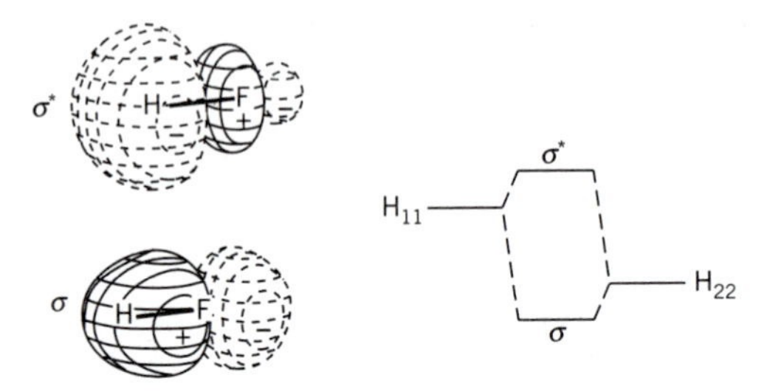

Figure 10.11. Asymmetric energy levels and MOs in HF, a polar bond (compare to the symmetric nonpolar bond in Fig. 10.4). The bonding MO is built predominantly from the AO of the more electronegative atom (the $2p_z$ of F) the nonbonding MO from the AO of the less electronegative atom (the 1*s* of H). This asymmetry gives the polar molecule an electric dipole moment, related to the "fractional ionic character" of the bond.

MO will spend more time near the fluorine ion than near the hydrogen ion. The bond will be *polar*, or "partly ionic."

For the HF molecule, LCAO theory yields $c_2/c_1 = 1.8$ for the bonding MO. This represents a shift of negative electron charge from H to F, and this polar molecule will have an *electric dipole moment.* This moment has been measured to be about 6×10^{-30} C-m. The bond strength is measured to be 5.9 eV, the bond length R_0 to be 0.092 nm. From the dipole moment and the bond length, we can make an estimate of the *fractional ionic character* of the bond. We make the simplistic assumption that a net charge of δe was transferred from the site of the H nucleus to the site of the F nucleus, which would give a dipole moment of

$$p = \delta e R_0 \tag{10.13}$$

If the bonding MO had *no* contribution from the H ($c_1 = 0$), that would correspond to $\delta = 1$, a complete transfer of one electron from H to F, a fully ionic bond. From the measured electric dipole moment and bond length of the HF molecule, the fractional ionic character δ of the H-F bond is about 0.41.

In contrast, consider the molecule LiH. The bonding and anti-bonding MOs are constructed from the 1*s* of H (at -13.6 eV) and the 2*s* of Li (at -5.4 eV). In this case, however, the bonding MO will be weighted toward the hydrogen, which will have a net negative charge. (Although F is more electronegative than H, H is more electronegative than Li.) Calculation indicates that $c_2/c_1 = 2.4$. Bond strength is 2.4 eV, bond length 0.16 nm, and the dipole moment is nearly 2×10^{-29} C-m. The fractional ionic character δ of LiH is about 0.78, substantially more ionic than HF.

Finally, consider LiF. The bonding and antibonding MOs are built from the 2*s* of Li and the $2p_z$ of F. The bonding MO is weighted heavily toward F, which is *much* more electronegative than Li. Bond length is 0.16 nm, electric dipole moment is about 2.1×10^{-29} C-m, and the fractional ionic character $\delta \approx 0.82$. The bond strength of LiF is 5.9 eV, considerably larger than that of either Li_2 or F_2. The ionic character of the bond contributes substantially to its strength.

The three polar molecules considered above (HF, LiH, LiF) were all held together by single bonds (two bonding electrons in one MO) that were sigma bonds. More complex polar bonds are of course possible. For example, a polar molecule isoelectronic to N_2 is CO, which, like N_2, has a *triple bond* made up of one sigma bond and two pi bonds. CO has a fractional ionic character of only a few percent, but it has a bond strength of 11.1 eV, the strongest bond of all diatomic molecules. (However, breathing a lot of it is *not* recommended.) CO, like N_2, has no net electron spin and is therefore diamagnetic. On the other hand, NO, because it has an odd number of electrons, must have a net spin and is therefore paramagnetic.

Pauling and others have devised semiquantitative methods of estimating the fractional ionic character of polar bonds, but present-day quantum-mechanical calculations make it clear that simple assignment of a single *electronegativity* number to an element cannot fully characterize the nature of its bonding to all other elements. With the LCAO-MO and related approximation methods, chemists can treat the full continuum of bond types from pure covalent (H_2) to predominantly ionic (LiF). Chemical bonds can *all* be viewed as the sharing of electrons in bonding MOs, which can be approximated by a linear combination of appropriate AOs. The sharing can be very unequal, but even in alkali halides like LiF, it is inappropriate to consider such a bond as purely ionic, with complete transfer of an electron from Li to F.

SUMMARY

In this chapter, we considered the quantum mechanics of interatomic bonding, starting with the simplest molecule, the hydrogen molecular ion H_2^+. We used the Born-Oppenheimer approximation, which allowed us to consider the comparatively heavy nuclei to be stationary at a fixed separation R while we solved Schrödinger's equation for the much lighter electron, using the LCAO-MO approximation $\psi = c_1\phi_1 + c_2\phi_2$ (a linear combination of atomic orbitals yielding a molecular orbital). This gave us two solutions, a symmetric ($c_1 = c_2$) bonding MO and an antisymmetric ($c_1 = -c_2$) antibonding MO. Calculation of the electron energy of each MO primarily involved the calculation of the bond integral H_{12}. The corresponding electron energies $E_{el}(R)$ were added to the energy term representing the repulsion between the protons to yield the total energy of the molecule $E_t(R)$.

Although the $E_t(R)$ curve for the antibonding MO increases monotonically with decreasing R, the curve for the bonding MO goes through a minimum, indicating a stable molecule with an equilibrium internuclear spacing of R_0 and a bond energy of E_b. The curvature of the $E_t(R)$ curve at $R = R_0$ represents the bond stiffness K. In the neutral H_2 molecule, *two* electrons occupy the bonding MO, and full calculation requires use of the SCF method to account for the repulsion between electrons.

The additional electron makes the bond in H_2 stronger, shorter, and stiffer than the bond in H_2^+ (the 3S rule).

Calculated values of the three quantities E_b, R_0, and K for these simple diatomic molecules agree with experiment and are analogous to three important properties of a solid—cohesive energy, lattice spacing, and bulk elastic modulus. Both in simple molecules and in solids, such properties of interatomic bonds are determined by a balance between attractive Coulomb forces produced by bonding electrons and repulsive Coulomb forces between neighboring ions.

In H_2^+ and H_2, the LCAO-MO method used $1s$ AOs to build sigma MOs ($1s + 1s \rightarrow 1\sigma + 1\sigma^*$). Two AOs gave us two MOs, our first example of the *conservation of states*. We extended the approach to other homonuclear diatomic molecules, producing further MOs: $2s + 2s \rightarrow 2\sigma + 2\sigma^*$, $2p_z + 2p_z \rightarrow 3\sigma + 3\sigma^*$, $2p_x + 2p_x \rightarrow 1\pi_x + 1\pi_x^*$, $2p_y + 2p_y \rightarrow 1\pi_y + 1\pi_y^*$. With the $2p$ AOs, six AOs gave us six MOs, another example of the conservation of states. Whereas sigma MOs have circular symmetry around the molecular z-axis, pi MOs do not—they reverse in sign with a half rotation about the z-axis. The threefold degeneracy of $2p$ levels in spherically symmetric atoms is replaced by a twofold degeneracy of 1π levels in cylindrically symmetric diatomic molecules—an example of reduced symmetry leading to reduced degeneracy.

The net bond order depends on the number of bonding electrons minus the number of antibonding electrons. Bond order and bond strength increase from B_2 (single pi bond) to C_2 (double pi bond) to N_2 (triple bond—two pi bonds and one sigma bond). However, since the bonding MOs are fully occupied in N_2, further electrons must occupy antibonding MOs. Bond order therefore decreases to a double bond in O_2, which has two antibonding electrons, and to a single bond in F_2, which has four antibonding electrons. The degeneracy of the 1π and $1\pi^*$ energy levels and Hund's rule result in a net electron spin in B_2 and O_2, making them paramagnetic. Molecular spectra, including photoelectron spectra, can be used to confirm the energy levels of various MOs.

In *homo*nuclear diatomic molecules, MOs are built from matching AOs, and the resulting MOs are either symmetric or antisymmetric with respect to the molecular midplane. In *hetero*nuclear (polar) molecules, on the other hand, MOs are built from different AOs from the different atoms, with symmetry and overlap considerations determining the appropriate AOs. The decreased symmetry of heteronuclear molecules makes the mathematics of the LCAO approximation a bit more complex. Now the MOs are asymmetric, weighting one AO more heavily than the other and giving the molecule a polar bond and an electric dipole moment. The bonding MO is built primarily from the AO of the more electronegative atom, the anti-bonding MO from that of the less electronegative atom. A fractional ionic character δ of the bond can be calculated from measured dipole moments and bond lengths. Although some bonds have a highly ionic character, all MOs involve some contribution from AOs of both atoms. All bonds involve electron sharing, and the LCAO method can be applied to the full continuum of bond types, from pure covalent (H_2) to predominantly ionic (LiF).

PROBLEMS

10-1. At the equilibrium spacing of 0.108 nm, the total energy of the H_2^+ molecular ion, -2.7 eV, is the sum of an attractive and a repulsive energy term. What are the values of those two separate energy terms?

10-2. From the data shown in Fig. 10.6, calculate the net attractive force produced by the two electrons in the neutral H_2 molecule.

10-3. Assume that, in the vicinity of the equilibrium spacing R_0, the electron energy for the H_2^+ molecular ion can be approximated by the formula $E_{el} = -BR^{-5/6}$. (a) Using the experimental value of R_0 (0.108 nm), calculate the constant B. (b) Using this value of B, calculate the bond strength and compare with experiment. (c) Using this value of B, calculate the spring constant K. (An experimental value is 140 N/m.)

10-4. At the equilibrium internuclear spacing of H_2^+ (0.108 nm), calculate the numerical values of the normalized bonding and anti-bonding MOs at the four points described by the following values of (r_1, r_2): (0.054 nm, 0.054 nm), (0.108 nm, 0), (0.108 nm, 0.108 nm), (0.108 nm, 0.216 nm). Use $1/\sqrt{2}$ for the normalization constant. What is the electron probability density at each of these locations?

10-5. N_2 has a larger binding energy than N_2^+, but O_2^+ has a larger binding energy than O_2. Explain briefly.

10-6. The following homonuclear diatomic molecules are bonded by MOs formed from p AOs: P_2, S_2, Ge_2, As_2, Br_2, Te_2, I_2. (a) For each, give the number of electrons in bonding MOs, the number of electrons in antibonding MOs, and the bond order. (Consider only the MOs formed from the p AOs of the outer shell.) (b) Which of these molecules do you expect to have the strongest bond? the longest bond? the stiffest bond? (c) Which of these molecules do you expect to be paramagnetic? For each paramagnetic molecule, show the occupation of MOs with the standard energy-level sketch, and identify from which AOs the various MOs are formed. (As before, consider only the MOs formed from p AOs in the outer shell.)

10-7. The energy of a diatomic molecule as a function of the internuclear distance R is given by $E(R) = AR^{-2} - BR^{-1}$. In terms of the constants A and B, derive (a) the equilibrium bond length R_0, (b) the bond energy E_b, (c) the spring constant K, and (d) the spacing at which the molecule will break if the atoms are pulled apart with an external force (i.e., the spacing at which the restoring force no longer increases with increasing spacing).

10-8. The vibrational states of diatomic molecules are quantized, and transitions between different vibrational states can be induced by incident photons. Would the photon emitted by the H_2 molecule when falling from its first excited vibrational state to its vibrational ground state be of higher or lower energy than the corresponding photon for the H_2^+ molecular ion? Why?

10-9. The B_2 molecule has a bond energy of 3.0 eV, a bond length of 0.16 nm, and a spring constant of 350 N/m. (a) Assuming that the quadratic approximation is valid, calculate the energy of this molecule at spacings of 0.15 and 0.17 nm. (b) Calculate the restoring force at each of these spacings. (c) What is the classical frequency of molecular vibrations? (For a homonuclear diatomic molecule, the effective mass is one-half the atomic mass.) (d) What is the energy separation between quantized vibrational energy levels for this molecule?

10-10. The CO molecule has a bond energy of 11.1 eV, a bond length of 0.11 nm, and an electric dipole moment of 3.7×10^{-31} C-m. (a) What is the fractional ionic character of this bond? (b) If an electric field of 2 V/m is directed perpendicular to the molecular axis, what is the torque and net force on the molecule? (c) By how much does the energy of interaction between the molecule and the field decrease if the molecule rotates until its axis is parallel to the field?

10-11. From what AOs is the major bonding MO of the NaCl molecule built? (Define the molecular axis as the z-axis.) Which of these AOs is dominant in this MO? Which AO is dominant in the corresponding antibonding MO? Sketch an energy-level diagram showing the relative positions of these two MO and two AO levels and label each level.

10-12. Write the total electronic configuration of the molecular ion Be_2^+. Do you expect Be_2^+ to be stable? If so, is the bond a double bond, a single bond, or a half-bond?

Chapter 11

From Bonds to Bands (and Why Grass Is Green)

11.1 ADDING MORE ATOMS

An ancient Chinese adage says that even the longest journey must start with a single step. In Chapter 9 we considered the electron energy levels and wave functions in isolated atoms, and in Chapter 10 we considered diatomic molecules—we went from one atom to two atoms. Our ultimate goal is to understand electron energy levels and wave functions in solids, so in this chapter we'll cover the remaining gap between two atoms and 10^{23} atoms! That's a big jump, but it's possible because the most important bonds in a solid are between an atom and its nearest neighbors, and we've already introduced most of the important concepts related to interatomic bonds in Chapter 10. However, since atoms in most solids have more than one nearest neighbor, we must first consider some additional aspects of bonding that we encounter when atoms have more than one neighbor. One such aspect is *hybridization*. To introduce hybridization and one or two other necessary concepts, we'll approach solids starting with small polyatomic molecules, gradually adding more atoms as we go.

In discussing diatomic molecules with the linear combination of atomic orbitals (LCAO) approach, we considered for simplicity only molecular orbitals (MOs) that were built from *one* atomic orbital (AO) from each atom. We remind you, however, that LCAO is only an approximation method and that greater accuracy can be obtained by using trial wave functions that include more adjustable parameters. In particular, use of trial wave functions incorporating two or more AOs from individual atoms—forming *hybrid* AOs—can lead to MOs of lower energy through increased overlap and increased bond integrals. Building MOs from hybrid AOs can produce increased accuracy even in many diatomic molecules, but it becomes an absolutely necessary approach to get reasonable quantitative results for most polyatomic molecules, and even to understand their geometries.

Consider, for example, beryllium dihydride (BeH_2). This is known to be a linear molecule (H-Be-H), with an angle of 180° between the two Be-H bonds. MOs to describe the Be-H bonds must be built from the $1s$ AOs of hydrogen; there we have no choice. But which Be AOs should we use? The ground state of Be atoms is $1s^2 2s^2$, but constructing MOs for the two Be-H bonds using only the $2s$ AOs of Be yields results for the bond strength and bond length that disagree strongly with experiment. Suppose we instead construct our MOs for BeH_2 with the following hybrid AOs that also contain $2p$ AOs (which have energies only slightly higher than the $2s$ AOs):

$$\phi_1 = \frac{\phi_{2s} + \phi_{2p_z}}{\sqrt{2}} \quad \text{and} \quad \phi_2 = \frac{\phi_{2s} - \phi_{2p_z}}{\sqrt{2}} \tag{11.1}$$

(We have chosen p_z because we define the molecular axis as the z-axis, and the $\sqrt{2}$ normalizes ϕ_1 and ϕ_2, presuming that ϕ_{2s} and ϕ_{2p_z} are normalized.) Because the $2p$ orbitals have a slightly higher energy than the $2s$ orbitals, these hybrid AOs have a slightly higher energy than pure $2s$ AOs. But these hybrids extend much farther from the nucleus (Fig. 11.1). This produces more overlap with the hydrogen $1s$ orbitals, resulting in a bond integral H_{12} that is more negative, resulting in a lower total energy—a stronger bond. The slight energy cost of producing the hybrid AOs is more than compensated by the increased overlap, which results in a lower energy for the molecule. Calculations based on these hybrids yield bond strengths and bond lengths in good agreement with experiment. The electrons in BeH_2 do their best to form wave functions (MOs) that minimize their energy, and apparently equations (11.1) represent Be-based AOs that, combined by LCAO with hydrogen $1s$ AOs, can be used to describe to a good approximation the MOs of this molecule.

The hybrid atomic orbitals in (11.1), which involve equal mixing of s and p atomic orbitals, are called *sp* hybrids, or *digonal* hybrids. They are useful in describing bonding in many *linear* molecules, including various beryllium halides, such as $BeCl_2$.

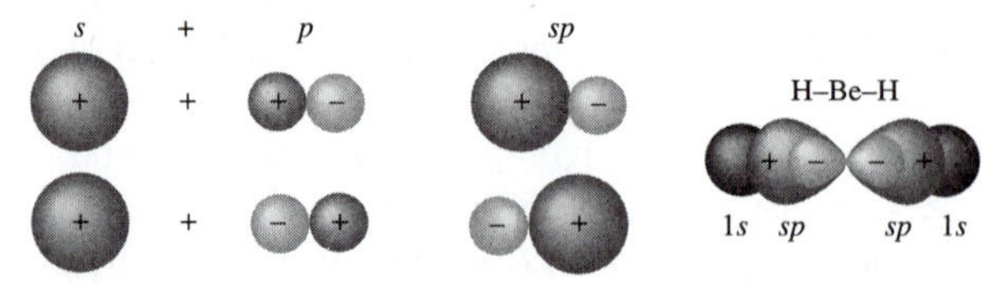

Figure 11.1. Digonal *sp* hybrid AOs for bond angles of 180°.

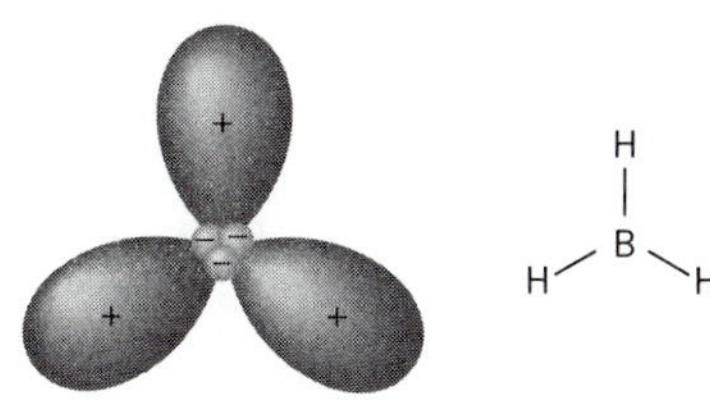

Figure 11.2. Trigonal sp^2 hybrid AOs for bond angles of 120°.

Analogously, consider boron trihydride (BH_3), which forms a *planar* molecule with three equivalent B-H bonds at angles of 120° to each other. The properties of this molecule cannot successfully be calculated, or the symmetry explained, by using only the $1s^2 2s^2 2p^1$ electronic ground state of B. However, we can construct (Fig. 11.2) three hybrid AOs for boron using one s AO and two p AOs. These hybrid AOs extend far from the B nucleus in directions at 120° to each other, as needed. As in the previous molecule, the energy cost of creating the hybrid AOs is more than offset by the increased overlap and more negative bond integrals, thereby producing a more stable BH_3 molecule than is possible without hybridization. Such hybrids are called sp^2 or *trigonal* hybrids, and are useful in explaining the structure and properties of many *planar* molecules.

An interesting and useful molecule is methane (CH_4), the chief constituent of natural gas. In methane, there are four equivalent C-H bonds arranged in tetrahedral symmetry, with an angle of 109.5° between bonds. These bonds are interpreted as arising from the overlap between $1s$ AOs of hydrogen with four hybrid AOs of carbon constructed as follows:

$$
\begin{aligned}
\phi_1 &= \frac{\phi_{2s} + \phi_{2p_x} + \phi_{2p_y} + \phi_{2p_z}}{2} \qquad & \phi_2 &= \frac{\phi_{2s} + \phi_{2p_x} + \phi_{2p_y} - \phi_{2p_z}}{2} \\
\phi_3 &= \frac{\phi_{2s} + \phi_{2p_x} - \phi_{2p_y} + \phi_{2p_z}}{2} \qquad & \phi_4 &= \frac{\phi_{2s} - \phi_{2p_x} + \phi_{2p_y} + \phi_{2p_z}}{2}
\end{aligned}
\tag{11.2}
$$

These hybrids, constructed from one s and three p orbitals, are called sp^3 or *tetrahedral* hybrids (Fig. 11.3). These hybrids are able to explain the tetrahedral bonding of carbon in many organic compounds, as well as the bonding in diamond and in silicon, the dominant semiconductor in today's technology. In such crystals, the hybrids extend in each of the four ⟨1 1 1⟩ directions, as can be seen by the variation in signs between the various terms in (11.2).

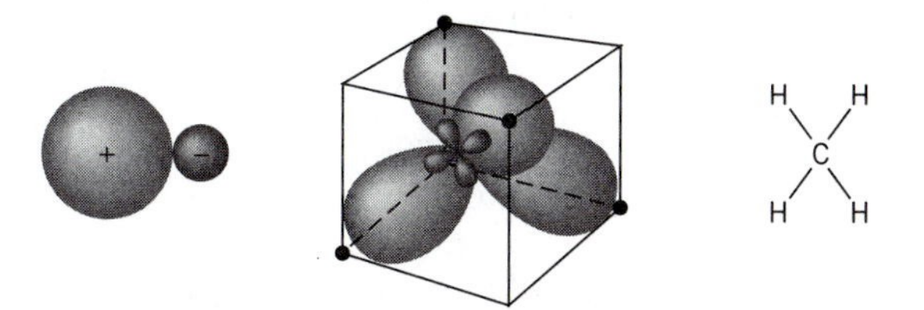

Figure 11.3. Tetrahedral sp^3 hybrid AOs for bond angles of 109.5°.

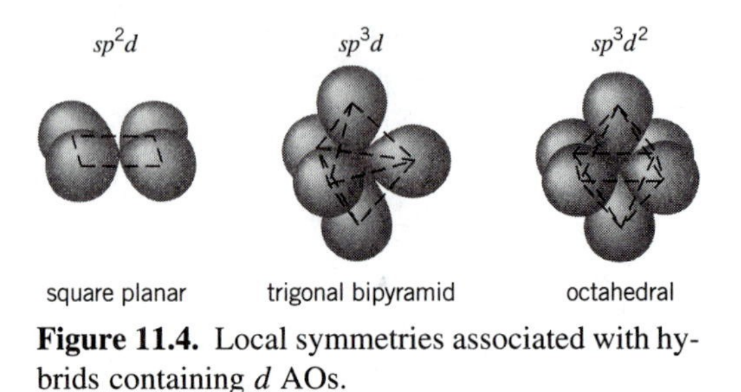

Figure 11.4. Local symmetries associated with hybrids containing *d* AOs.

Bonding in many molecules and solids can be explained, and accurate calculations made, by assuming MOs constructed from either *sp*, sp^2, or sp^3 hybrid AOs. In other molecules, intermediate percentages of *s-p* hybridization may produce minimum energy. Consider a common molecule that is vitally important to all of us—H_2O. In water molecules, the two O-H bonds make an angle of 104.5° with each other. Constructing MO bonds with unhybridized *p* AOs of oxygen would predict an angle of only 90°. Assuming sp^3 hybridization, and using two of the hybrids for O-H bonds and the other two to accommodate the nonbonding *lone pairs* of oxygen's valence electrons (in the *electron domain* theory used by chemists to predict molecular shapes), would predict a bond angle of 109.5°—better, but still not quite right. A detailed model assuming hybrids with 80% *p* and 20% *s* for the O-H bonds, a hybrid with 60% *p* and 40% *s* for a lone pair in the plane fixed by the bonds, and pure *p* orbitals out of that plane for the other lone pair, yields a lower energy than the model based on pure sp^3 hybridization. This model also accurately predicts the experimentally observed bond angle and the experimentally measured electric dipole moment. A similar model applied to ammonia (NH_3), which has three equivalent N-H bonds and one lone pair, accurately predicts the experimentally observed bond angle of 107° and other properties of the molecule.

For molecules containing heavier atoms than the ones we've considered so far, hybrids that contain *d* AOs in addition to *s* and *p* AOs may be necessary to understand molecular geometries and calculate bond properties (Fig. 11.4). For example, the octahedral symmetry of sulfur hexafluoride (SF_6), an important insulating liquid, can be explained with the use of sp^3d^2 hybrids built from one 3*s*, three 3*p*, and two 3*d* AOs of sulfur. With the creative use of appropriate hybrid AOs, the properties of interatomic bonds in even more complex molecules can be calculated using LCAO theory.

11.2 THE VERSATILE BONDING OF CARBON

The four valence electrons of carbon ($Z = 6$), combined with various degrees of hybridization, give it great versatility in bonding. This versatility is responsible for the richness of organic chemistry and for the richness of carbon-based life forms like us.

In many molecules, carbon uses sp^3 hybridization to form bonds with four other atoms, as seen earlier in methane. Such molecules are said to be *saturated.* Another

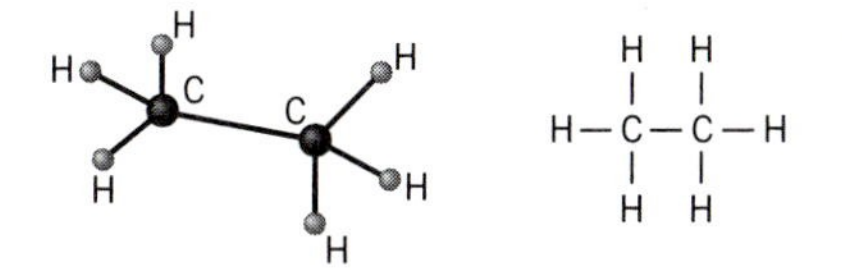

Figure 11.5. Ethane, a saturated molecule using sp^3 hybridization of carbon.

example is ethane (C_2H_6), in which each carbon atom has three C-H bonds and one C-C bond, with bond angles of approximately 109.5° (Fig. 11.5). All the bonds are single bonds (two electrons participating) and are also *sigma bonds*, as are all the bonds in BeH_2, BH_3, and CH_4. All other hydrocarbons of the *alkane* (C_nH_{2n+2}) series—propane ($n = 3$), butane ($n = 4$), pentane, hexane, etc.—are saturated molecules containing only single sigma bonds with tetrahedral symmetry.

In contrast, consider ethylene (C_2H_4). This molecule (Fig. 11.6) is *planar*, and each carbon has only three neighbors, with two C-H bonds and one C-C bond. Hybridization is sp^2, with bond angles of approximately 120°. That leaves one $2p$ AO on each carbon, extending out of the molecular plane, which forms a pi bond with the corresponding $2p$ AO on the other carbon. The C-C bond is therefore a *double bond*, consisting of one sigma bond (sp^2-sp^2) and one pi bond (p-p). Molecules like ethylene, in which one or more carbon atoms are bonded to less than four atoms and have multiple bonds, are called *unsaturated* compounds.

Unsaturated carbon compounds are ripe for further chemical bonding, and ethylene can serve as a "monomer" for *polymer formation*. Breaking the C-C double bond and attaching an extensive series of CH_2 groups produces the familiar polymer *polyethylene* (Fig. 11.7). Polyethylene is a saturated compound containing only single sigma bonds. It can contain many thousands of carbon atoms in each molecular chain, and can be thought of as the ultimate alkane. Because sigma bonds have circular symmetry about the interatomic axis, each C-C bond can be rotated, allowing a wide variety of configurations. This flexibility of the molecular chain is basic to the mechanical properties of such a polymer.

Next consider the highly flammable molecule acetylene (C_2H_2). This is a linear molecule (Fig. 11.8) in which each carbon atom has only two neighbors. With digonal (sp) hybrids, each carbon can form two sigma bonds, one with hydrogen, one with the other carbon. This leaves two unused orbitals on each carbon. Calling the molecular axis the z-axis, as usual, the p_x and p_y AOs can form π_x and π_y MOs with the neighboring carbon atom. The C-C bond in acetylene is therefore a *triple bond*, consisting of one sigma bond and two pi bonds, as in N_2. It is *very* unsaturated.

Acetylene can also serve as a monomer for polymerization, and the result is *polyacetylene*. Breaking the triple bond and attaching multiple CH groups yields a

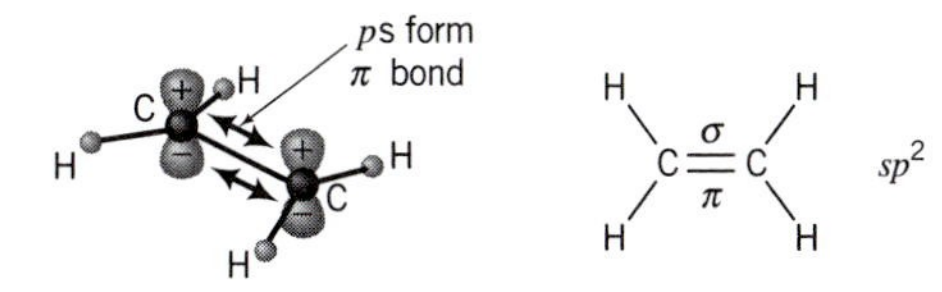

Figure 11.6. Ethylene, a planar molecule with a double C-C bond.

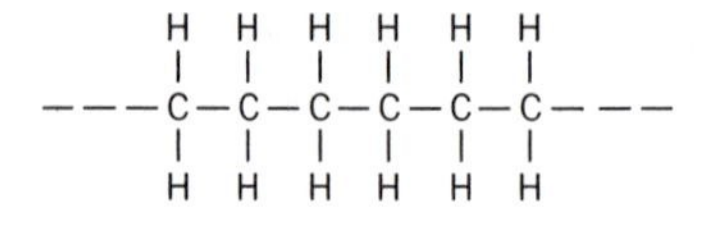

Figure 11.7. Polyethylene, a saturated polymer with sp^3 hybridization.

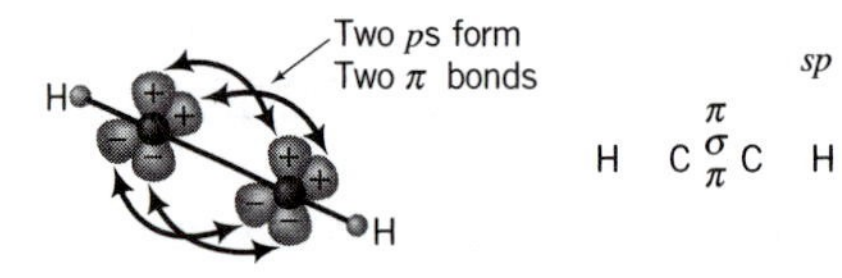

Figure 11.8. Acetylene, a linear molecule with a triple C-C bond.

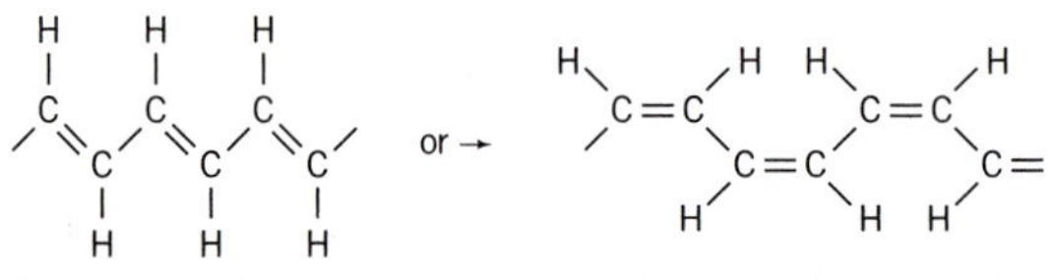

Figure 11.9. Two forms of polyacetylene, a conjugated polymer with sp^2 hybridization.

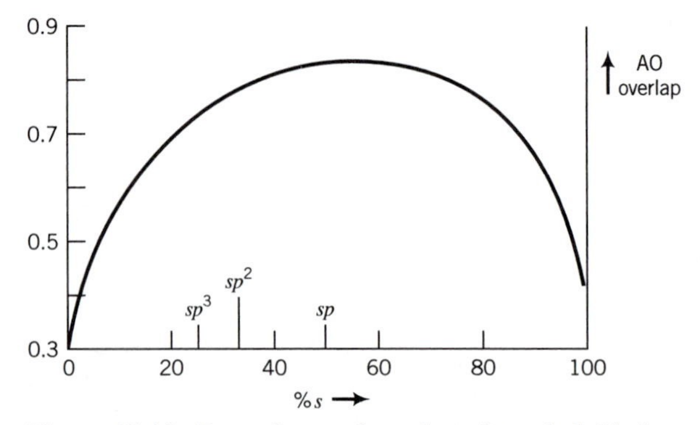

Figure 11.10. Dependence of overlap of two hybrids in a C-C bond on the percentage of *s*-character of the hybrids.

polymer chain (Fig. 11.9) that can be drawn with alternating single and double bonds. Hybridization in polyacetylene is sp^2, and bond angles are approximately 120°. The multiple C-C bonds cannot be rotated like the single sigma bonds in polyethylene, and the mechanical properties are therefore very different. Such molecules are called *conjugated.* More about conjugated compounds soon.

With varying types of hybridization, carbon can form a variety of interatomic bonds with other atoms as well as with other carbon atoms. Figure 11.10 shows the overlap integral S_{12} between two carbon atoms at fixed separation as a function of the percentage of s character in the hybrid AO. Whereas the overlap for pure p is 0.3 and for pure s it is less than 0.5, with hybridization the overlap can exceed 0.8. It is this increased overlap and the associated increase in bond energy (more negative bond integral) that favor hybridization.

The figure indicates that overlap is greatest for sp hybridization (50% s), less for sp^2 hybridization (33% s), and still less for sp^3 (25% s), which suggests that single sp bonds should be stronger than single sp^2 bonds, which in turn should be stronger than single sp^3 bonds. This is consistent with measured values of C-H bond strengths in acetylene (sp, 5 eV), ethylene (sp^2, 4.4 eV), and methane (sp^3, 4.1 eV). Comparative values of bond stiffness (640, 613, and 539 N/m) and bond length (0.1061, 0.1086, and 0.1093 nm) are consistent with the comparative bond strengths and with the 3S rule (stronger, stiffer, shorter). It is also of interest to compare the bond strengths and bond lengths of various C-C bonds: single (sigma)—3.6 eV and 0.145 nm, double (sigma plus pi)—6.3 eV and 0.134 nm, and triple (sigma plus two pis)—8.7 eV and 0.12 nm.

Sample Problem 11.1

In the acetonitrile (H_3C-CN) molecule, the nitrogen atom is triple-bonded to one of the carbons. What hybrid AOs of the two carbon atoms are used to build the MOs in this molecule?

Solution

One carbon is bonded to four atoms (3 H's and one C). It therefore is saturated and forms four σ bonds with four hybrid sp^3 AOs. The other carbon is bonded to only two atoms (one N and one C). It forms two σ bonds with two hybrid sp AOs. Two remaining p AOs form two π bonds with the nitrogen to complete the triple bond.

11.3 DELOCALIZATION

In most polyatomic molecules, bonding can be described in terms of MOs that are essentially localized on individual bonds (built from AOs of two neighboring atoms). This approach, however, breaks down for conjugated compounds, for which a new concept must be introduced. A prime example of this is *benzene* (C_6H_6). Diffraction and spectroscopic evidence make it clear that the six atoms of the benzene ring form a regular plane hexagon (Fig. 11.11), with six C-H bonds and six *equivalent* C-C bonds, and with bond angles of approximately 120°. This implies trigonal sp^2 hybridization of the carbon AOs, and the formation of sigma MOs with the hydrogens and with each other. This leaves six carbon valence electrons to be accounted for—one $2p$ AO from each carbon, directed out of the molecular plane. These *could* form three pi bonds, and benzene rings are often drawn with alternating single and double bonds. However, experiments make it clear that all C-C bonds in benzene are equivalent, and the strengths and lengths of the C-C bonds are intermediate between single and double bonds.

With LCAO-MO theory, we can understand the benzene molecule by constructing *delocalized* pi MOs that extend around the entire molecule. We write our MOs as

$$\psi = c_1\phi_1 + c_2\phi_2 + c_3\phi_3 + c_4\phi_4 + c_5\phi_5 + c_6\phi_6 \tag{11.3}$$

In (11.3), each of the ϕ_i are $2p$ AOs, centered on each of the six carbon atoms and directed out of the molecular plane. The ring nature of benzene requires that ϕ_6 overlaps with ϕ_1, as well as with ϕ_5. As we discussed originally with H_2^+, our goal is to find what combinations of the coefficients c_i give us approximate solutions to Schrödinger's equation with the appropriate Hamiltonian, and what electron energy levels these MOs will correspond to. This is a daunting task, however, since the potential energy function for electrons in benzene is pretty complicated, and we have a lot of terms in (11.3). For complex molecules, it is common to take a short cut and use the semi-empirical approximation method introduced in 1931 by Hückel. Hückel made the following simplifying assumptions:

1. The Coulomb integrals H_{ii} for each ϕ_i are all the same—call them α.
2. The bond integrals H_{ij} for ϕ_is from adjacent carbon atoms are all the same—call them β—and the bond integrals from nonadjacent AOs (say between ϕ_2 and ϕ_4) are all zero.
3. The overlap integrals $S_{ij} = \int \phi_i\phi_j dV$ are all zero. (This is the only assumption that seems unreasonable at first glance, but it greatly simplifies the mathematics and can be made legitimate with modest adjustments in the AOs.)

Applied to benzene, the Hückel theory yields six MOs—three bonding and three antibonding (conservation of states—six p AOs yield six pi MOs). The lowest-

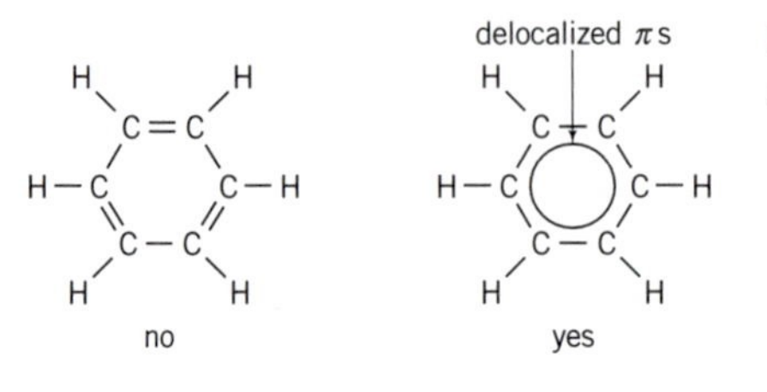

Figure 11.11. Benzene, a ring molecule with sp^2 hybridization.

energy MO corresponds to all c_i being equal (like the bonding MO for H_2^+). The highest-energy MO corresponds to c_i with alternating signs, with nodal planes between each atom pair (like the antibonding MO for H_2^+). Figure 11.12 shows, schematically, the relative orientation of the AOs from each atom that form each MO. The form of the lowest-energy MO is like two doughnuts, one resting on the other, with a node in the plane of the molecule but continuity around the ring. In addition to the node in the molecular plane, the next two MOs have two nodal planes perpendicular to the ring, the next two have four, and the highest-energy MO has six.

The figure also shows the corresponding energy levels, which include two pairs of degenerate levels. In terms of the integrals defined above, the energy of the bottom level is found to be $\alpha + 2\beta$ (as before, the bond integral β is negative). There are two degenerate bonding MOs with $E = \alpha + \beta$ (these are the highest occupied MOs—the "HOMO" levels), two degenerate antibonding MOs with $E = \alpha - \beta$ (these are the lowest unoccupied MOs—the "LUMO" levels), and the highest antibonding MO level with $E = \alpha - 2\beta$. The Coulomb energy α represents the reference state for the $2p$ electrons, and the difference in energy between the highest and lowest level is four times the bond integral β. In the ground state of the benzene molecule, the three bonding levels are occupied, and the three antibonding levels are unoccupied.

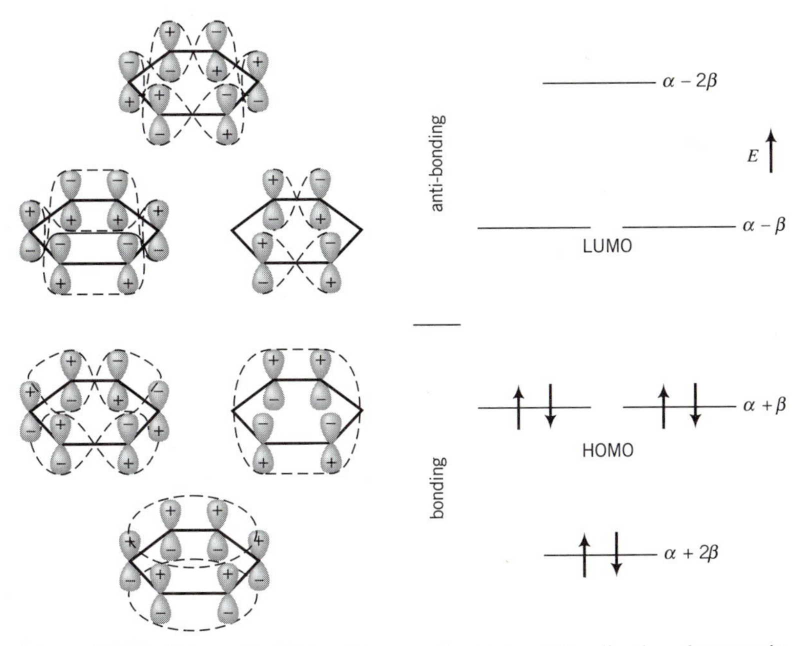

Figure 11.12. Schematic MOs and energy levels for delocalized π electrons in benzene molecule from Hückel model.

Sample Problem 11.2
From the knowledge that benzene is transparent to visible light, what conclusion can you draw about the magnitude of the bond integral β that is associated with the formation of delocalized pi MOs from p AOs in benzene?

Solution
Since benzene cannot absorb photons of visible light, we know that those photons do not have enough energy to excite electrons from the highest-occupied MO (HOMO) energy level to the lowest-unoccupied MO (LUMO) energy level. From Fig. 11.12, this HOMO-LUMO energy gap is $(\alpha - \beta) - (\alpha + \beta) = -2\beta$, which must be greater than the energy of photons at the violet end of visible light, which is about 3.1 eV. Thus we know that $-\beta > 1.55$ eV.

A simple example of a conjugated *chain* molecule is *butadiene* (C_4H_6, Fig. 11.13). As with benzene, each of the carbon atoms has three neighbors, and sp^2 hybrids form sigma bonds. The remaining $2p$ AOs, directed out of the molecular plane, form delocalized pi MOs that extend along the carbon chain, and can be treated in the Hückel approximation. Figure 11.13 schematically shows the relative orientation of the p AOs in each of the four MOs (two bonding, two antibonding), with dashed lines suggesting the outline of the various MOs. The four energy levels are $\alpha + 1.62\beta$, $\alpha + 0.62\beta$, $\alpha - 0.62\beta$, and $\alpha - 1.62\beta$.

There are many conjugated chain molecules that are much longer than butadiene.

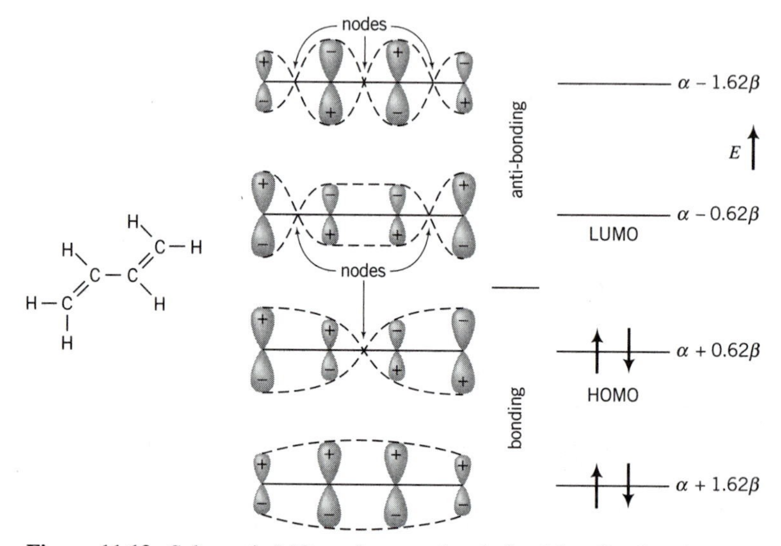

Figure 11.13. Schematic MOs and energy levels for delocalized π electrons in butadiene molecule from Hückel model.

Figure 11.14, for example, shows the nine energy levels and four of the nine MOs generated by the Hückel approximation for a simple conjugated chain containing nine carbon atoms. In the ground-state electronic configuration of this molecule, the four bonding MOs are fully occupied, the four antibonding MOs are unoccupied, and the intermediate nonbonding MO is half-occupied. As in benzene and butadiene (and H_2^+), the c_i are all of the same sign for the lowest-energy MO but of alternating sign, with nodal planes between each atom pair, for the highest-energy MO. It can be seen from the four MOs shown that, as in Figs. 11.12 and 11.13, increasing energy corresponds to more oscillations in the envelope containing the MO and more interatomic nodal planes (increasing $\nabla^2\psi$).

More generally, the energy levels of the delocalized pi MOs for a conjugated chain of N carbon atoms are given by:

$$E_j = \alpha + 2\beta \cos\left(\frac{j\pi}{N+1}\right) \qquad (j = 1, 2, 3, \ldots N) \qquad \textbf{(11.4)}$$

(Note that for $N = 2$, this would correspond to $E_j = \alpha \pm \beta$, equivalent to equations (10.7) and (10.8) for a diatomic molecule.) There are N energy levels, with the difference between the highest and lowest level never exceeding 4β. *Thus with increasing N, the levels become more and more closely spaced.* At *very* large N, we would have a nearly continuous *band* of energy levels of width 4β (Fig. 11.15). In the ground state, the lower half of the *energy band* would be occupied bonding MOs, the upper half unoccupied antibonding MOs. The energy band associated with the delocalized pi electrons of the carbon chain would be half filled with electrons. (Or, if you're a pessimist, it would be half empty.)

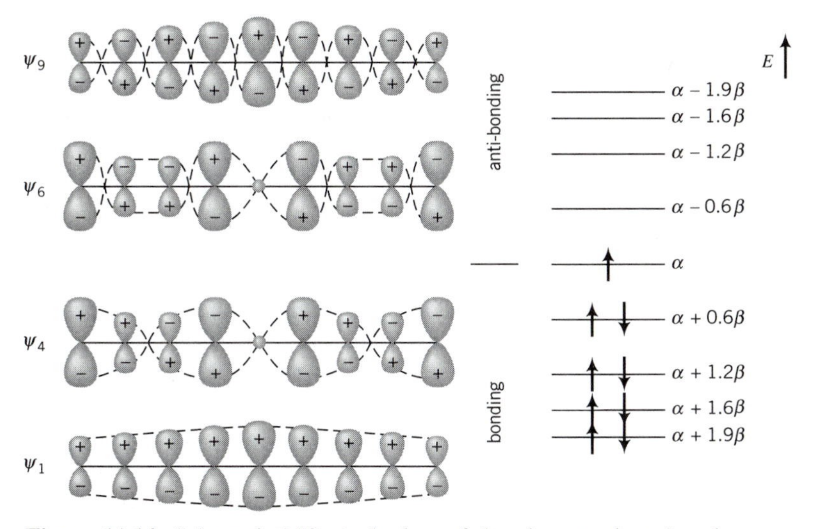

Figure 11.14. Schematic MOs (only four of the nine are shown) and energy levels for delocalized π electrons in C_9H_{11}, a conjugated chain molecule.

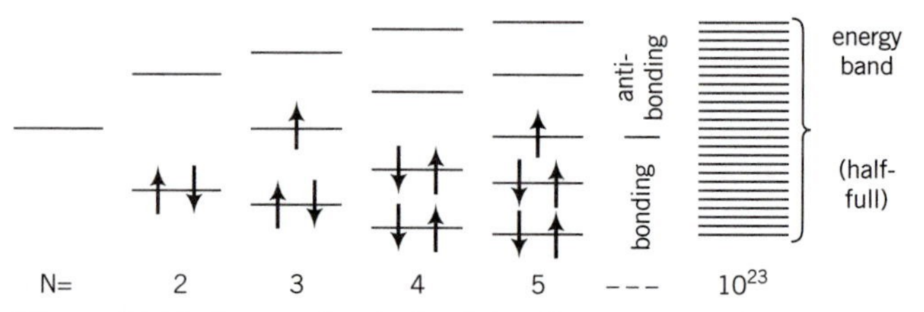

Figure 11.15. Energy levels for delocalized π MOs of conjugated chain molecules become closer together with increasing length of the molecule, eventually merging into a *band* of closely spaced levels.

Sample Problem 11.3

If a conjugated chain with $N = 8$ is transparent but a conjugated chain with $N = 10$ is colored, estimate the value of the bond integral β.

Solution

For $N = 8$, the HOMO is $j = 4$ and the LUMO is $j = 5$. From (11.4), the HOMO-LUMO energy gap is $2\beta \cos(5\pi/9) - 2\beta \cos(4\pi/9) = -0.69\beta$. For $N = 10$, the gap is $2\beta \cos(6\pi/11) - 2\beta \cos(5\pi/11) = -0.57\beta$. Apparently -0.69β is greater than 3.1 eV (making the molecule transparent), but -0.57β is not. Thus $-\beta$ must be between 4.5 and 5.4 eV.

(Warning: For problems like this, be sure the angle setting on your calculator is for radians, not degrees!)

All these pi MOs are delocalized, extending along the entire length of the molecule. The Hückel model also yields an expression for the N coefficients of the N terms in each of the N MOs. With the general expression for the jth MO, analogous to (11.3) for benzene, given by $\psi_j = \Sigma c_{ji} \phi_i$, the coefficients are

$$c_{ji} = \sqrt{\frac{2}{N + 1}} \sin\left(\frac{\pi j i}{N + 1}\right) \tag{11.5}$$

For *very* large N, we could think of (11.5) as a series of nearly continuous functions, representing the envelope functions (like the dashed lines in Figs. 11.13 and 11.14) that modulate the $2p$ AOs, the ϕ_i. These envelope functions are reminiscent of the sinusoidal standing-wave solutions (8.24) for the one-dimensional box, with i substituted for z/a, j for the quantum number n, and $(N + 1)$ for L/a. With increasing j in (11.5), the MOs have more and more interatomic nodes, and more and more kinetic energy, *just like the wave functions of the electrons in the box (Chapter 8).* More on this analogy later, but first a brief consideration of the optical properties of some important conjugated molecules that add color to our world.

11.4 LIVING COLOR (WHY GRASS IS GREEN)

Some natural color results from a dependence on frequency and wavelength of classical optical effects we discussed in Chapter 4. Rainbows result from dispersive (frequency-dependent) refraction of light in raindrops. The blue sky and red sunsets result from a strong dependence of Rayleigh scattering on wavelength. Bragg diffraction from periodic structures can separate colors, as from the surfaces of CDs and some snakes and insects, and from natural and artificial opals.

Much of the color of materials, however, cannot be explained classically, and results from electron transitions between quantized energy levels. Light *emission* can result from electron transitions from excited states to lower-energy states, the energy of the emitted photons corresponding to the energy difference between the initial and final states. We saw this in the emission of red photons from neon lights and helium-neon lasers (Chapter 9). However, the color that we see in most objects in our daily lives results instead from the excitation of electrons from lower to higher-energy states by the *absorption* of photons. Red glassware is red because it absorbs blue and transmits red. Red ties are red because they absorb blue and reflect red. (Opaque objects like ties and grass contain many internal surfaces that reflect and scatter light, so that unabsorbed light is reflected rather than transmitted.)

The subtractive nature of the color of most objects was brought to my attention some years ago when I left work one winter evening and went out to the parking lot to locate my red Dodge Dart. It wasn't there. In its place was a black Dodge Dart—but it had the same license plate as my car (and even had my gloves on the front seat where I had left them). The former parking-lot lights had been removed and replaced with mercury-vapor lamps, which emit strongly in the blue but negligibly in the red. Since the light incident on my car contained essentially no red light, it could reflect none, and my car lost its distinctive color. It turned red again once I drove it out under the better-balanced streetlights.

Pure saturated hydrocarbons like methane and polyethylene, with only sigma bonds, are transparent to visible light because their HOMO-LUMO separation is nearly 9 eV, much greater than the photon energies of visible light (1.8 to 3.1 eV). Small conjugated hydrocarbon molecules, like benzene and butadiene, also have HOMO-LUMO separations greater than 3.1 eV and are transparent. *But equation (11.4) indicates that the HOMO-LUMO separations of conjugated chain molecules decrease as the molecules get longer and longer. When the molecules are long enough, they can absorb visible light and produce color.* For example, the β-carotene molecule contains a long conjugated chain (Fig. 11.16) and absorbs blue light (Fig. 11.17), thereby producing the orange color of carrots. It is used to color margarine and other food and cosmetic products. A closely-related molecule is crocin, derived from the saffron flower (a relative of the spring crocus) and used as a yellow dye.

Other molecules that color our natural world with energy transitions of delocalized pi electrons are *indigo*, *chlorophyll*, and *heme* (Fig. 11.18), which contain conjugated rings. Indigo absorbs preferentially in the yellow-green part of the spectrum (Fig. 11.17) and is a natural blue dye, often used to dye blue jeans. Chlorophyll, the molecule that plants use to draw energy from sunlight, has absorption bands in both

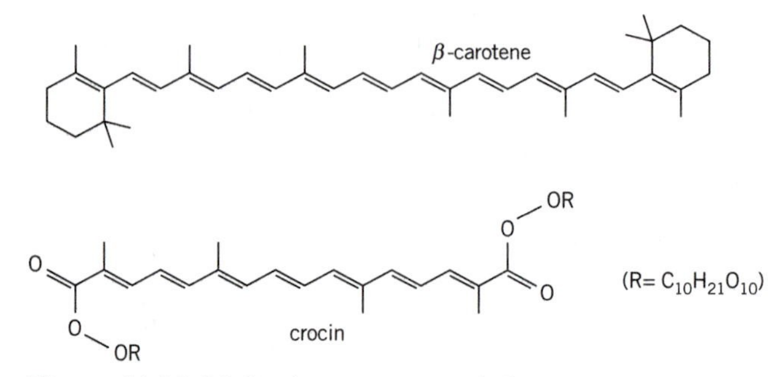

Figure 11.16. Molecular structures of β-carotene and crocin, conjugated chain molecules. For simplicity, the Cs and Hs are omitted.

the blue and red ends of the visible spectrum (Fig. 11.19), and preferentially reflects the center of the visible spectrum, green. Heme, as part of the hemoglobin molecule, serves to transport oxygen from the lungs throughout our bodies and makes our red blood cells red. (The structural similarity of chlorophyll and heme molecules suggests the common evolutionary origin of plants and animals.)

Many other conjugated organic compounds color plants and animals (including melanin, which colors our skin, hair, and eyes). Many others have been produced artificially for dyes. All produce color with HOMO-LUMO transitions of the delocalized pi electrons. (Color produced by photon absorption and electron transitions in *in*organic materials will be discussed in Chapter 14.)

As a child, you may have asked your parents why grass is green, and you probably got an unsatisfactory answer. You now are prepared to give a detailed and accurate answer the next time a child asks you the same question. You can simply say: "When conjugated carbon-based chain or ring molecules are sufficiently large, the MO levels

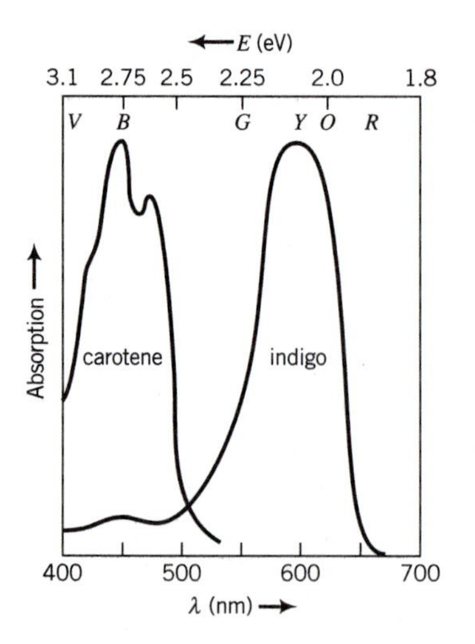

Figure 11.17. Absorption spectra of β-carotene and indigo. β-carotene absorbs in the blue-violet, and colors carrots orange. Indigo absorbs in the yellow-green, and colors blue jeans blue. (Adapted from K. Nassau, *The Physics and Chemistry of Color*, John Wiley & Sons, New York, 1983.)

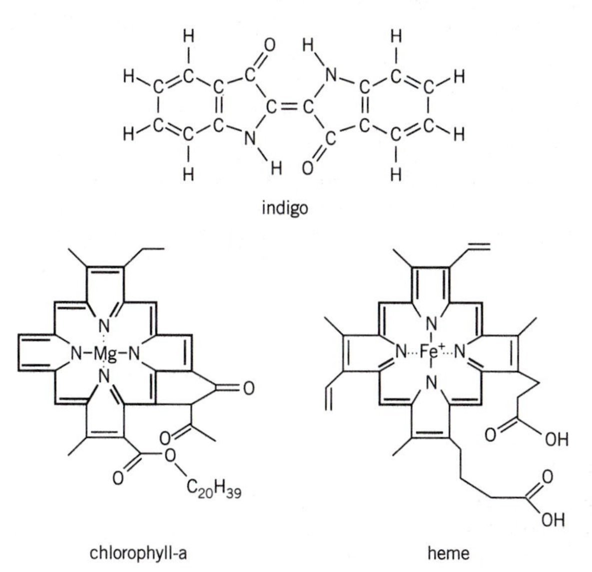

Figure 11.18. Examples of conjugated ring molecules—indigo, chlorophyll (the molecule responsible for energy production in plants), and heme (part of hemoglobin, responsible for oxygen delivery in blood).

of the delocalized pi electrons are close enough together to absorb visible photons and thereby produce color. Grass contains chlorophyll, a complex conjugated molecule with an MO structure that absorbs both red and blue light, thereby producing green by preferential reflection." After that, you probably won't get any more questions!

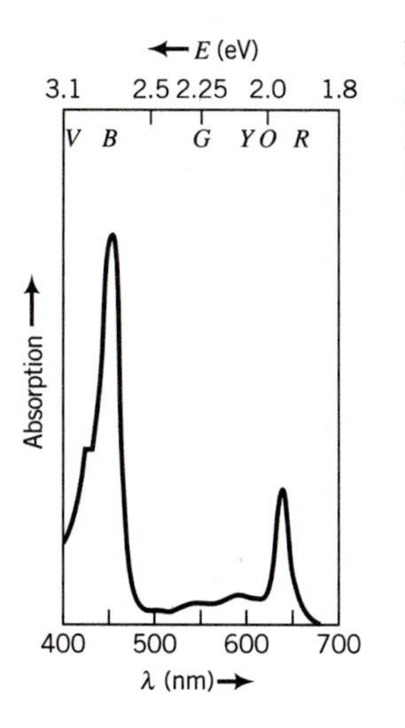

Figure 11.19. Absorption spectrum of chlorophyll, which makes grass green (by absorbing both red and blue, but reflecting green). (Adapted from K. Nassau, *The Physics and Chemistry of Color*, John Wiley & Sons, New York, 1983.)

11.5 FROM TWO TO MANY—ENERGY BANDS

With conjugated ring and chain carbon molecules, we were forced to consider the concept of delocalization—that some MOs were not localized on individual interatomic bonds, but extended over entire molecules. Examples included benzene, butadiene, many biological molecules (like chlorophyll and heme), and polymers like polyacetylene. The concept of delocalized MOs is more general, and can be especially appropriate when applied to conduction electrons in metals.

In the previous chapter, we considered the diatomic Li_2 molecule, bound by a single sigma bond constructed from 2*s* AOs. Figure 11.20 shows, schematically, the two energy levels and two MOs built from the Li 2*s* valence electrons ($2s + 2s \rightarrow 2\sigma + 2\sigma^*$). If we add a third Li atom to form a three-atom chain, the LCAO approach can be used to construct *three* MOs, with *three* energy levels, also shown in the figure. When occupying these Li_3 MOs (in the ground state, the bottom level will be fully occupied, the middle level half-occupied), the electrons cannot be considered as localized in one particular interatomic bond. They are shared by all three atoms—as with the pi electrons in benzene, they are delocalized.

With four Li atoms in the chain, there are four delocalized MOs and four energy levels. No matter how many Li atoms you add to the chain, the *lowest*-energy MO corresponds to equal coefficients on each of the AOs in the LCAO equation $\psi_j = \Sigma c_{ji}\phi_i$, and the *highest*-energy MO corresponds to coefficients of alternating sign, with nodal planes between each pair of atoms. *The energy difference between the lowest and highest levels thus remains essentially unchanged.* As in the conjugated carbon chains, an increasing number of atoms leads to energy levels that are more and more closely spaced, approaching a band of closely spaced energy levels for very large N (like in Fig. 11.15). A total of N atoms (N 2*s* AOs) will lead to N MOs and N energy levels in the band—conservation of states. Each level can be occupied by two electrons, so the full band could accommodate $2N$ electrons. However, with

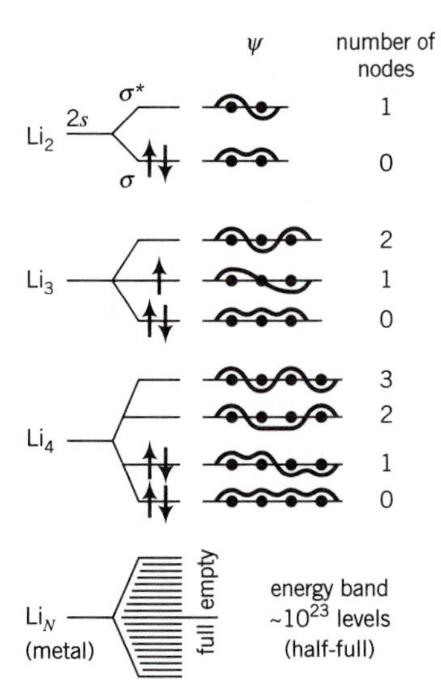

Figure 11.20. Evolution of MO energy levels into an energy band as more and more Li atoms are added, and evolution of MOs to approximate sine waves.

only one valence electron contributed by each Li atom, the energy band built from $2s$ AOs will be only *half full*.

If we extend these concepts to a bulk crystalline metal, with $N \approx 10^{23}$, the delocalized MOs can be thought of as *crystal orbitals* (COs). The coefficients of the terms in the LCAO equation, as in equation (11.5), are such that the envelopes of the COs approximate sine waves, much like the wave functions of electrons in a box; see Fig. 8.7. The 10^{23} CO energy levels built from the $2s$ valence electrons of Li are so many and so close that they form an essentially *continuous band of allowed energy levels*. At midband, there are unoccupied levels immediately above occupied levels (*the HOMO-LUMO separation is vanishingly small*), so that the delocalized $2s$ electrons can absorb not only all visible photons (making the metal opaque) but even photons with much less energy—infrared photons, even radio waves. And if you apply an electric field, electrons can pick up energy from the field—be accelerated and carry electric current—for the same reasons. The Li $2s$ electrons occupying delocalized COs are thus *conduction electrons*, the quantum-mechanical counterparts to our free electrons of Chapter 1.

This half-filled *conduction band* of lithium and other monovalent metals can therefore explain both the opacity and the conductivity of metals. Of course, we are also able to explain those properties of metals with the classical free electron theory, as we did in Chapters 1 and 2. But we'll find in the ensuing chapters that quantum mechanics can explain some properties of metals that we couldn't explain classically, such as electron heat capacity and x-ray and photoelectron spectra. And we'll find quantum mechanics absolutely essential to explain semiconductors.

If our arguments about energy levels of the delocalized pi electrons in conjugated chain molecules (Fig. 11.15) could be extended to polyacetylene, then it should, like lithium, have a half-filled energy band and be metallic and opaque. It comes pretty close, but a subtle structural instability of one-dimensional structures (called a *Peierls distortion*) makes pure polyacetylene a semiconductor instead of a metal, and it needs impurity doping to make it a decent conductor. However, even undoped polyacetylene has an electrical conductivity many orders of magnitude higher than that of polyethylene, a saturated sp^3-bonded polymer that is a transparent insulator (see Fig. 1.1). As we'll see in the next section, there's a similar huge difference between the properties of sp^2-bonded graphite and sp^3-bonded diamond.

Figure 11.21 reminds you of the variation of the total energy of a diatomic molecule with interatomic spacing R for both the bonding and antibonding MOs. We

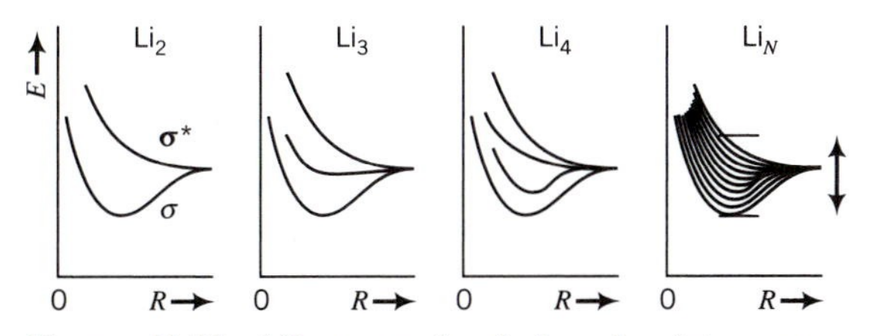

Figure 11.21. AO energy levels broaden into energy bands as atoms come closer and closer together.

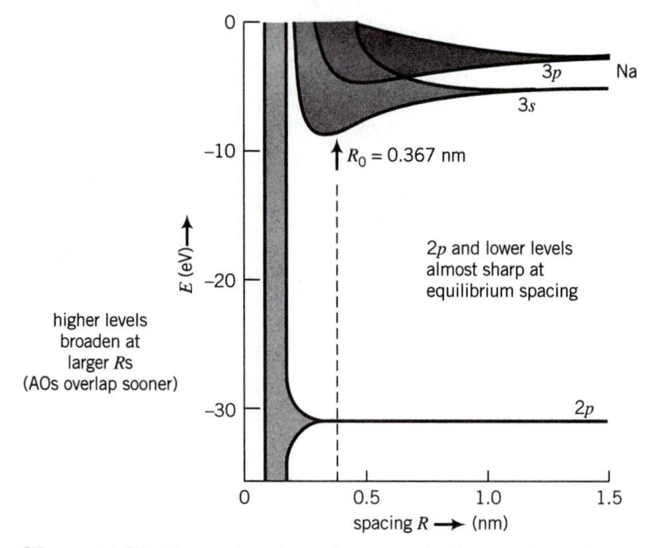

Figure 11.22. Energy levels vs. interatomic distance for sodium. (Adapted from J. C. Slater, *Phys. Rev.* 45, 1934, p. 794.)

arrived at such a curve first for H_2^+ (Fig. 10.3), but the concept applies equally well for the 2 MOs of Li_2 built from 2*s* AOs. The same figure shows energy vs. separation curves for Li_3, Li_4, and Li_N. The figure for Li_N shows how the energy band associated with the 2*s* atomic orbitals of Li broadens from the initially discrete AO energy levels as the many atoms in a long chain are brought closer and closer together.

At the equilibrium interatomic distance determined by the 2*s* bonding electrons, the 1*s* wave functions of Li would have little overlap and little broadening of the 1*s* energy level. At closer separations, the 1*s* energy levels would also broaden into an energy band (in this case a *full* energy band, since each of the *N* Li atoms has *two* 1*s* electrons to occupy the 2*N* energy levels). But at the equilibrium interatomic separation, the 1*s* electrons in Li metal are little changed from those in isolated Li atoms.

Figure 11.22 shows calculated energy vs. interatomic separation curves for sodium, another alkali metal. Just as the 1*s* electrons of Li had little overlap and little energy broadening in metallic Li, the 2*p* and lower levels of Na have little overlap and little energy broadening at the equilibrium interatomic spacing of Na (0.367 nm), which is determined primarily by the COs built from 3*s* AOs. For a given interatomic spacing, the higher AO levels broaden more in energy because those AOs extend further and overlap more in space (and thus have a more negative bond integral β).

11.6 SECONDARY BONDS

We've discussed the three types of primary bonds that hold atoms together in solids—covalent, ionic (polar covalent), and metallic (delocalized). In some solids, weak secondary bonds based on dipole-dipole interactions play an important role.

For example, when inert gases like neon solidify, the atoms are held together only by van der Waals (dipolar) forces. Since dipolar forces are larger for larger atoms (because of increased polarizability), the melting and boiling points of inert gases increase with increasing Z, a trend opposite to that observed for covalent bonding.

When molecules like N_2 and benzene (C_6H_6) solidify, strong covalent and ionic bonds bind the atoms together within the individual molecules, but only weak van der Waals forces provide the intermolecular bonds. Similarly, the atoms within polymer molecules are held together by strong covalent and ionic bonds, but often the bonds between different polymer molecules are provided only by weak van der Waals forces (unless the molecules are cross-linked). Like the proverbial chain, solids are only as strong as their weakest links, so all such solids have relatively low melting points and low elastic moduli (see Chapter 7). However, some polymers can be processed into fibers within which the polymer chains are aligned along the length of the fiber, producing a high elastic modulus in the fiber direction.

Some *layered* solids have strong bonds within planar layers but only weak van der Waals bonds between the layers. The classic example is graphite (Fig. 11.23). Within the layers, carbon atoms form a strongly bonded two-dimensional hexagonal lattice. Each carbon is covalently bonded to three neighbors with sigma bonds formed from sp^2 hybrid AOs. The remaining p AOs, directed normal to the plane, form delocalized pi MOs that allow electrical conduction in the layer planes and have closely spaced energy levels that can absorb visible light. Whereas diamond (sp^3 hybridization, saturated bonds) is a transparent insulator, graphite (sp^2 hybridization, conjugated bonds) is an opaque conductor, at least within the layers. However, the bonding *between* layers is provided only by weak van der Waals forces, which allows the layers to slide over each other easily under the application of low stresses. Thus graphite is a much better lubricant and a much better "lead" for pencils than diamond. (Nevertheless, most people seem to prefer diamonds.)

In recent years, carbon atoms have been shown to form in arrangements other

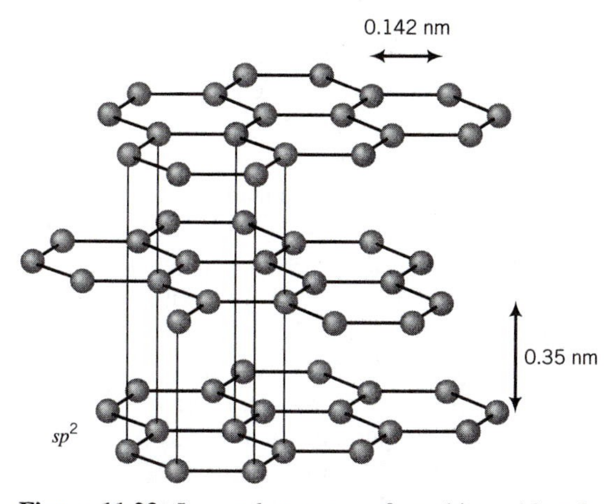

Figure 11.23. Layered structure of graphite, with only secondary van der Waals bonds between layer planes.

than graphite and diamond. The simplest of these are the *buckeyballs* of 60 carbon atoms forming a sphere, with a pattern of hexagons and pentagons like that of a soccer ball. The covalent bonds in this case are intermediate between sp^2 and sp^3, and can be approximated by $sp^{2.2}$. Individual buckeyballs exert attractive forces on other buckeyballs by secondary van der Waals forces and can crystallize into face-centered cubic arrays of C_{60} balls.

SUMMARY

The bonding and geometry of many polyatomic molecules cannot be understood in terms of MOs built simply from one pure AO from each atom, as we got away with in the previous chapter for diatomic molecules. *Hybrid* AOs must be used. Although hybrid AOs are of higher energy than pure AOs, they produce greater overlap between orbitals, more negative bond integrals, and stronger bonds—bonds and molecular geometries consistent with experiment. Commonly used hybrid AOs include *sp* (digonal) AOs for linear molecules, sp^2 (trigonal) AOs for planar molecules with 120° angles between bonds, and sp^3 (tetrahedral) AOs for tetrahedrally bonded molecules and crystals (including diamond and silicon) with 109.5° angles between bonds. Bonding and structure of some molecules, like H_2O and NH_3, are best explained by intermediate mixtures of *s* and *p* AOs, and bonding and structure in some molecules containing heavier atoms, like SF_6, are best understood in terms of hybrids that also contain *d* AOs.

Carbon is especially versatile in its bonding, producing the richness of organic chemistry and biological molecules. With four nearest neighbors, carbon bonds tetrahedrally with sp^3 hybrids, forming *saturated* compounds. With three nearest neighbors, it forms sigma bonds with sp^2 hybrid AOs, allowing the remaining $2p$ electron to form a pi bond, creating a double bond. With only two nearest neighbors, it can form sigma bonds with each, leaving two $2p$ electrons that can form two pi bonds, creating a triple bond.

Unsaturated carbon molecules like ethylene (containing a double bond) and acetylene (containing a triple bond) can serve as monomers for the formation of polymers such as polyethylene (saturated, all single sigma bonds) and polyacetylene (conjugated, alternating single and double bonds).

In conjugated organic molecules like benzene, butadiene, and many biological molecules, sigma bonds are built from sp^2 hybrids, and the remaining $2p$ electrons of the carbon atoms form *delocalized* pi molecular orbitals that extend over the entire molecule. In large chain or ring molecules, there are many delocalized pi MOs with closely spaced energy levels. If these levels are sufficiently close, electron transitions between the highest-occupied MO (HOMO) and the lowest-unoccupied MO (LUMO) can absorb visible photons, producing color (such as the orange of carrots, the green of grass). In some conjugated molecules, the energy levels can be so close that they approximate a continuous *band* of energy levels, half-filled with electrons, leading to the absorption of *all* visible photons (opacity) and the possibility of electrical conduction.

The concept of delocalized molecular orbitals is also applicable to the MOs built from *s* valence electrons in metals like Li. In bulk metal crystals, these MOs (now COs—crystal orbitals) also form a nearly continuous band of closely spaced energy levels, which can be viewed as arising from a broadening of the AO energy level as the many Li atoms are brought closer and closer together. Because Li is monovalent, this energy band is also only half-filled, since N atoms produce N energy levels in the band, which can accommodate up to $2N$ electrons. With unoccupied energy levels adjacent to occupied levels, conduction electrons can absorb even low-energy photons, and can be accelerated by electric fields (a quantum-mechanical view of why metals are opaque and conductive).

In some materials, weak secondary bonds based on dipole-dipole interactions play an important role. In solids where van der Waals forces provide the intermolecular bonds, melting temperatures and elastic moduli are relatively low. Linear polymers have strong covalent and ionic bonds within each molecule, but often have only van der Waals bonds between molecules. Layered solids like graphite have strong bonds within each layer but only van der Waals bonds between layers. In solids with both strong and weak bonds, it is usually the weaker van der Waals bonds that dominate the mechanical properties.

PROBLEMS

11-1. From the data given in this and the previous chapter on interatomic bonds, how much energy would be released by the reaction $C_2H_2 + H_2 \rightarrow C_2H_4$?

11-2. In terms of the Hückel bond integral β, what is the HOMO-LUMO energy separation in benzene? In butadiene? In a conjugated chain molecule with $N = 16$?

11-3. For a conjugated chain molecule containing 6 carbon atoms, calculate the 36 coefficients c_{ji}. (That's not as bad as it sounds—many will be identical.) Sketch the corresponding six MOs in the fashion of Fig. 11.14. Calculate the six energy levels in terms of α and β. Identify the HOMO and LUMO.

11-4. Explain the difference between the electrical, optical, and mechanical properties of alkanes (C_nH_{2n+2}) and alkenes (e.g., C_nH_{n+2}) in terms of the different MOs and AOs involved in their bonding.

11-5. If $\alpha = -15$ eV and $\beta = -6$ eV, which is the shortest conjugated carbon chain molecule capable of absorbing some visible light (and thereby becoming colored)?

11-6. In terms of α and β, what are the energies of (a) the lowest bonding MO and the highest antibonding MO of the conjugated chain molecule $C_{10}H_{12}$? (b) of the HOMO and the LUMO? (c) of the lowest-energy photon capable of inducing an electron transition in this molecule?

11-7. (a) Derive a simple expression for the HOMO-LUMO separation in an alkene chain (e.g., C_nH_{n+2}) as a function of n in the limit of very large n. (b) Using this expression and $\beta = -6$ eV, at what n will the HOMO-LUMO separation be equal to k_BT at room temperature? at 77 K?

11-8. Equation (11.4) is an expression for the energy levels of the delocalized MOs for a conjugated chain of N carbon atoms. In the limit of a very long chain of length L, with $a = L/N$ the distance between neighboring carbon atoms, this equation becomes $E = \alpha + 2\beta \cos(ka)$, with $k = \pi/L, 2\pi/L, 3\pi/L, \ldots N\pi/L$ ($= \pi/a$). (Yes, k is the wave number of the "waves" represented by equation (11.5)—see discussion at the end of Section 11.3.) (a) Assuming that N is so large that the energy levels can be considered to form a continuous energy band, and that k is a continuous rather than a discrete variable, derive an expression for the group velocity of electrons in this band. (b) Sketch the variation of energy and group velocity with k from the bottom ($k = 0$) to the top ($k = \pi/a$) of this band.

11-9. From Fig. 11.22, what is the lowest energy of electrons in the conduction band of sodium? Would this energy increase or decrease with application of hydrostatic pressure?

11-10. The melting points and boiling points of inert gases increase with increasing atomic number. So does the dielectric constant of these gases at standard temperature and pressure. Explain the connection between these two sentences in terms of an atomic property discussed in Chapter 3.

Chapter 12

Free Electron Waves in Metals

12.1 ELECTRON WAVES IN A BIG BOX—THE FERMI ENERGY

In Chapters 9–11, we approached solids the way that quantum chemists usually do. We started with individual atoms (Chapter 9), considered atoms bonding with other atoms to form diatomic molecules (Chapter 10), and then gradually added more and more atoms (Chapter 11) until we approached bulk solids. The energy levels of atomic orbitals (AOs) were replaced by molecular orbital (MO) energy levels, which, when we approached bulk solids with as many as 10^{23} atoms, broadened into *bands* of nearly continuous energy levels.

Near the end of Chapter 11, we began to picture the wave functions of conduction electrons in metals as *delocalized MOs* (or crystal orbitals—COs) that approximated AOs in the vicinity of each atom but were added together in such a way (LCAO) that they began to look more and more sinusoidal. The shape of our COs (Fig. 11.20) looked rather like the shape of the sinusoidal wave functions that we got in Chapter 8 when we considered an electron in a box (Fig. 8.7), except that there we were picturing a small box containing only one or a few electrons and only a few wavelengths. Compared to the quantum wells we mentioned in Chapter 8, a bulk metal is certainly a very big box, one containing an awful lot of electrons—and many, many wavelengths.

There's another common approach to thinking about the quantum mechanics of electrons in solids—one that many physicists often prefer. They prefer to start where Drude started with his classical theory of electrons in metals that we discussed way back in Chapters 1 and 2. They start, as Drude did, by ignoring the presence of an atomic structure and considering a metal simply as a big box containing *free electrons*. However, they now add something that Drude didn't know—that electrons are *waves*, waves that satisfy Schrödinger's wave equation. We'll find that the ideas of quantum mechanics can carry us a bit further toward understanding the properties of metals than Drude's classical theory could. We'll add some atomic structure in Chapter 13, but in this chapter we'll tackle the *quantum-mechanical free electron theory of metals*.

As we hinted at earlier, the quantum-mechanical free-electron theory of metals is simply an extension of the model problem we treated in Chapter 8—the electron in a box. We treat a macroscopic piece of metal as simply a volume in which the potential energy $V = 0$ for electrons inside the metal and $V > 0$ outside. This is a pretty simple model of a bulk metal—we essentially ignore the presence of the ionic cores (the nuclei and the many bound inner-shell electrons) and consider them only as providing the energy well that keeps the free conduction electrons from escaping.

ONE-DIMENSIONAL BOX

Instead of one electron in a small box or energy well, we consider a metal as a large box or energy well containing *many* free electrons. To keep things even simpler, we'll first limit our consideration to a *one-dimensional* metal, so that our electron wave functions will simply be the same ones that we had in equation (8.19):

$$\Psi(z, t) = e^{-i\omega t}\psi(z) = e^{-i\omega t}\{Ae^{ikz} + \mathrm{B}e^{-ikz}\} \qquad \textbf{(12.1)}$$

where

$$E = \hbar\omega = \frac{\hbar^2k^2}{2m} \qquad \textbf{(12.2)}$$

The next step is to apply boundary conditions to determine the allowable values of k. In Chapter 8, where we assumed $V = \infty$ outside the box, our boundary conditions were $\psi = 0$ at $z = 0$ and $z = L$. The resulting wave functions, expressed in (8.24), were $\psi_n = A_n \sin(n\pi z/L)$, *standing waves* with $k = \pi n/L$ $(L = n\lambda/2)$. This led to the energy levels described by equation (8.25): $E_n = n^2h^2/8mL^2$.

We could do the same again, but for the problem of bulk metals it is more conventional to apply instead what are called *periodic boundary conditions*:

$$\psi(0) = \psi(L) \qquad \textbf{(12.3)}$$

The mathematical difference is minor: we can satisfy (12.3) with $k = 2\pi n/L$ ($L = n\lambda$). The major difference is that we now can envision *traveling waves* instead of standing waves:

$$\psi_n = A_n e^{i2\pi nz/L} \tag{12.4}$$

Whereas we considered only positive values of n for standing waves, we can consider here both positive values of n (waves traveling in the positive-z direction) and negative values of n (waves traveling in the negative-z direction). The kinetic energy of these traveling waves is now given by

$$E_n = \frac{\hbar^2 k^2}{2m} = \frac{n^2 h^2}{2mL^2} \tag{12.5}$$

Periodic boundary conditions are really only a mathematical trick (after all, we recall that a standing wave can be constructed from two traveling waves traveling in opposite directions). And it seems unlikely that the properties of free electrons in a bulk metal would really depend on the details of the boundary conditions at the external surface. But since one of the goals of the model is to describe the transport of charge and energy across the metal by the free electrons, physicists find it more satisfying to think about a metal full of traveling electron waves than one full of standing electron waves. You have to admit that it does seem more dynamic.

When we dealt with an electron in a box in Chapter 8, our major result was that we had discrete allowed energy levels. For quantum wells, the widths and the energy depths are so small that only a few energy levels are available. Now that we have our electron in a *big* box, with side L of macroscopic dimensions, we recognize that we now have *many* allowed energy levels for our electrons (about $10^{29}/\text{m}^3$ in a typical 3-D free-electron metal), and that they are *very closely spaced* in energy (less than 10^{-7} eV apart). So for mathematical purposes we can think of the distribution of energy levels versus energy as a *continuum*—a band—of allowed energy levels. However, we still keep in the back of our minds that these are discrete levels, and recall that each of the allowed energy levels can accommodate only *two* electrons (one with spin up, one with spin down). For a given number of electrons N, we are now interested in knowing what our simple model tells us about their energies. Let's delay considering the effect of temperature, and assume that $T = 0$.

We have a lot of allowed and quantized energy levels in our box, described by (12.5). Pauli tells us that we can put only two electrons in the lowest-energy level, two in the next, and so on. We now have to do a bit of accounting to calculate how far up we're going to fill those levels—what the maximum energy of our electrons in this energy band will be. It will depend, of course, on how many free electrons we have.

For this accounting task, we will find it convenient to consider the allowed states in our one-dimensional free-electron metal geometrically. We therefore imagine a line (Fig. 12.1) in k-space, with a point for each allowed energy level at the values

$$k = \frac{2\pi n}{L} \tag{12.6}$$

as noted above. The points in our one-dimensional k-space that correspond to allowed electron energy levels are spaced uniformly along the line, separated by a constant spacing of $2\pi/L$. *Per unit length* along this line in k-space, we therefore have $L/2\pi$ allowed energy levels.

We imagine putting two electrons in each allowed energy level (each of the points along the line), starting from $k = 0$ and building to higher and higher energies until we have used up all of our N electrons. This process fills $N/2$ of the lowest energy levels, which will include points from $-k_F$ to $+k_F$, a length along the line in k-space of $2k_F$. Since this line contains $L/2\pi$ allowed levels per unit length, $N/2 = (L/2\pi)2k_F$, and therefore the maximum k value reached with these N electrons in our one-dimensional metal is

$$k_F = \frac{N\pi}{2L} \quad \text{(1-D)} \tag{12.7}$$

We have used the subscript F because we will call this, in honor of the Italian physicist Enrico Fermi, the *Fermi wave number.* And we will call the corresponding maximum energy the *Fermi energy*:

$$E_F = \frac{\hbar^2 k_F^2}{2m} = \frac{h^2}{32m}\left(\frac{N}{L}\right)^2 \quad \text{(1-D)} \tag{12.8}$$

The Fermi energy is an important concept in considering electrons in solids. Although we will be broadening the definition later in this chapter when we consider finite temperatures, and in Chapters 14 and 15 when we consider semiconductors and insulators, at this point the Fermi energy E_F simply represents the *maximum kinetic energy of conduction electrons at $T = 0$.*

As a result of the Pauli principle, which limits the number of electrons per energy level to two, we have free electron waves with kinetic energies up to the Fermi energy E_F in our *one-dimensional* metal even though the temperature $T = 0$. And that maximum electron energy varies with the square of the number of free electrons per unit length (N/L). We already have a free-electron theory of metals that is *very*

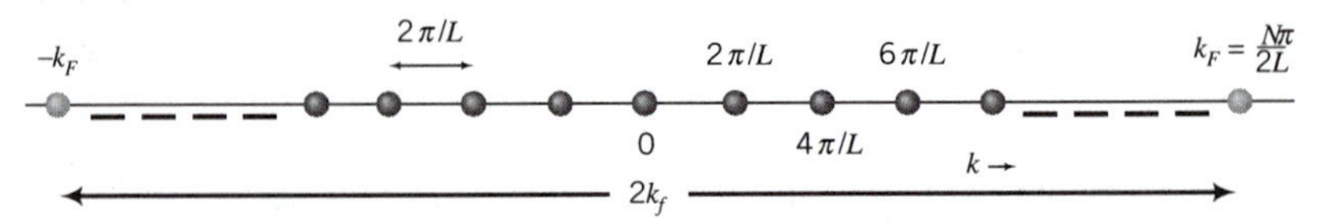

Figure 12.1. Allowed electron states in one-dimensional free-electron metal represented as points along a line in k-space.

different from Drude's classical theory. In Drude's classical theory, conduction electrons have *no* kinetic energy at absolute zero, but the quantization of electron waves and Pauli's principle tell us that they actually do.

TWO-DIMENSIONAL BOX

Let's extend the same geometric reasoning to a *two-dimensional* metal. (We're really more interested in three-dimensional metals, but this gradual approach has pedagogical advantages.) We'll assume it's a square L on a side and apply periodic boundary conditions like (12.3) in both y and z directions. Instead of electron waves defined by *one* quantum number n, we'll now have *two* quantum numbers n_y and n_z. Our wave number k will now instead be a *wave vector* $\mathbf{k}$ with two components, k_y and k_z. If we define position in our 2-D metal by the vector $\mathbf{r}$ (with components y and z), our wave function can be written as

$$\psi = Ae^{i\mathbf{k}\cdot\mathbf{r}} = Ae^{ik_y y}e^{ik_z z} \qquad \textbf{(12.9)}$$

The periodic boundary conditions in y and z will require $k_y = 2\pi n_y/L$ and $k_z = 2\pi n_z/L$. The allowed energy levels will now correspond to points in two-dimensional k-space (Fig. 12.2) that make up a square lattice with spacing $2\pi/L$, so that *per unit area* in two-dimensional k-space there are $(L/2\pi)^2$ allowed levels. The energy of each level will be

$$E(n_y, n_z) = \frac{\hbar^2 k^2}{2m} = \frac{(n_y^2 + n_z^2)h^2}{2mL^2} \qquad \text{(2-D)} \qquad \textbf{(12.10)}$$

where k is the absolute value of the wave vector $\mathbf{k}$, and $k^2 = k_y^2 + k_z^2$.

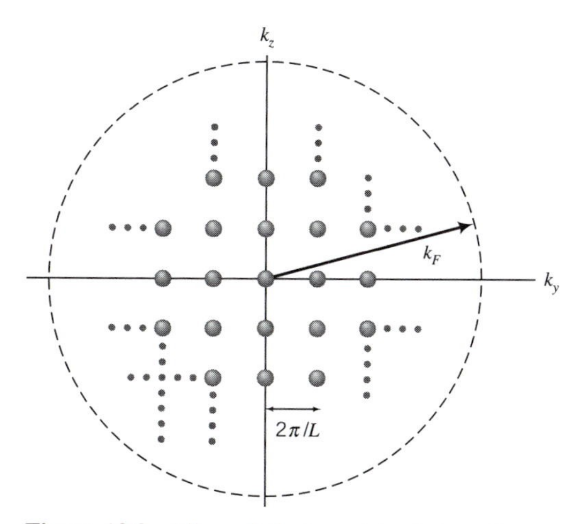

Figure 12.2. Allowed electron states in two-dimensional free-electron metal represented by points in a square lattice in two-dimensional k-space.

As in one dimension, we now imagine that our two-dimensional metal contains N free electrons. At $T = 0$, we fill up $N/2$ of the lowest-energy allowed levels. This corresponds to a *circle* around the origin in k-space, which will have an area of πk_F^2 and contain $\pi k_F^2(L/2\pi)^2$ levels. Thus $N/2 = \pi k_F^2(L/2\pi)^2$,

$$k_F^2 = 2\pi\left(\frac{N}{L^2}\right) \qquad \text{(2-D)} \qquad \textbf{(12.11)}$$

and

$$E_F = \frac{\hbar^2 k_F^2}{2m} = \frac{h^2}{4\pi m}\left(\frac{N}{L^2}\right) \qquad \text{(2-D)} \qquad \textbf{(12.12)}$$

Sample Problem 12.1

Calculate the Fermi energy, Fermi wave vector, and Fermi velocity for a two-dimensional monovalent free-electron metal in which the atoms form a square lattice with a nearest-neighbor separation of 3×10^{-10} m.

Solution

Each atom contributes one free electron and occupies an area of $(3 \times 10^{-10})^2 = 9 \times 10^{-20}\ \text{m}^2$. Thus $(N/L^2) = 1.11 \times 10^{19}\ \text{m}^{-2}$.
From (12.11), $k_F = 8.35 \times 10^9\ \text{m}^{-1}$.
$v_F = \hbar k_F/m = (1.05 \times 10^{-34})(8.35 \times 10^9)/9.11 \times 10^{-31}) = 9.62 \times 10^5$ m/s.
From (12.12), $E_F = (3.83 \times 10^{-38})(1.11 \times 10^{19}) = 4.25 \times 10^{-19}\ \text{J} = 2.66$ eV

THREE-DIMENSIONAL BOX

Finally, we apply similar geometric arguments to calculate the Fermi energy of a *three-dimensional* metal, in the shape of a cube L on a side. Now we have three quantum numbers, n_x, n_y, and n_z, and our allowed energy levels correspond to points in a cubic lattice with spacing $2\pi/L$ in k-space (Fig. 12.3). Thus *per unit volume* in k-space, we have $(L/2\pi)^3$ allowed levels. Our N electrons will fill $N/2$ energy levels, which will occupy a *sphere* around the origin with a volume of $4\pi k_F^3/3$. Thus $N/2 = (4\pi k_F^3/3)(L/2\pi)^3$,

$$k_F^3 = 3\pi^2\left(\frac{N}{L^3}\right) \qquad \text{(3-D)} \qquad \textbf{(12.13)}$$

and

$$E_F = \frac{\hbar^2 k_F^2}{2m} = \frac{h^2}{2m}\left(\frac{3N}{8\pi L^3}\right)^{2/3} \qquad \text{(3-D)} \qquad \textbf{(12.14)}$$

Note that the Fermi energy has a different dependence on free-electron concentration in 1-D (12.8), 2-D (12.12), and 3-D (12.14). Let's plug in some numbers for

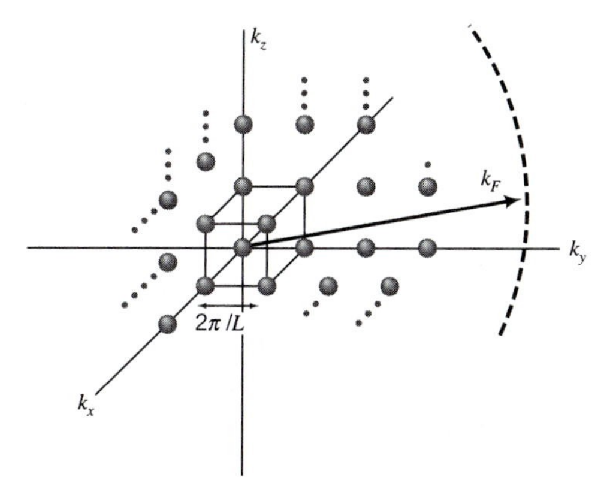

Figure 12.3. Allowed electron states in three-dimensional free-electron metal represented by points in a cubic lattice in three-dimensional k-space.

a 3-D metal, for which the Fermi energy varies with the 2/3 power of N/L^3, the number of free electrons per unit volume (the quantity we labeled N_e in Chapters 1 and 2). For alkali metals, we can assume one free electron per atom, so that N/L^3 is just the atom density. Considering Li, for example, equation (12.14) gives 4.72 eV as the maximum electron kinetic energy. That corresponds to $1.11 \times 10^{10}\ \text{m}^{-1}$ for k_F, the maximum wave vector, and 1.29×10^6 m/s for the maximum electron velocity, the *Fermi velocity.*

In Chapter 1, with the *classical* free-electron theory of metals, we calculated a thermal electron velocity *at room temperature* of about 10^5 m/s. And since thermal velocity varied as $\sqrt{T}$, it would be zero at $T = 0$. Quantum mechanics tells us instead that, *even at* $T = 0$, some electrons have velocities and kinetic energies much higher than predicted by classical theory for room temperature! Classical theory assumed that the probability of occupation of an energy level E_n was proportional to $\exp(-E_n/k_BT)$, but quantum mechanics tells us that each level can be occupied by no more than two electrons. Some electrons have high kinetic energies at $T = 0$ not because they want to, but because all of the lower-energy levels are already full.

Thankfully, the velocities of conduction electrons are still much smaller than the velocity of light, so we don't need to worry about relativity. Absorbing quantum mechanics is probably enough for one course. We should warn you, however, that relativity can no longer be neglected when considering the inner-shell electrons of heavy atoms, which have very high kinetic energies.

WHY *k*-SPACE?

Most people seem content to think about the normal three-dimensional space in which we live, where lengths are in meters, areas in m^2, and volumes in m^3. Scientists thinking about electrons in solids instead often think in terms of the k-space of Figs.

12.1, 12.2, and 12.3. We first introduced k, the wave number, way back in Chapter 2. Since $k = 2\pi/\lambda$, lengths in k-space are measured in reciprocal or inverse meters, areas in m^{-2}, and volumes in m^{-3}. Another term for k-space is therefore *reciprocal space*. Since k is directly proportional to electron momentum ($p = \hbar k$), another commonly used term is *momentum space*.

Believe it or not, k-space is not used because scientists enjoy thinking in abstractions. (At least that's not the main reason.) The quantization condition for free-electron waves, expressed in (12.6) for one dimension, resulted in allowed electron energy levels that were *equally spaced in k-space*, as seen in Figs. 12.1, 12.2, and 12.3. Thus in three dimensions, for example, a given volume of k-space corresponding to a given number of electron energy levels, simplifying our calculation of the Fermi vector and Fermi energy. This is analogous to our use of n-space in Fig. 8.1 when we were considering classical electromagnetic waves in a cavity, where we could also use geometrical thinking to help our calculations.

We'll also find k-space very useful in Chapter 13 for picturing the complex effects of the interaction of electron waves with the crystal lattice (which we're ignoring in this chapter). As in our discussion of sound waves in Section 7.4, we'll use the Bragg diffraction condition, equation (4.13), to divide k-space into Brillouin zones and then study how $E(k)$—the relation between electron energy and momentum—is affected as k approaches the zone boundaries.

12.2 THE DENSITY OF STATES

Equation (12.14) gives the maximum electron energy at $T = 0$ for a three-dimensional metal with N/L^3 free electrons per unit volume. Of course, most electrons will have lower energies than E_F, and it will be useful to define a function that describes the *distribution of allowed electron states as a function of energy*. We define $Z(E)dE$ as the number of allowed electron states with energies between E and $E + dE$, and we can treat it as a mathematically continuous function because there are so many closely spaced states. (An electron *state* is defined both by an allowed energy *level* and by its *spin*. Thus the number of "states" between E and $E + dE$ is *twice* the number of energy levels between E and $E + dE$.)

In 3-D, equation (12.14) was derived by filling N states ($N/2$ levels) with N electrons, and calculating the resulting maximum energy E_F. We can turn that relation around and solve for how many states $N(E)$ correspond to energies up to a given energy E:

$$N(E) = \frac{8\pi L^3}{3h^3}(2mE)^{3/2} \qquad \text{(3-D)} \tag{12.15}$$

The derivative of this function will then tell us how many states are in each energy interval:

$$Z(E) = \frac{dN(E)}{dE} = \frac{4\pi L^3(2m)^{3/2}}{h^3}E^{1/2} = CE^{1/2} \qquad \text{(3-D)} \tag{12.16}$$

This important function $Z(E)$ is traditionally called the *density of states*. Here *density* has the unusual meaning of *per unit energy*, and the units for $Z(E)$ are therefore J^{-1}. We see that in a three-dimensional free-electron metal, the density of states increases as $E^{1/2}$.

By similarly inverting equations (12.12) for 2-D and (12.8) for 1-D to derive $N(E)$, and taking $dN(E)/dE$ to get $Z(E)$, the density of states is seen to be independent of E for 2-D and to vary as $E^{-1/2}$ for 1-D (see Fig. 12.4). In 1-D, E_n is proportional to n^2, so that the energy levels in 1-D become more and more widely spaced with increasing n, resulting in a density of states—states per unit energy—that decreases with increasing energy. But E_n is proportional to $(n_y^2 + n_z^2)$ in 2-D and to $(n_x^2 + n_y^2 + n_z^2)$ in 3-D. So with increasing dimensions, there are more combinations of integers that correspond to a given energy or nearly the same energy (more degeneracy or near-degeneracy), which results in a different variation of the density of states with energy.

Sample Problem 12.2

In 1-D, 2-D, and 3-D free-electron metals at $T = 0$, what fraction of the total free electrons have energies above half the Fermi energy?

Solution

By inspection of Fig. 12.4, this fraction must be 0.5 in 2-D, >0.5 in 3-D, and <0.5 in 1-D.

In 3-D, from (12.15), $(0.5)^{3/2} = 0.35$ will have $E \leq E_F$, so 0.65 will have $E \geq E_F$.

In 1-D, since $N \propto E^{1/2}$, $(0.5)^{1/2} = 0.71$ will have $E \leq E_F$ and 0.29 will have $E \geq E_F$.

(For those who are bothered by the seeming arbitrariness of the periodic boundary conditions used earlier, as I sometimes am, please be assured that if you instead stick with $\psi = 0$ at the surfaces and standing waves, you get exactly the same equations for Fermi energy and for density of states that we got above. You would use $k =$

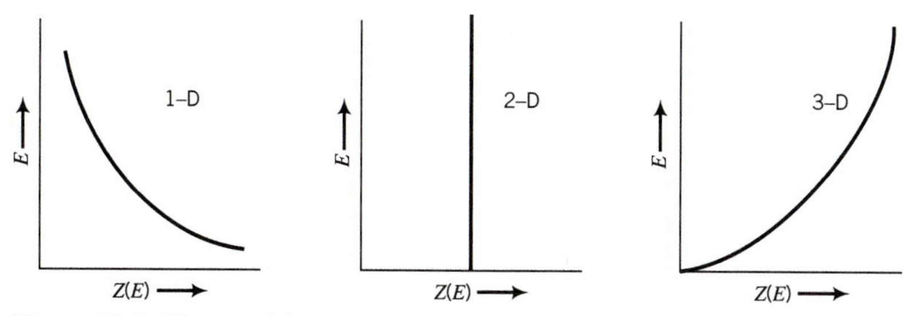

Figure 12.4. Shapes of density-of-states function $Z(E)$ in 1-D, 2-D, and 3-D free-electron metals.

$\pi n/L$ instead of $k = 2\pi n/L$, but then limit yourself to positive values of k and it all works out to give the same final results. That approach is close to the classical treatment of electromagnetic waves in a cavity and Fig. 8.1.)

Among other things, the density of states can be used to average various quantities over the assembly of free electrons. For example, since $EZ(E)$ is the total energy of the electrons between E and $E + dE$, to calculate the *total* energy of the full assembly of free electrons, you integrate $EZ(E)$. At $T = 0$, the range of integration will be only from $E = 0$ to $E = E_F$. To calculate the *average* energy of the free electrons, you must divide the total energy by the number of electrons, which is given by $N(E)$, that is, by the integral of $Z(E)$. In a 3-D free-electron metal, since $Z(E)$ is described by (12.16),

$$E_{\text{avg}}(T = 0) = \frac{\int EZ(E)dE}{\int Z(E)dE} = \frac{\int CE^{3/2}dE}{\int CE^{1/2}dE} = \frac{3E_F}{5} \tag{12.17}$$

Similarly, using the appropriate forms of $Z(E)$, you can show that the average electron energy at $T = 0$ is $E_F/2$ for a 2-D free-electron metal and $E_F/3$ for a 1-D free-electron metal. Check it out. Other average quantities, like speed or momentum, can be similarly calculated.

Sample Problem 12.3

What is the average speed of electrons in a 1-D free-electron metal at $T = 0$?

Solution

$v = (2E/m)^{1/2}$ and $Z(E) = CE^{-1/2}$, so $vZ(E) = (2/m)^{1/2}C$

and $v_{\text{avg}} = \dfrac{\int vZ(E)dE}{\int Z(E)dE} = (2/m)^{1/2}\dfrac{\int CdE}{\int CE^{-1/2}dE} = (2/m)^{1/2}\dfrac{E_F}{2E_F^{1/2}} = \dfrac{v_F}{2}$

(integrating from $E = 0$ to $E = E_F$)

12.3 HEATING THINGS UP

Although $T = 0$ is convenient mathematically, it is very inconvenient experimentally, and we are really more interested in the distribution of electron energies at temperatures greater than absolute zero, for example, at room temperature. The density of states $Z(E)$ only describes the density of *allowed* states. At $T = 0$, the allowed states are fully occupied below E_F, and completely unoccupied above E_F. At temperatures greater than zero, some electrons will be excited to higher energy states, so that some states below E_F will be unoccupied and some above E_F will be occupied. (At $T > 0$, entropy enters the game, and it is free energy, not energy, that is minimized.) The probability $F(E)$ that an electron state at energy E will be occupied is determined by *Fermi-Dirac statistics*, applicable to a collection of energy levels whose maximum

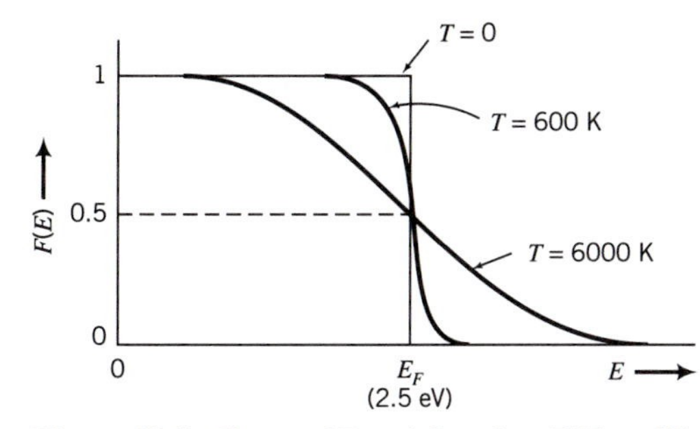

Figure 12.5. Shape of Fermi function $F(E)$ at different temperatures (for $E_F = 2.5$ eV). (Adapted from L. Solymar and D. Walsh, *Lectures on the Electrical Properties of Materials*, Oxford, 1993.)

occupation is limited by the Pauli exclusion principle. The result is the *Fermi function*

$$F(E) = \frac{1}{1 + e^{(E-E_F)/k_BT}} \tag{12.18}$$

This is an important function, and it will pay to spend some time becoming familiar with it. The function $F(E)$ is shown in Fig. 12.5 for several different temperatures (and an assumed Fermi energy of 2.5 eV). It can be seen that

$$F(E) = 1 \quad \text{for} \quad E < E_F \quad \text{and} \quad F(E) = 0 \quad \text{for} \quad E > E_F \quad (\text{at } T = 0) \tag{12.19}$$

Thus, as we assumed earlier, at $T = 0$ all the available electron states are occupied up to the Fermi energy E_F, and all the electron states are empty above that energy. For *any* $T > 0$, the exponential in (12.18) becomes equal to 1 at $E = E_F$, so that

$$F(E_F) = \frac{1}{2} \tag{12.20}$$

At *any* temperature, *the probability of occupation of electron states at the Fermi energy is 1/2.*

Sample Problem 12.4
What is the probability of occupation of electron states located k_BT above and below the Fermi energy?

Solution
For $E - E_F = k_BT$, $F(E) = 1/(1 + e) = 0.269$
For $E_F - E = k_BT$, $F(E) = 1/(1 + e^{-1}) = 0.731$

Next, consider states far above the Fermi energy, at which $(E - E_F) \gg k_BT$. In this case, the exponential term dominates the denominator, and the Fermi function becomes approximately

$$F(E) \approx e^{-(E-E_F)/k_BT} \qquad \textbf{(12.21)}$$

Far above the Fermi energy, the Fermi function decreases exponentially toward zero. Since this is reminiscent of the classical Maxwell-Boltzmann probability of occupation $\exp(-E/k_BT)$, this limit of the Fermi function at high energies is called the *Boltzmann tail.*

For energies far *below* the Fermi energy, where $(E_F - E) \gg k_BT$, the exponential in the denominator becomes $\ll 1$, and using the approximation that $1/(1 + \varepsilon) \approx 1 - \varepsilon$ for $\varepsilon \ll 1$, the Fermi function approaches one exponentially:

$$F(E) \approx 1 - e^{(E-E_F)/k_BT} \qquad \textbf{(12.22)}$$

This is like an *inverted* Boltzmann tail (see Fig. 12.5). If $F(E)$ is the probability of occupation, $1 - F(E)$ is the probability that a state will *not* be occupied, and in this region

$$1 - F(E) \approx e^{(E-E_F)/k_BT} \qquad \textbf{(12.23)}$$

However, at normally accessible temperatures and for typical Fermi energies, $F(E)$ drops rather abruptly from $F = 1$ to $F = 0$ in the immediate energy vicinity of E_F. In particular, it drops from 0.9 to 0.1 between $(E_F - 2.2k_BT)$ and $(E_F + 2.2k_BT)$, that is, an energy range of $4.4k_BT$. Typical Fermi energies for metals are several eV, and at room temperature $k_BT \approx 0.025$ eV $\ll E_F$. So *at ordinary temperatures, thermal activation affects only a small fraction of the free electrons.* Most of them are well below the Fermi energy in fully occupied states, sandwiched between fully occupied states above and below, and their energies and velocities at room temperature are essentially unchanged from what they were at $T = 0$.

The density of *allowed* electron states is given by $Z(E)$, and the probability that states are occupied is given by $F(E)$, so the product $Z(E)F(E)$ defines the *density of occupied states.* It describes the *actual* energy distribution of the electrons at any temperature, and will be the function used to calculate various properties of the solid. The remainder of this chapter is devoted to applying the quantum-mechanical free-electron theory of metals and the concepts we have introduced—Fermi energy E_F, density of states $Z(E)$, and Fermi function $F(E)$—to explain a number of the experimental properties of metals that classical free-electron theory cannot explain.

12.4 ELECTRONIC HEAT CAPACITY

Our quantum-mechanical model of a free-electron metal already allows us to explain one experimental result that classical theory could not explain—why free electrons contribute so little to the heat capacity of metals. Classical free-electron theory predicted (Section 1.8) a contribution to the heat capacity per unit volume of $3k_BN_e/2$, where N_e was the density of free electrons. Experimental values are much smaller, and this discrepancy was one of the important failures of Drude's theory. We now can see the source of this discrepancy. It's Pauli's principle.

As temperature increases, most of the free electrons are little affected, because they are sitting in completely filled energy levels and have completely filled energy levels above and below them. The only electrons that can take advantage of $T > 0$ are those within a few k_BT of the Fermi energy E_F, and thus the electronic heat capacity is drastically reduced. To get a ballpark figure, assume that only a fraction $2.2k_BT/E_F$ of the total of N_e electrons are affected by temperature (those between E_F and $E_F - 2.2k_BT$, where $F = 0.9$), and assign them the classical energy per electron of $3k_BT/2$. This gives an energy of $(3k_BT/2)$ times $(2.2N_ek_BT/E_F)$, and, taking d/dT to get the specific heat, yields an approximate value for the electronic heat capacity per unit volume of

$$C_e = 6.6N_ek_B \frac{k_BT}{E_F} \qquad \textbf{(12.24)}$$

Since typical Fermi energies are a few eV (we'll soon see direct experimental evidence of that), and k_BT at room temperature is only about 0.025 eV, we can see that the quantum-mechanical prediction for electronic heat capacity at room temperature is smaller than the classical prediction by about two orders of magnitude. (That's promising, since the *experimental* value was also about two orders of magnitude smaller than the classical prediction.) To do a more accurate job, we should calculate the total energy of the free electrons by integrating over the full density of occupied states, that is, the density of allowed states $Z(E)$ times the Fermi function $F(E)$:

$$E_{\text{tot}} = \int EZ(E)F(E)dE \qquad \textbf{(12.25)}$$

and then take d/dT of that. It's a bit messy mathematically, but for a 3-D free-electron monovalent metal, the more accurate result of quantum-mechanical theory for electronic heat capacity per unit volume is

$$C_e = \frac{\pi^2}{2} N_ek_B \frac{k_BT}{E_F} \qquad \textbf{(12.26)}$$

This is just a bit less than our rough estimate in (12.24).

Sample Problem 12.5

What is the electronic heat capacity per unit volume at $T = 300$ K of a metal with a free-electron density of $10^{29}\ \text{m}^{-3}$? What would it would be according to classical free-electron theory?

Solution

From (12.14), $E_F = 1.25 \times 10^{-18}$ J $= 7.87$ eV, so $k_BT/E_F = 3.31 \times 10^{-3}$.
From (12.26), $C_e = 0.0163\ N_e k_B = 2.25 \times 10^4\ \text{JK}^{-1}\text{m}^{-3}$.
Classically, from (1.17), $C_e = 1.5\ N_e k_B = 2.07 \times 10^6\ \text{JK}^{-1}\text{m}^{-3}$.

Equation (12.26) is a great improvement over the classical result, equation (1.17), which was too large by two orders of magnitude. But how do we compare (12.26) more precisely with experiment? It's not that easy to measure the electronic heat capacity accurately at elevated temperatures, where the lattice heat capacity is so much larger than the electronic contribution. But at very low temperatures, the lattice heat capacity C_L (see Fig. 7.4) varies as T^3, while the electronic heat capacity C_e varies linearly with T. The coefficient of the electronic term is usually written as γ, and that of the lattice term as A:

$$C = C_L + C_e = AT^3 + \gamma T \qquad (\text{near } T = 0) \tag{12.27}$$

The different temperature dependence of the two terms allows them to be separated experimentally. Experimental data for low-temperature specific heat is usually plotted as C/T vs. T^2. From (12.27), $C/T = AT^2 + \gamma$, so the intercept for $T \rightarrow 0$ gives γ. The experimental value of γ for sodium is 1.3 mJ/mol-K^2, while equation (12.26) gives 1.0. In the same units, the experimental value for copper is 0.7 and equation (12.26) gives 0.5.

Sample Problem 12.6

For the material in Sample Problem 12.5, what is the electronic heat capacity coefficient γ per mole?

Solution

$C_e = 2.25 \times 10^4\ \text{JK}^{-1}\text{m}^{-3}$ at 300 K
With $N_e = 10^{29}/\text{m}^3$, there are $6.02 \times 10^{-6}\ \text{m}^3/\text{mol}$
$\gamma = (2.25 \times 10^4) \times (6.02 \times 10^{-6})/300 = 4.52 \times 10^{-4}$ J/mol-K^2

In general, experimental values of γ for simple metals are in fair quantitative agreement with (12.26). That's a great improvement over classical theory, and impressive considering how simple our model was—a metal modeled simply as a box of constant potential energy. Agreement between theory and experiment is, however, much worse for more complex metals like transition metals. For them we'll need a more sophisticated theory—the band theory of metals, which we'll be getting to in Chapter 13. It turns out that real metals are a bit more complicated than simple boxes.

12.5 MEASURING FERMI ENERGY WITH X-RAY SPECTROSCOPY

The agreement between theory and experiment for the electronic heat capacity coefficient γ of many metals supports the simple quantum-mechanical free-electron model. But it would be nice to get more direct evidence of this strange concept that some free electrons have high kinetic energies—up to the Fermi energy E_F—even at low temperatures. It turns out that we can directly measure the Fermi energy by several techniques. The simplest technique to explain is x-ray emission spectroscopy. (*Spectroscopy* is a general term referring to measurement as a function of frequency or, equivalently, as a function of energy.)

You've probably learned elsewhere how x-rays are generated in solids. You bombard the solid with high-energy electrons, which knock electrons out of some of the inner shells of the atoms. Then electrons from the atoms' outer shells "fall" from higher-energy states into the vacated lower-energy states in the inner shells, emitting an x-ray photon with energy $h\nu$ corresponding to the energy difference between the two states.

For example, consider the effect of knocking a $2p$ electron out of a sodium atom. In gaseous sodium, $3s$ electrons would then fall into the vacated $2p$ state and emit an x-ray photon with $h\nu = E_{3s} - E_{2p}$. In the atom, the $3s$ and $2p$ energy levels are sharp (Fig. 12.6a), and the photons given off by this transition will all have the same energy. Not so in sodium metal! Here, instead of a single frequency, the x-ray

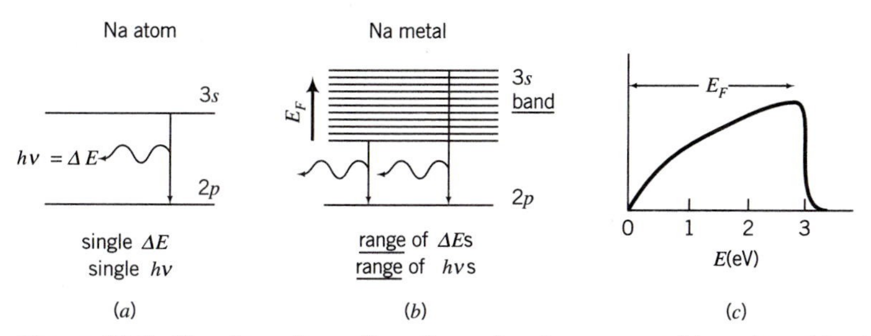

Figure 12.6. Showing why sodium $3s \rightarrow 2p$ electron transition gives (a) a single x-ray energy in an isolated atom, but (b) a range of energies (equal to the Fermi energy) in solid sodium. (c) X-ray emission data from $3s \rightarrow 2p$ transition in solid sodium, setting $E = 0$ for transition from bottom of $3s$ band. (Adapted from F. Seitz, *Modern Theory of Solids*, McGraw-Hill, New York, 1940.) The width of the x-ray emission band yields an experimental value for the Fermi energy.

Table 12.1. Experimental and Calculated Values of Fermi Energy (in eV)

Element	Li	Na	K	Cu	Au	Be	Mg	Zn	Al
E_F(calculated)	4.7	3.2	2.6	7.1	5.5	14.3	7.2	9.5	12.8
E_F(experimental)	3.9	2.8	1.9	6.5	5.4	13.8	7.6	11.5	11.8

photons corresponding to the $3s \rightarrow 2p$ transition have a *range* of frequencies (Fig. 12.6b). Although the $2p$ energy level of sodium remains fairly sharp in sodium metal (Fig. 11.22), its $3s$ electrons, the conduction electrons, have a *range* of energies. From our model of free electrons in a box, that range of energies corresponds essentially to the Fermi energy. Thus *the energy width of the x-rays corresponding to the $3s \rightarrow 2p$ transition (Fig. 12.6c) can be used to measure E_F.* The experimental value for sodium is 2.8 eV, while equation (12.14), the prediction of simple free-electron theory, gives 3.2 eV.

Again, considering the simplicity of the model, quantitative agreement for E_F of sodium is pretty good. In addition, the variation of x-ray intensity with energy in Fig. 12.6c is roughly parabolic, consistent with the parabolic density of states of the conduction electrons. The increase in x-ray intensity with increasing energy results from the increase in the density of states with increasing energy. (There are more $3s$ electrons per unit energy at higher energies than at lower energies.)

Table 12.1 shows experimental data for a number of monovalent, divalent, and trivalent metals. In each case, the experimental value is taken from the width of the x-ray band corresponding to conduction or valence electrons falling into the next-lowest energy level, and the theoretical value is from (12.14). X-ray emission spectroscopy gives strong direct evidence that the conduction electrons in metals have, instead of the single energy level they would have in an isolated atom, a *range* of energy levels, or an *energy band.* And that range of energies measured experimentally is in rough quantitative agreement with (12.14), the Fermi energy predicted by simple quantum-mechanical free-electron theory. More sophisticated theories are required to improve the quantitative agreement, but x-ray spectroscopy alone is strong proof that conduction electrons in many metals aren't *very* different from free electron waves in a big box.

12.6 PHOTOELECTRON SPECTROSCOPY

In x-ray emission spectroscopy, you bombard a solid with high-energy electrons and measure the energy distribution of the photons given off. In photoelectron spectroscopy, you do the opposite—you bombard a solid with high-energy *photons* and measure the energy distribution of the *electrons* given off. If the high-energy photons used are x-rays, it is called XPS (x-ray photoelectron spectroscopy), and if the photons used are in the ultraviolet, it is called UPS.

We discussed the photoelectric effect in Chapter 8 as one of the early experiments that led to the realization that electromagnetic waves were quantized in particles called photons. In particular, equation (8.5) gave the maximum energy of the electrons knocked out of a solid by high-energy photons, that is, by the photoelectrons, as equal to $h(\nu - \nu_c)$.

Our free-electron model so far assumed that the potential energy V of the electrons was zero inside the metal and, implicitly, that V was infinite outside the metal. We assumed that the electrons just couldn't get out of the box. More realistically, the metal should be considered as a *finite* energy well. This wouldn't change much of what we've done so far, as long as the potential barrier for leaving the metal is larger than the highest energy of the conduction electrons, which at $T = 0$ is E_F. In fact, equation (8.5) tells us just how much higher the potential barrier is than the highest energy of the conduction electrons. It shows that a minimum energy of $h\nu_c$ is needed to knock electrons out of a metal. It seems reasonable to assume that it is the conduction electrons with the highest initial energy—the electrons at the Fermi energy—that can be knocked out of the metal with the minimum photon energy. Thus our energy box with finite walls can be modeled (Fig. 12.7) with an external potential energy that is $h\nu_c$ higher than the Fermi level.

For historical reasons, this minimum energy barrier $h\nu_c$ for conduction electrons to leave a metal is called the *work function* and is usually given the symbol Φ. (In some texts the work function Φ is instead expressed in volts, with the energy barrier $h\nu_c$ given by $e\Phi$.) The corresponding quantity for isolated atoms (the minimum energy required to remove an electron) is called the first ionization energy. One of the elements with the lowest first ionization energy is cesium, and cesium metal also has one of the lowest work functions—1.8 eV. Most metals have significantly higher work functions, such as tungsten (4.5 eV), silver (4.8 eV), and platinum (5.3 eV).

Returning to the photoelectron experiment, suppose we bombard the metal with monochromatic (single frequency) photons with energy greater than $\Phi + E_F$. We now can not only knock electrons at the Fermi energy out of the metal, but we have enough photon energy to knock out even the lowest-energy conduction electrons—the electrons at the bottom of the energy band of conduction electrons, called the *conduction band*. And by measuring the range of kinetic energies of the photoelectrons (analogous to our measurement of the range of x-ray energies in Fig. 12.6), we would be able to measure the Fermi energy—the energy width of the conduction band.

Photoelectron spectroscopy involves measuring how many photoelectrons are produced at each electron energy by a photon of fixed energy. Electrons of each energy can be separated from the others in a *spectrometer* (Fig. 12.8), which measures how many photoelectrons travel in a circular path of a particular radius as

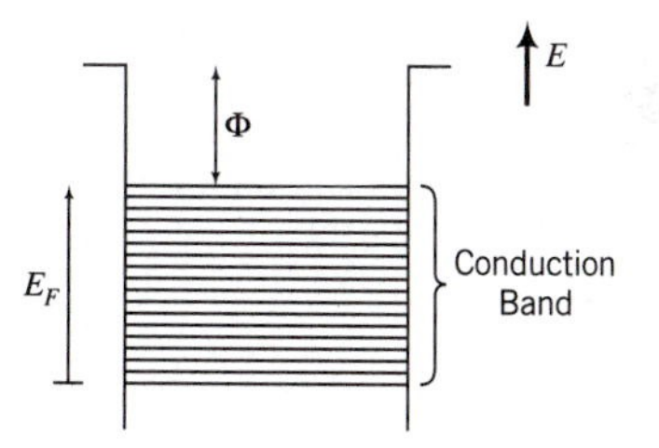

Figure 12.7. The work function Φ of a metal.

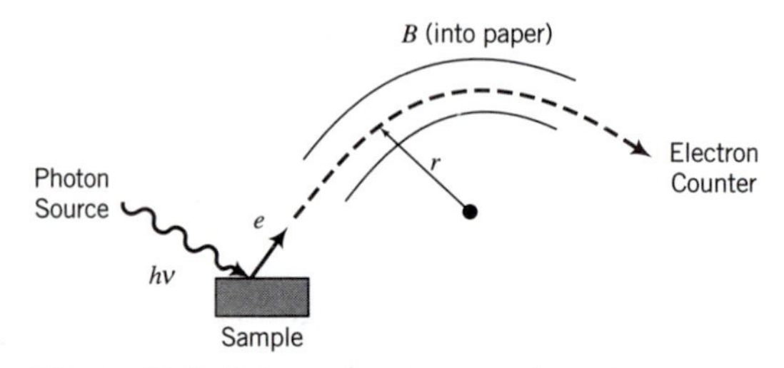

Figure 12.8. Schematic representation of spectrometer used to select photoelectrons of specific energies.

function of applied magnetic field. (From the magnetic field **B** and the radius r, the photoelectron velocity—and hence its energy—can be determined, since $evB = mv^2/r$, and therefore the electron velocity $v = Bre/m$.) It is assumed that each photon gives *all* of its energy ($h\nu$) to each of these electrons. The electron requires a certain amount of energy, its *binding energy* E_b, to escape the metal, and therefore the remaining kinetic energy E_K of the photoelectron can be used to measure the electron's binding energy through

$$E_b = h\nu - E_K \quad \textbf{(12.28)}$$

Experimental data for the distribution of binding energies of conduction electrons in aluminum metal generated from (12.28) are shown in Fig. 12.9, with the zero set for electrons at the Fermi energy (maximum E_K, minimum E_b). We see (subtracting the experimental background) a range of binding energies of conduction electrons in aluminum of about 12 eV. The Fermi energy of aluminum, calculated from (12.14) assuming three free electrons per atom (aluminum is trivalent) is 12.8 eV. And the shape of the distribution curve suggests the parabolic $Z(E)$ predicted from (12.16). *Thus both x-ray spectroscopy and photoelectron spectroscopy present direct evidence in support of the general predictions of the quantum-mechanical free-electron model.* These and other forms of spectroscopy are among the most powerful tools used to probe electron energy levels in atoms, molecules, and solids.

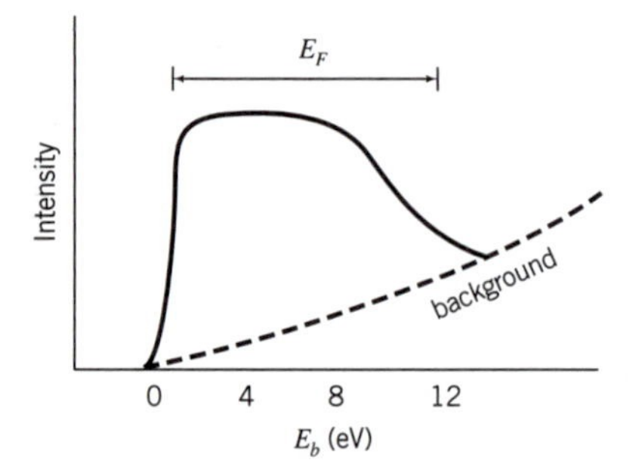

Figure 12.9. Photoelectron spectrum for aluminum from which the distribution of binding energies of conduction electrons can be determined (setting $E_b = 0$ at the Fermi level, i.e., for the maximum photoelectron energy E_K). Results indicate $E_F \approx 12$ eV. (Adapted from P. A. Cox, *The Electronic Structure and Chemistry of Solids*, Oxford University Press, Oxford 1987.)

Sample Problem 12.7
A metal is bombarded with ultraviolet photons of 12 eV energy. The velocities of the photoelectrons emitted from the conduction band range from 1.19×10^6 m/s to 1.74×10^6 m/s. What is the Fermi energy of the metal? What is the work function?

Solution
Those minimum and maximum velocities correspond to minimum and maximum photoelectron energies ($E_K = mv^2/2$) of 4.03 and 8.62 eV.
The Fermi energy E_F (the range of E_K, thus the range of E_b) is $8.62 - 4.03 = 4.59$ eV.
The work function $\Phi = h\nu_c = 12 - 8.62 = 3.38$ eV.
(Φ is the minimum E_b, and is calculated from the maximum E_K.)

12.7 EMITTING HOT ELECTRONS

In television tubes, metallic filaments are raised to high enough temperatures by resistive heating that they emit quantities of electrons (which are then driven to the screen to excite phosphors and produce the images that many of us spend too much time watching). This *thermionic emission* is further evidence of the finiteness of the energy barrier holding conduction electrons in metals, and offers another way to measure the work function Φ.

To escape from the metal, an electron must have an energy at least Φ greater than the Fermi energy. At any temperature T, the number of electrons with energies greater than $E_F + \Phi$ is given by the appropriate integral over the density of occupied states $Z(E)F(E)$:

$$N(E > E_F + \Phi) = \int_{E_F+\Phi}^{\infty} Z(E)F(E)dE = \int_{E_F+\Phi}^{\infty} CE^{1/2}e^{-(E-E_F)/k_BT}dE \quad \textbf{(12.29)}$$

We have substituted $CE^{1/2}$ for $Z(E)$ from (12.16) and, since $(E_F + \Phi) \gg k_BT$, we have used approximation (12.21) for F(E). (It is only the high-energy Boltzmann tail of the electron distribution that can escape the barrier.) It is not difficult to show that this integral is proportional to $e^{-\Phi/k_BT}$.

Not all of these electrons will escape from the metal (it will depend on the direction of their momentum vector **p**). And the emission current from some particular surface (say $x = 0$), is proportional to p_x. So proper calculation of the emission current requires integrating over the different components of momentum rather than simply over energy, which yields the Richardson equation

$$J = \frac{4\pi em}{h^3}(k_BT)^2e^{-\Phi/k_BT} \quad \textbf{(12.30)}$$

But equation (12.29) already suggested that this current was likely to include the

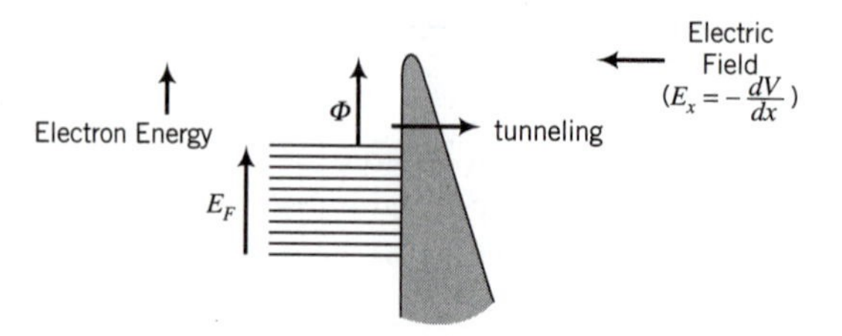

Figure 12.10. Field emission. High electric field, represented by high *slope* of electron energy $e(dV/dx)$, allows tunneling through a narrow energy barrier at surface.

term $e^{-\Phi/k_BT}$. This exponential term dominates the variation of thermionic emission current with temperature, and plotting experimental values of log J against $1/T$ allows measurement of the work function.

If we wish to increase the emission of electrons from the metal filament beyond what simple thermionic emission and equation (12.30) would give, we can apply a very high electric field **E**. In terms of the $V(x)$ diagrams that we have been using, an electric field corresponds to a sloping line, since $E_x = -dV/dx$. Thus a large electric field will produce a narrowing of the energy barrier for electron escape (Fig. 12.10). With a high enough field, the barrier will become thin enough that electrons have a reasonable probability of *tunneling* through the barrier (as discussed in Chapter 8), and escaping from the metal despite the fact that their energy is insufficient to escape thermally. This process is called *field emission*, and it is used to increase electron emission from filaments in many devices, including electron microscopes. It's perhaps intuitively obvious that applying a large electric field would help "pull" electrons out of a metal filament, but it's *not* intuitively obvious that the process involves tunneling—a quantum-mechanical effect that allows electrons to penetrate regions from which they are classically forbidden, within which classical theory would say they have negative kinetic energy.

12.8 CONTACT POTENTIAL

What happens if we bring two metals with different work functions, Φ_A and Φ_B, into electrical contact? The energy scales for the two metals must line up at the energy of escape (Fig. 12.11a), so that if Φ_A is less than Φ_B, it is clear that electrons at the Fermi level in metal A have a higher energy than electrons at the Fermi level in metal B. That's an unstable situation, and electrons will flow from metal A into metal B.

This electron flow will give metal B a net negative charge, which will increase electron energy there, and will leave metal A with a net positive charge, which will decrease electron energy there. Charge will flow until the resulting energy shifts are sufficient to bring the two Fermi levels to the same energy (Fig. 12.11b). In equilib-

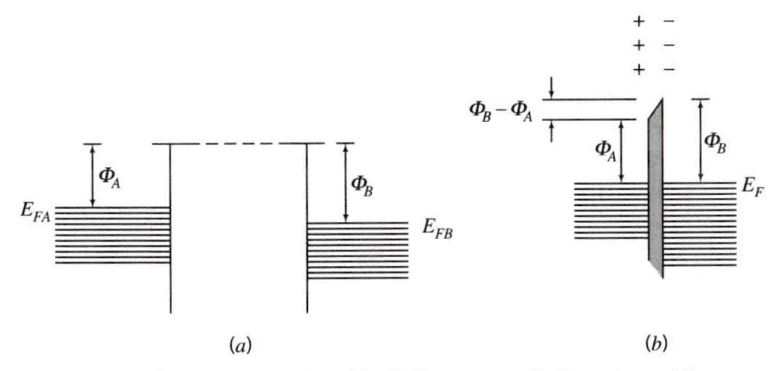

Figure 12.11. Two metals with different work functions (a) separated, (b) in contact, and after electron flow has produced a double charge layer and a contact potential.

rium, there will be a voltage difference, a *contact potential* V_{AB}, between the two metals equal to

$$V_{AB} = (\Phi_B - \Phi_A)/e \tag{12.31}$$

(Here V_{AB} is in volts, and the electronic charge appears because we have defined work functions as energies.) There will be an associated double layer (dipole layer) of electric charge at the contact and an associated electric field between the two layers of charge. As in Fig. 12.10, the electric field appears in Fig. 12.11b as a steeply sloped line.

The contact potential is an equilibrium effect, and it cannot be measured by attaching a voltmeter across the junction. If, for example, the voltmeter is connected to each of the metals with a wire of metal C, this produces two more contact potentials: $\Phi_B - \Phi_C$ where the wire contacts metal B, and $\Phi_C - \Phi_A$ where the wire contacts metal A. With the circuit completed, the various contact potentials cancel, and there is no net voltage in the circuit.

The contact potential results from the electrons in one metal coming into equilibrium with the electrons in another metal. Although contact potentials between materials in electrical contact cannot be measured directly with a voltmeter, they can sometimes have important effects. In particular, we will find in Chapter 16 that the contact potential between two semiconductors can greatly influence the charge flow across the *p-n junction* between them. Such junctions provide the basis for many modern devices, including light-emitting diodes (LEDs), photodetectors, solar cells, and transistors.

SUMMARY

Armed with some basic principles of quantum mechanics, including the wave nature of electrons and the Pauli principle, we returned to the material we considered way back in Chapter 1—a metal containing free electrons. We chose boundary conditions that put the solutions to Schrödinger's equation in the form of traveling electron waves. Because the "box" is so large, the energy levels corresponding to the various electron waves are very closely spaced, and can be considered as a continuous *band* of allowed energy levels. Because each of the allowed levels can accommodate only two electrons, even at $T = 0$ some of the free electrons have very high velocities and kinetic energies. We called the maximum electron energy at $T = 0$ the *Fermi energy* E_F, and found relations between the free-electron concentration and Fermi energy for 1-D, 2-D, and 3-D metals.

The density of states function $Z(E)$ defines how many allowed electron states (up and down spin allowing two "states" per energy level) exist *per unit energy*. $Z(E)$ varies as $E^{1/2}$ in 3-D, is independent of E in 2-D, and varies as $E^{-1/2}$ in 1-D. Although $Z(E)$ defines the density of *allowed* states, the probability that any state is actually occupied by an electron at a nonzero temperature is defined by the Fermi function $F(E)$, which differs from the classical Boltzmann distribution because of the Pauli principle. The density of *occupied* states is therefore represented by the product $Z(E)F(E)$.

The quantum-mechanical free-electron model explains why the free electrons contribute so little to the specific heat of metals, and calculations of the electronic heat capacity coefficient γ are in fair agreement with low-temperature measurements for simple metals. X-ray emission spectroscopy and photoelectron spectroscopy provide direct experimental evidence that conduction electrons have a range of energies, and experimental measurements of the width of that range—the Fermi energy—are in fair agreement with theory.

Although the simplest free-electron model assumes an infinite energy barrier at the metal surface, a more realistic model assumes a finite energy barrier. The threshold photon energy $h\nu_c$ required to eject photoelectrons corresponds to the difference between the electron escape energy and the Fermi energy—the work function Φ. At elevated temperatures, the Boltzmann tail of the electron energy distribution can exceed the energy barrier and produce thermionic emission of electrons from the metal. Measurement of the temperature dependence of thermionic emission offers another means of measuring the work function. Application of a high electric field to the metal can thin the energy barrier opposing electron emission and allow electrons to tunnel through, a process known as *field emission*.

Electrical contact between two metals with different work functions produces charge flow across the junction and equilibration of the two Fermi levels. In the equilibrium junction there is a dipole charge layer, an electric field, and a contact potential equal to the difference between the two work functions.

PROBLEMS

12-1. In a 3-D free-electron metal with a Fermi wave vector of 10^{10} m^{-1}, (a) what are the Fermi energy and Fermi velocity? (b) What is the mean speed of the electrons at $T = 0$? (c) What is the root-mean-square speed of the electrons at $T = 0$?

12-2. From the experimental value of the Fermi energy for sodium determined by x-ray emission spectroscopy (2.8 eV), (a) what are the corresponding Fermi velocity and Fermi wave vector? (b) What is the average speed of free electrons in sodium? (c) What are their root-mean-square speed and average kinetic energy?

12-3. A metal with a work function of 4 eV and a Fermi energy of 3 eV is exposed to electromagnetic radiation with a wavelength of (a) 1 micron and (b) 0.1 micron. For each case, what will be the range of energies and wavelengths of the photoelectrons ejected from the conduction band?

12-4. (a) For a one-dimensional free electron metal, calculate the ratio of the density of occupied electron states at the Fermi energy to that at one-third the Fermi energy. Assume $k_BT \ll E_F$. Do the same calculation for (b) 2-D and (c) 3-D free-electron metals.

12-5. A three-dimensional free-electron metal has a Fermi velocity of 1×10^6 m/s. You heat this metal until the density of occupied states at an energy 0.3 eV above the Fermi energy equals one percent of the density of occupied states at the Fermi energy. What is the temperature?

12-6. Metals A and B, with work functions of 5 eV and 2 eV, respectively, are brought into electrical contact. (a) Which metal now has a net negative charge on its surface? (b) In which direction does the electric field point? (c) Assume that the electron displacement that created the surface charges was 0.3 nm. What is the magnitude of the resulting electric field? (d) If the metals were 1-cm cubes and the contact area 1 cm^2, how many electrons were transferred?

12-7. For a three-dimensional free-electron metal, at what fraction of the Fermi energy will the density of occupied states be a maximum for $k_BT = 0.1\ E_F$? For $k_BT = 0.01\ E_F$?

12-8. (a) From quantum-mechanical free-electron theory, derive expressions for the Fermi energy and for the plasma frequency of alkali metals as a function of the body-centered cubic (bcc) lattice parameter a. (Reminder: There are two atoms per unit cell in a bcc crystal.) (b) The lattice parameters of Li and Cs are, respectively, 0.351 and 0.614 nm. Calculate the Fermi energy and plasma frequency for these two metals. (c) Would either of these metals be transparent to any visible light?

12-9. According to quantum-mechanical free-electron theory, what fraction of the free electrons have kinetic energies more than 75% of the Fermi energy at $T = 0$ in (a) a 3-D metal? (b) A 2-D metal? (c) A 1-D metal?

12-10. Equations (12.15) and (12.16) give expressions for $N(E)$ and $Z(E)$ in a three-dimensional free-electron metal. Derive analogous expressions for 1-D and 2-D free-electron metals.

12-11. From the experimental value for E_F of copper given in Table 12.1, calculate the electronic heat capacity at room temperature predicted by equation (12.26). By what factor does this differ from the value predicted by classical free-electron theory (Chapter 1)?

12-12. (a) In a three-dimensional free-electron metal at $T = 300$ K with $E_F = 3$ eV, what is the ratio of the density of occupied states at 2.9 and 3.1 eV to that at 3 eV? (b) Calculate the same ratio for energies of 2.95 and 3.05 eV.

12-13. Calculate the current density produced by thermionic emission from tungsten heated to (a) 1000 K, (b) 2000 K, (c) 3000 K.

12-14. Will a beam of yellow light produce photoelectrons from silver? From platinum? From cesium?

Chapter 13

Nearly Free Electrons—Bands, Gaps, Holes, and Zones

13.1 LESS THAN FREE

In 1858, just before the American Civil War, Abraham Lincoln said, "I believe this government cannot endure half slave and half free." But the electrons that hold solids together and determine most of the properties of engineering materials—the valence electrons—can! They are not as completely "slave" as the core electrons in the inner shells, which are bound to their nuclei in orbitals little different from the AOs of isolated atoms, or as completely "free" from the local Coulomb attraction of the ion cores as we assumed for the valence electrons in Chapter 12.

In Chapter 11, we considered bonding in larger and larger molecules, and eventually discussed solids as very large molecules. We specifically considered *delocalized* pi electrons in conjugated carbon chains and delocalized sigma electrons in lithium crystals, and found that molecular orbitals (MOs) evolved into crystal orbitals (COs), with the valence electrons shared by as many as 10^{23} atoms. The envelopes of the resulting wave functions in Figs. 11.14 and 11.20 resemble the sine waves of the simple free-electron waves of Chapter 12, their sinusoidal nature represented explicitly in the form of the coefficients of the LCAO equation—see equation (11.5). However, in the vicinity of each atom they retained much of the character of the atomic orbitals (AOs) from which the MOs and COs were constructed. In a sense, these valence electrons were half slave, half free.

This LCAO → MO → CO approach to valence electrons in solids starts with AOs—electrons confined to specific atoms—and observes their gradual increase in "freedom" as molecules grow into solids. An alternate approach starts with the completely free electrons-in-a-box of Chapter 12 and then recognizes reality by putting atoms in the box. This is called the *nearly free electron* (NFE) model, which we consider next. It is comforting that both approaches lead to qualitatively similar pictures of the wave functions and energy levels of valence electrons in solids. Which approach is more quantitatively accurate varies from case to case. (For example, in the first transition-metal series, the 4s electrons are best treated by the NFE approach, while the 3d electrons, which have less overlap, are best treated with the LCAO approach.)

The presence of atoms in the box will change, at least slightly, both the free-electron wave functions and the simple parabolic $E(k)$ relation that we had in Chapter 12. However, in this chapter we'll focus only on the change in $E(k)$ produced by the atoms, and continue to assume that our wave functions are simple $e^{i\mathbf{k}\cdot\mathbf{r}}$ functions as in equation (12.9). More advanced treatments modify those simple free-electron wave functions into *Bloch functions*, expressed as $u(r)e^{i\mathbf{k}\cdot\mathbf{r}}$, where $u(r)$ is a function with the periodicity of the lattice, allowing the $e^{i\mathbf{k}\cdot\mathbf{r}}$ waves to adjust a bit near each atom. Bloch functions, in fact, have the form of the COs we developed with the LCAO method—the $e^{i\mathbf{k}\cdot\mathbf{r}}$ part replacing the c_i coefficients and the $u(r)$ part replacing the series of AOs. But for our present purposes we can let $u(r)$ remain a constant in our nearly free electron picture and just see how the presence of a periodic lattice of atoms (ion cores, to be more precise) will influence $E(k)$ for simple $e^{i\mathbf{k}\cdot\mathbf{r}}$ wave functions.

13.2 ATOMS IN THE BOX—DIFFRACTION AND THE GAP

What happens to those free-electron waves of Chapter 12 when we admit that there are also some atoms in the box? Limiting ourselves for now to a *one-dimensional* solid, the simple assumption that $V = 0$ throughout the box must be replaced by a periodic potential energy function $V(z)$ that recognizes the attraction between the ions (the atoms minus the valence electrons) and the valence electrons (Fig. 13.1). For electrons with a wavelength λ that is very long compared to the interatomic

Figure 13.1. Periodic electron potential energy produced by lattice.

spacing a (a wave vector $k = 2\pi/\lambda$ very small compared to $2\pi/a$), the effect will be small. These low-k electrons will simply experience some average potential energy, which we can set equal to zero for convenience. The kinetic energy of such electrons will be little changed from that of a completely free electron:

$$E = \frac{\hbar^2 k^2}{2m} \tag{13.1}$$

However, as electron kinetic energy increases (k increases, λ decreases) and the wavelength approaches a few interatomic spacings, the effect of the periodic potential will begin to be felt. Consider the extreme case where $\lambda = 2a$ ($k = \pi/a$). You might recognize this as the *Bragg diffraction condition*—equation (4.13)—for this one-dimensional lattice. Electron waves traveling in either direction will be reflected, and the sum of incident and reflected waves will produce *standing waves*.

Consider two standing electron waves (Fig. 13.2), one with the maxima of ψ^2 centered on the atoms, the other with the maxima of ψ^2 centered midway between the atoms. The average potential energy felt by the first wave will be much less than the average potential energy (defined as $V = 0$) felt by the long-wavelength (low k) wave considered above, so its total energy will be *less* than expressed in the simple parabolic expression (13.1). The second electron wave, on the other hand, will be sampling primarily those regions of high potential energy between the atoms. It will feel an average potential energy *greater* than the $V = 0$ felt by low-k waves, so its total energy will be *greater* than expressed by (13.1).

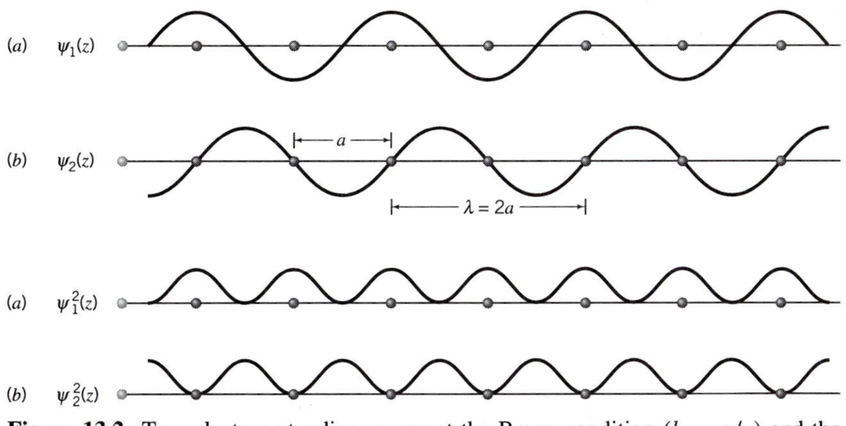

Figure 13.2. Two electron standing waves at the Bragg condition ($k = \pi/a$) and the associated probability distributions. Since one wave (a) is located mostly on the atoms and the other wave (b) is located mostly between the atoms, they experience different average potential energies.

Sample Problem 13.1

Suppose the periodic electron potential energy produced by the lattice in a one-dimensional crystal (Fig. 13.1) consists simply of a series of square energy wells with $V = -V_0/2$, of width $a/2$, centered on each atom, separated by regions of width $a/2$ with $V = +V_0/2$. Calculate the difference in the average potential energies (the band gap) experienced by the two wave functions $\psi_1(z)$ and $\psi_2(z)$ shown in Fig. 13.2.

Solution

$\psi_1(z) = A\cos(\pi z/a)$ and $\psi_2(z) = A\sin(\pi z/a)$

From Section 8.7, we know that the normalization constant $A = \sqrt{(2/a)}$.

So the two expectation values we want are: $V_1 = (2/a)\int V(z)\cos^2(\pi z/a)dz$ and $V_2 = (2/a)\int V(z)\sin^2(\pi z/a)dz$, where $V(z) = -0.5\ V_0$ from 0 to $a/4$ and from $3a/4$ to a, and $V(z) = +0.5\ V_0$ from $a/4$ to $3a/4$.

The results are $V_1 = -0.33\ V_0$ and $V_2 = +0.33\ V_0$, so the gap $V_2 - V_1 = 0.66\ V_0$. (The average of V_1 and V_2 is 0, the average of $V(z)$.)

Electron waves with $k = \pm\pi/a$ will therefore have a total energy either greater than or less than the simple parabolic expression (13.1) that is appropriate for completely free electrons (Fig. 13.3). There will be an *energy gap* at the value of k corresponding to Bragg diffraction. Electron waves with wave vectors slightly lower or slightly higher than $\pm\pi/a$ will also be strongly affected by the periodic potential energy $V(z)$, and will have total energies different from (13.1). The difference between the $E(k)$ behavior for completely free electrons—equation (13.1)—and the $E(k)$ behavior of these nearly free electrons is shown in Figure 13.3. The magnitude of the energy gap at $k = \pm\pi/a$ will depend on both the depth and the shape of the energy well at each atom site.

Of course, we have encountered gaps of forbidden energies between allowed energy levels before—the gaps between the allowed energy levels (1*s*, 2*s*, 2*p*, etc.) of atoms (Chapter 9). In Chapters 10 and 11, the LCAO → MO → CO approach

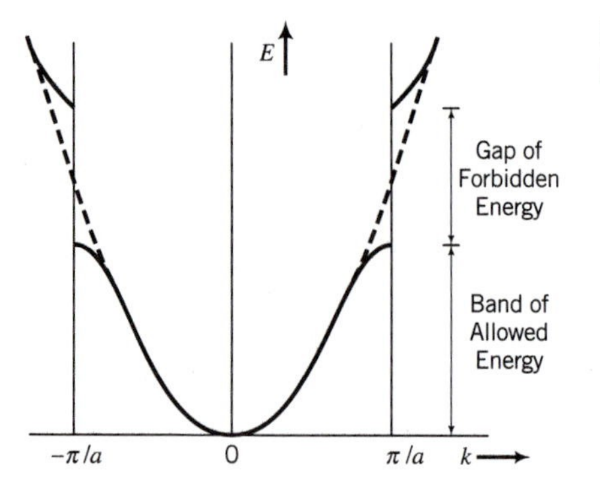

Figure 13.3. Energy gaps in $E(k)$ produced by Bragg diffraction.

showed us that the energy levels of the AOs led to many MO levels. As N, the number of atoms, became very large, the N MO levels corresponding to one AO energy level became very closely spaced and could be considered as a continuous *band* of allowed energies—an energy band of N closely spaced levels capable of accommodating $2N$ electrons. But energy gaps usually remained, although reduced by the broadening of AO energy levels into bands, as in Fig. 11.22.

In Chapter 12, we considered instead quantum-mechanical *free* electron theory. In a one-dimensional crystal of length L, we had a series of allowed energy levels for the valence electrons defined (for periodic boundary conditions) by $k = 2\pi n/L$. With L large, the many allowed energy levels were very closely spaced, and could be treated essentially as a continuous band of allowed energy levels. However, as long as we ignored the presence of the atomic lattice, we had *no gaps* of forbidden energy. We now have seen that Bragg diffraction creates such a gap, and can calculate how many energy levels are available in the allowed energy band between $k = 0$ and $k = \pm\pi/a$. (With periodic boundary conditions, we have traveling-wave solutions, and we must consider both positive and negative k, that is, waves traveling in either direction.)

The band extends from $k = -\pi/a$ to $k = +\pi/a$, a range of $2\pi/a$, with each allowed level separated along the k-axis by $2\pi/L$ from its neighbors. Thus the number of energy levels in this band is $(2\pi/a)$ divided by $(2\pi/L)$, or L/a. But L/a is just N, the number of atoms in our long one-dimensional crystal. So our NFE (nearly free electron) approach leads to the same conclusion as our LCAO approach—*one AO from each of N atoms leads to an energy band with N energy levels (which can accommodate 2N electrons)*. This is the extension to macroscopic solids of the *conservation of states* principle that we applied earlier to small molecules.

That's no coincidence. The two approaches look very different—one starts with atoms and builds bands from interatomic bonds, while the other starts with electron waves in an otherwise empty box and adds the atoms later. Yet both lead to energy bands (containing $2N$ states) separated by energy gaps. And further consideration of Fig. 13.2 can show that the connection is greater than that.

Suppose that, with the LCAO approach applied to this same one-dimensional crystal, you build a band of COs (crystal orbitals) with s-s overlap, as we did for lithium in Fig. 11.20. The highest-energy wave function in that s-band will be, as seen in that figure, an antibonding CO with maxima at the atom centers and nodes between each pair of atoms. That looks a lot like the wave function in Fig. 13.2a, which was the highest-energy wave function in the lower NFE energy band.

Now suppose that you instead build COs with p_z-p_z overlap, as in Fig. 10.7. The p AOs have nodes at the center of the atoms, and the lowest-energy wave function built from the p_zs will have maxima *between* the atoms. That looks a lot like the wave function in Fig. 13.2b, which was the lowest-energy wave function in the upper NFE energy band. So if we identify our LCAO s-band with the lower NFE band and our LCAO p-band with the upper NFE band, the wave functions just below and just above the energy gap look very much the same with both LCAO and NFE approaches. Although the necessary calculations are quite different in the two approaches, there are many similarities in the final results.

Returning to our NFE model, the Bragg diffraction condition occurs not only at $k = \pi/a$, but also at $k = 2\pi/a$, $k = 3\pi/a$, and so on. At each of these k values, an energy gap appears, so that the $E(k)$ curve divides into a series of energy bands of increasing energy (Fig. 13.4a) corresponding to different segments along the k-axis. These segments are called *Brillouin zones* in honor of Louis Brillouin, a French scientist who studied the effects of periodic structures on the propagation of waves. (We introduced this concept in Section 7.4 when considering elastic waves.) The segment between $k = -\pi/a$ and $k = +\pi/a$ (corresponding to the band of lowest energy) is called the first Brillouin zone, the segments between $k = -2\pi/a$ and $k = -\pi/a$ and between $k = +\pi/a$ and $k = +2\pi/a$ (corresponding to the next energy band) are together called the *second Brillouin zone*, and so on.

The periodicity of the potential energy function produced by the crystal lattice (Fig. 13.1) leads to a periodicity of $E(k)$ in k-space such that

$$E(k) = E(k \pm 2n\pi/a) \tag{13.2}$$

This allows the segments of the $E(k)$ curve in the second and higher Brillouin zones to be replotted within the first zone as in Fig. 13.4b, a representation called the *reduced-zone scheme* that is often convenient. The representation in Fig. 13.4a is called the *extended-zone scheme*.

There is more to be learned by considering both the LCAO and NFE treatments of a one-dimensional crystal. We next consider the detailed form of the $E(k)$ curve, which will lead us to solutions of some of the mysteries we encountered many chapters ago, such as why electrons can sometimes behave as if they had effective masses different from 9.11×10^{-31} kg (as measured by cyclotron resonance; see Chapter 2), and why they can sometimes behave as if current were being carried by

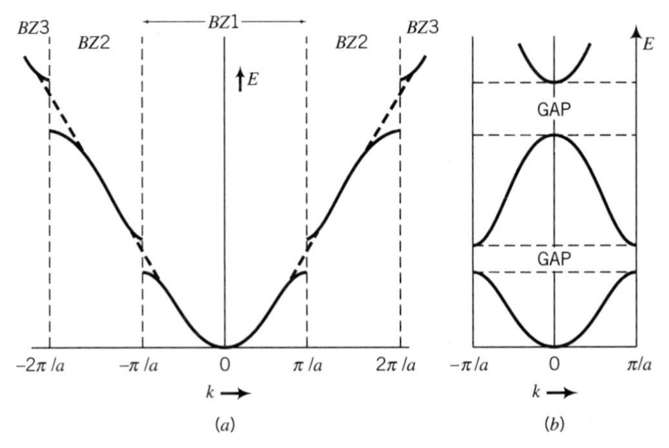

Figure 13.4. (a) $E(k)$ in extended-zone scheme, showing division of one-dimensional k-space into Brillouin zones. (b) $E(k)$ in reduced-zone scheme, with segments from higher zones replotted in first Brillouin zone, using equation (13.2).

positively charged particles (as determined from the sign of the Hall voltage; see Chapter 1).

13.3 EFFECTIVE MASS

In Figure 13.3, we drew the $E(k)$ curve with reverse curvature near the top of the first energy band, with the slope $\partial E/\partial k$ approaching zero as k approached π/a. (We have redrawn this $E(k)$ at the top of Fig. 13.5.) From the NFE point of view, this can be understood in terms of the interaction of electron waves traveling in a periodic potential as k nears the Bragg diffraction condition. We recall that if you want to apply particle-like thinking to traveling waves (in a wave packet), the appropriate velocity is the *group velocity* $\partial\omega/\partial k$. We defined group velocity in Section 7.4 during our discussion of sound waves, and discussed the group velocity of electron wave packets in Section 8.6. Recalling that for electrons $E = \hbar\omega$, the group velocity of electrons is

$$v_g = \frac{\partial\omega}{\partial k} = (1/\hbar)\,\frac{\partial E}{\partial k} \tag{13.3}$$

Near the bottom of the band, equation (13.1) applies, so the group velocity is simply $\hbar k/m$, or p/m, the classical velocity of a particle with momentum $p = \hbar k$. An electron that is increasing in energy, say, by drawing energy from an applied electric field, will increase in velocity and momentum. That seems right for a classical particle. But this will no longer be true in the upper half of the energy band. Here increasing energy corresponds to a decreasing slope of the $E(k)$ curve—a *decreasing* group velocity as the electron wave begins to feel increasing interference from the periodic lattice potential. The group velocity in fact approaches *zero* as the Bragg

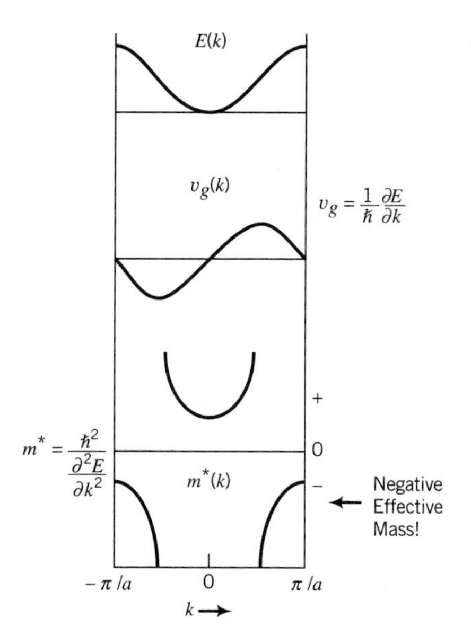

Figure 13.5. Energy, group velocity, and effective mass as a function of wave number k across energy band. Note that the effective mass is negative in the upper half of the band!

diffraction condition is approached. At $k = \pm\pi/a$, we now have a standing wave instead of a traveling wave, for which a group velocity of zero seems reasonable.

A traveling wave interacting with a periodic potential leads to strange concepts if we attempt to interpret it in terms of classical particle-like behavior, and define an effective mass m^* in Newtonian terms as $m^* = \mathbf{F}/\mathbf{a}$. From (13.3), the acceleration **a** is

$$a = \frac{dv_g}{dt} = (1/\hbar)\,\frac{\partial^2 E}{\partial k^2}\,\frac{dk}{dt} \qquad \textbf{(13.4)}$$

The force F can be related to the rate of change of momentum:

$$F = \frac{dp}{dt} = \hbar\,\frac{dk}{dt} \qquad \textbf{(13.5)}$$

Dividing (13.5) by (13.4), the effective mass of an electron wave is therefore

$$m^* = \frac{\mathbf{F}}{\mathbf{a}} = \frac{\hbar^2}{\partial^2 E/\partial k^2} \qquad \textbf{(13.6)}$$

Near the bottom of our NFE band, equation (13.1) is satisfied, and $m^* = m$, the normal electron mass. That's Newtonian particle-like behavior. But the curvature of $E(k)$ *reverses* in the upper half of the band. According to (13.6), the electron waves have *negative effective mass* in the upper half of the band! As a result of the interaction of the wave with the periodic potential, the induced acceleration near the top of the band is in the *opposite direction* to the force that causes it. (The vector **a** is opposite to the vector **F**.) The effect of diffraction as k approaches π/a leads to a *decrease in group velocity of the electron wave with increasing k and increasing energy*. Figure 13.5 shows group velocity and effective mass as a function of wave number k across an energy band. The weird result of a negative effective mass is simply the result of attempting to apply Newtonian particle-like concepts to the more complex problem of a wave traveling in a periodic potential.

The energy band resulting from NFE theory at least had classical behavior near the bottom of the band. How about the energy band in our one-dimensional crystal resulting from LCAO theory? By applying the Hückel theory to a conjugated chain of N carbon atoms, we reached expression (11.4) for the energy levels of the delocalized pi electrons, which we repeat here:

$$E_j = \alpha + 2\beta\cos\left(\frac{j\pi}{N+1}\right) \qquad (j = 1, 2, 3, \ldots N) \qquad \textbf{(13.7)}$$

In the limit of very large N, and with the upper level of the band ($j = N$) corresponding to $k = \pi/a$, we can transform this expression into an expression for $E(k)$:

$$E(k) = \alpha + 2\beta\cos(ka) \qquad \textbf{(13.8)}$$

(We remind you that in the LCAO treatment, α is the Coulomb integral and β is the bond integral.) This function has the same shape as the $E(k)$ curve in Fig. 13.5.

The energy varies from $\alpha + 2\beta$ at the bottom of the band ($k = 0$) to $\alpha - 2\beta$ at the top of the band ($k = \pi/a$). (Recall that the bond integral β is negative.) Thus *the total energy width of the LCAO-derived band is 4β.* The greater the overlap of the AOs, the larger the bond integral, and the wider the resulting energy band in the solid.

Applying (13.3) to (13.8), the group velocity in this LCAO energy band is

$$v_g = -2a\beta\hbar^{-1}\sin(ka) \qquad \textbf{(13.9)}$$

Consistent with the curve in Fig. 13.5, the group velocity is zero at both the top and bottom of the energy band. Applying (13.6) to (13.8), the effective mass is

$$m^* = \frac{\hbar^2}{\partial^2 E/\partial k^2} = \frac{-\hbar^2}{2a^2\beta\cos(ka)} \qquad \textbf{(13.10)}$$

Since the bond integral β is negative, the effective mass is positive in the lower half of the band, infinite at midband ($k = \pi/2a$), and negative in the upper half of the band, consistent with the curve in Fig. 13.5. Near the bottom of the band, the effective mass is $m^* = -\hbar^2/2a^2\beta$. The electron here behaves *qualitatively* like a free electron but not quantitatively, since its effective mass may be different from the free-electron mass.

Sample Problem 13.2

A one-dimensional crystal with an interatomic spacing of 0.3 nm has a bond integral $\beta = -7$ eV. The relation between energy and wave number is given by (13.8). (a) Calculate the group velocity and effective mass of electrons with a wavelength of $8a$. (b) For what value of β would the effective mass at $k = 0$ equal the free-electron mass?

Solution

(a) $\lambda = 8a$ corresponds to $k = 2\pi/\lambda = \pi/4a$

$v_g = -2(3 \times 10^{-10})(-1.12 \times 10^{-18})\sin(\pi/4)/(1.05 \times 10^{-34}) = 4.53 \times 10^6$ m/s

$m^* = -(1.05 \times 10^{-34})^2/2(3 \times 10^{-10})^2(-1.12 \times 10^{-18})\cos(\pi/4) = 7.74 \times 10^{-32}$ kg

(b) At $k = 0$, the effective mass for $\beta = -7$ eV is 5.47×10^{-32} kg, only 6% of the free electron mass. The effective mass at $k = 0$ would equal the free electron mass for $\beta = -\hbar^2/2a^2m = -0.42$ eV.

We see again that many of the qualitative features of energy bands are similar in the LCAO and NFE approaches. Which approach is more suitable for calculations depends on the specific material and the specific energy band of interest. There are also numerous other theoretical treatments, some of which are intermediate between NFE and LCAO. In LCAO bands, it can be seen from (13.10) that the *effective electron mass is inversely proportional to the bond integral β, while the energy*

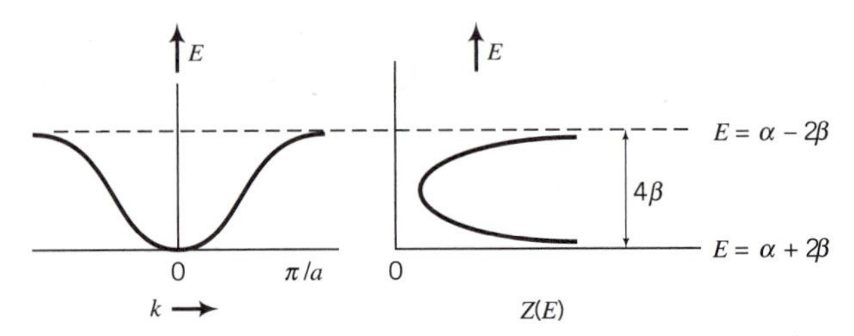

Figure 13.6. Expected density of states $Z(E)$ in an energy band of a one-dimensional crystal.

width of the band is proportional to β. Thus AOs with large overlap lead to *wide* bands with *low* effective mass ("light electrons"), while AOs with small overlap lead to *narrow* bands with *high* effective mass ("heavy electrons").

From (13.8) we can also derive $Z(E)$, the expected variation of the *density of states* with energy across the band of a one-dimensional crystal. We ask you to derive that function in Problem 13-8, but we can reach the qualitative result fairly easily. Near the bottom of the band, where $ka \ll 1$, (13.8) becomes approximately

$$E(k) \approx \alpha + 2\beta[1 - (k^2a^2/2)] = (\alpha + 2\beta) - \beta k^2 a^2 \qquad \textbf{(13.11)}$$

In this region, $E(k)$ is parabolic, as in the case of free electrons, except with an effective mass $m^* = -\hbar^2/2a^2\beta$. We learned in Chapter 12 that for free electrons in a one-dimensional crystal, with $E = 0$ at the bottom of the band, $Z(E)$ varied as $E^{-1/2}$ (Fig. 12.4). Thus in the present case, $Z(E)$ near the bottom of the band will also decrease as the inverse square root of the energy difference from the bottom of the band (Fig. 13.6). Near the top of the band, $E(k)$ is an inverted parabola, so $Z(E)$ will increase again, in a symmetric fashion.

There is experimental evidence for the general shape of the $Z(E)$ curve in Fig. 13.6. Figure 13.7 shows the x-ray photoelectron spectrum (XPS) of the long-chain alkane $C_{36}H_{74}$ corresponding to the C-C sigma bonds in this saturated one-dimensional molecule. The photoelectron spectrum reflects the density of states, which can be seen to have maxima near the top and bottom of the band, as expected from Fig. 13.6. The photoelectron spectra from other long-chain alkanes (C_nH_{2n+2}) and from polyethylene, the "ultimate alkane," are very similar. These energy bands corresponding to the sigma bonds in these saturated carbon-chain molecules are full. However, a band associated with the delocalized pi electrons in conjugated carbon chains would be only half-full, and would not show both peaks.

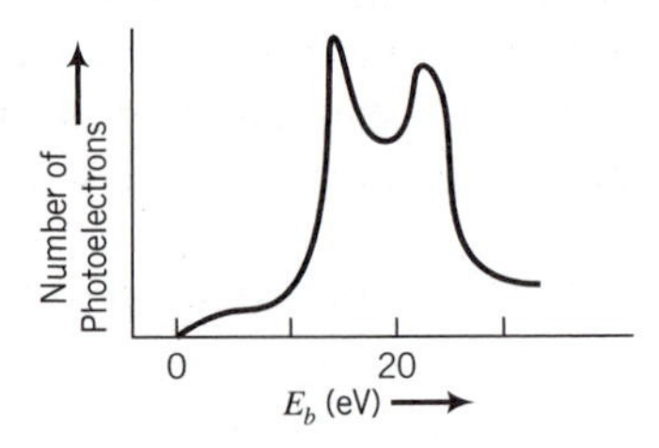

Figure 13.7. Photoelectron spectrum of $C_{36}H_{74}$, a long-chain alkane. (Adapted from P. A. Cox, *The Electronic Structure and Chemistry of Solids*, Oxford University Press, Oxford, 1987.)

13.4 THE HOLE TRUTH

In Section 11.5, we noted that when an energy band is half full, there are empty states near occupied ones, and electrons can draw energy from an electric field and move to higher energy states. Such a material could therefore conduct. (It could also accept energy from low-energy photons and therefore would also be opaque.) How about electrons in a completely full band? With all $2N$ states fully occupied—no empty states available—there's no way these electrons can carry current. From symmetry, for every electron wave moving in one direction, there's always an equivalent one moving in the opposite direction. *Full energy bands do not contribute to electrical conduction.*

But what about an energy band that's *almost* full, one that is occupied by a little *less* than $2N$ electrons? It has some empty states near the top of the band, but these empty states correspond to missing electrons with *negative effective mass.* The electrons in a nearly full energy band *can* carry current, but with some unusual consequences. Rather than considering the $(2N - x)$ electrons in the band, it is easier to focus on the x empty states—*x missing electrons with negative m^**. Detailed consideration of charge flow in this case shows that a missing electron with negative charge and negative m^* is equivalent to charge flow by a "particle" with *positive charge and positive m^*.*

Charge flow (electric current) in a nearly full band "really" occurs as a result of the motion of $(2N - x)$ electron waves interacting with the periodic lattice potential in a complicated, diffraction-related, way. But that's too hard for us mere mortals to visualize. It's much easier to visualize what turns out to be equivalent—the motion of x positively charged particles with positive mass. We call these invented (and inverted) particles *holes.*

The concept of holes in nearly full electron energy bands is somewhat similar to the concept of vacancies in a crystal lattice, with which you may already be familiar. Diffusion in solids can occur by the jumping of atoms into vacancies, and can be viewed equivalently as atoms diffusing in one direction or as vacancies diffusing in the opposite direction. The fundamentals of the concept of holes are actually a bit more complex—missing electron waves with negative effective mass and all that. But for many purposes you can get away with thinking of holes simply as vacancies, or missing electrons, in an otherwise full set of electrons in a specific set of interatomic bonds.

In particular, in ensuing chapters we will be especially interested in covalently bonded materials like silicon, which are held together in a tetrahedrally coordinated diamond lattice by sigma bonds (which, from the LCAO viewpoint, were built from sp^3 hybrid AOs). When all those sigma bonding MOs are fully occupied, the corresponding energy band (called the *valence band*) is completely full, and these electrons—four from each silicon atom, two in each interatomic bond—can carry no electric current. However, suppose that somehow one of these electrons is missing. We can view the location of the missing electron as a hole. Then electric current can be carried by an electron from a neighboring bond jumping into the original hole, another electron from a further bond jumping into the newly created hole, and so on. Much like diffusion via the vacancy mechanism, current carried by electrons moving in one direction can be viewed as resulting from positively charged holes moving in the opposite direction. From the viewpoint of energy bands and quantum mechanics, it's an overly simplified view, but we'll find it to be a helpful model when we get to understanding the physics of semiconductors and semiconducting devices like light-emitting diodes (LEDs) and solar cells.

Whenever electric current is dominated by holes (dominated by charge flow of electrons in nearly full energy bands), the sign of the Hall voltage will be opposite of that expected from simple electron flow. For reasons we will discuss in Chapter 15, electric current in many semiconductors can be dominated by holes. But what about those divalent metals like zinc that had positive Hall coefficients? With two valence electrons per atom, and $2N$ electron states per band, shouldn't the valence energy band of zinc be completely full? If so, how can those electrons conduct current? From what we've said so far, zinc shouldn't be a metal. To explain how zinc can be a metal even though it's divalent, we have to make a temporary excursion out of the one-dimensional world we've been considering so far in this chapter. Zinc, like most real materials, is usually not one-dimensional.

13.5 DIVALENT ZINC—OVERLAPPING BANDS AND HOLE CONDUCTION

We can tackle the question of how divalent materials can be metals, and see the effects of extra dimensions, without losing our heads and going all the way to three dimensions. Two dimensions are easier to think about, and much easier to draw. Instead of considering the three-dimensional hexagonal-close-packed lattice of zinc, let's just consider a simple *two*-dimensional *square* lattice of lattice parameter a (Fig. 13.8). Although it's not a very good approximation of 3-D zinc, this simple lattice will show us how additional dimensions may change some of our conclusions based on one-dimensional crystals.

The NFE approach concludes that energy gaps—regions of energy forbidden to electrons—develop where the Bragg diffraction condition is satisfied, which for our one-dimensional crystal was at $k = \pm\pi/a$. In our two-dimensional crystal, we'll also

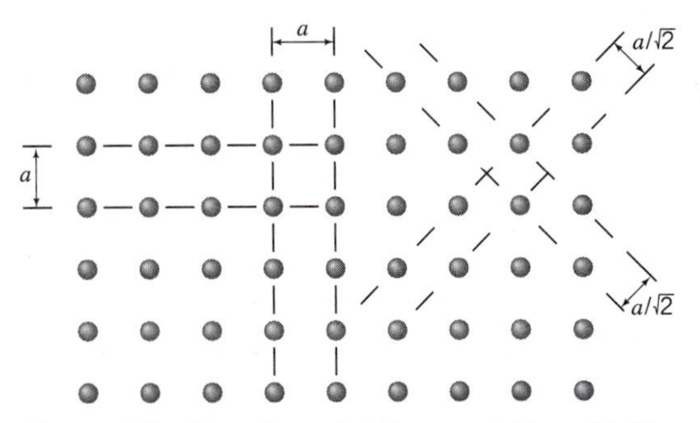

Figure 13.8. Two-dimensional square lattice of lattice parameter a. Spacing of (1 0) and (0 1) atom rows is a, but spacing of (1 1) atom rows is $a/\sqrt{2}$.

get diffraction at $k = \pm\pi/a$ for electron waves traveling in either the x-direction or the y-direction, since the spacing between rows of atoms equals a along either axis. (In a two-dimensional crystal, the diffracting units are atom rows, the equivalent of individual atoms in 1-D and atomic planes in 3-D.) But how about waves traveling at 45° to the x and y axes? In this direction, the spacing between atom rows is no longer a; it is only $a/\sqrt{2}$. With this smaller spacing, Bragg diffraction of electron waves traveling at 45° to the x and y axes will occur at a *larger* k (i.e., at $k = \pm(\sqrt{2})\pi/a$). *Waves in a two-dimensional crystal will diffract, and encounter energy gaps, at different values of k when traveling in different directions.*

To simplify simultaneous plotting of $E(k)$ curves for two different directions, we make use of the symmetry of the $E(k)$ curves seen in the earlier figures. Since $E(k) = E(-k)$, either the left half or the right half of each curve fully describes $E(k)$. So in Fig. 13.9, we plot $E(k)$ for waves traveling along the x or y axes—in the [1 0] or [0 1] directions—to the right, and $E(k)$ for waves traveling at 45° to the axes—in the [1 1] direction—to the left. The energy gaps occur at different values of k *and at different energies* in the two different directions. Thus some energies that are forbidden for [0 1] or [1 0] electron waves are possible for [1 1] waves. In fact, if the energy gaps are not too wide in energy, there may be *no* electron energies that are totally forbidden in our two-dimensional crystal. In particular, if (as drawn) the highest energy in the first band for [1 1] waves (A) is above the lowest energy of the *second* band for [1 0] waves (B), there will be no forbidden electron energies in this two-dimensional crystal, because the *energy range of the second energy band overlaps the energy range of the first energy band.*

The argument can of course be extended to three dimensions, with Bragg diffraction and energy gaps occuring at different values of k and different energies in

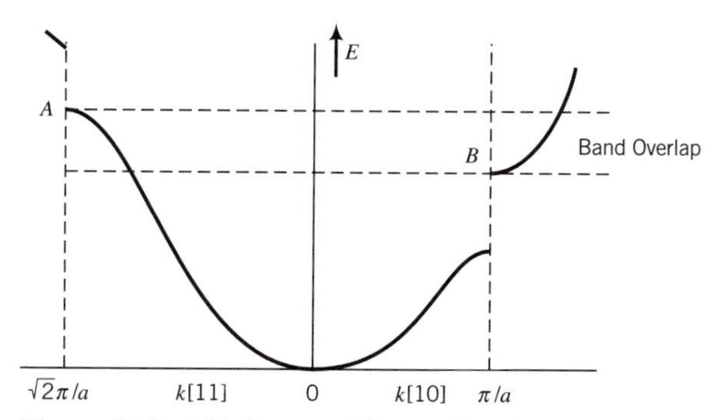

Figure 13.9. $E(k)$ for two different directions in the two-dimensional square lattice of Fig. 13.8. Energy gaps occur at different values of k and at different energies in the two directions. In the case shown, the highest energy for the first band for [1 1] waves (A) is higher than the lowest energy of the *second* band for [1 0] waves (B), so the two-dimensional crystal has no gap of totally forbidden electron energies between the first and second band (the first and second bands overlap).

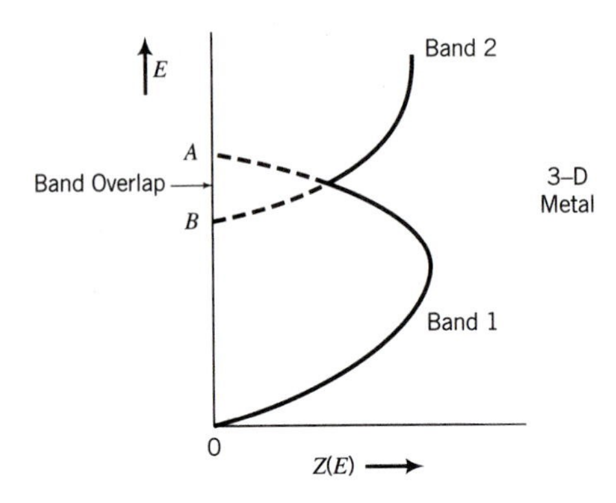

Figure 13.10. Schematic density of states $Z(E)$ for three-dimensional metal with overlapping energy bands. If metal is divalent, Fermi level is in the region of overlap, with both electrons (in the top band) and holes (in the bottom band) available for conduction.

the [1 0 0], [1 1 0], and [1 1 1] directions. Thus although a *one*-dimensional crystal would have a full energy band if it were divalent (and therefore had $2N$ electron states filled with $2N$ electrons), in two or three dimensions there may not be an overall energy gap because of *overlapping energy bands*. This is shown schematically in Fig. 13.10.

We can now see how zinc and other divalent metals can be metals despite having $2N$ valence electrons and $2N$ electron states per band. In these three-dimensional metals, the energy bands overlap, so that rather than filling one band, $2N - x$ electrons *almost* fill the first band, and x electrons occupy the lower part of the second band. The Fermi level is in the region of overlap of the energy bands, and metallic conduction can occur by both electrons (in the bottom of the higher band) and holes (in the top of the lower band). In zinc and some other divalent metals, hole conduction dominates, which explains one of the mysteries we encountered way back in Chapter 1. *With the Fermi level in the region of band overlap, that is, near the top of the lower band, holes sometimes dominate conduction, leading to positive Hall coefficients.*

13.6 FERMI'S SURFACE MEETS BRILLOUIN'S ZONE BOUNDARIES

Earlier in this chapter, we introduced the concept of diffraction-related energy gaps in the nearly free electron (NFE) picture, as well as the concepts of Brillouin zones, reduced-zone and extended-zone schemes, effective mass, and holes, all in terms of a simple one-dimensional crystal and one-dimensional k-space. However, as we learned in Section 13.5, some important effects require extending our thinking to more than one dimension. To become more comfortable with graphical methods of representing the effects of two or three dimensions on electron energies and momenta, we consider again the same problem we just dealt with—why divalent zinc is a metal—in terms of a diagram in two-dimensional k-space.

In Chapter 12 we showed that in a two-dimensional metal (a square L on a side), the periodic boundary conditions led to allowed electron states represented by a square lattice of points in k-space, with a lattice parameter of $2\pi/L$ (Fig. 12.2, reproduced in Fig. 13.11). In two dimensions, our wave number k is a *wave vector* $\mathbf{k}$,

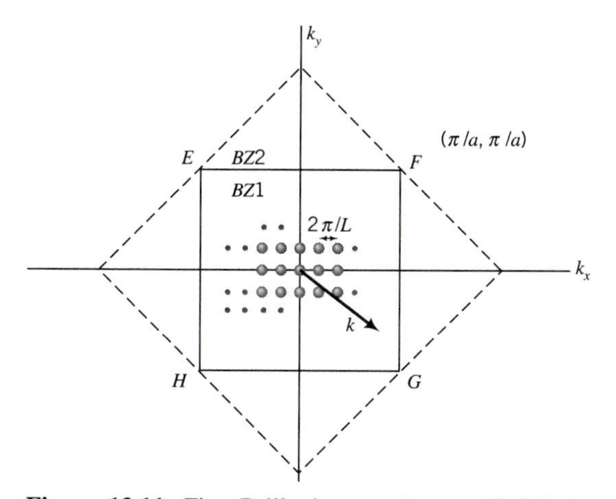

Figure 13.11. First Brillouin zone (square EFGH) in k-space of two-dimensional crystal with square lattice of parameter a (Fig. 13.8), showing a few of the allowed energy states as a lattice of dots in k-space. All allowed states in the first energy band—a total of $2N$—lie within square EFGH.

with two components, let's say k_x and k_y. We now, in the spirit of the NFE theory, recognize the existence of the atomic lattice (in "real" space), which we assume, as before, is a square lattice with lattice parameter a (Fig. 13.8). In the k_x and k_y directions, Bragg diffraction will occur at $k = \pm\pi/a$ (midpoints of edges EF, FG, etc. in Fig. 13.11). At 45° to the x and y axes (as discussed in the preceding section), Bragg diffraction will instead occur at $k = \pm(\sqrt{2})\pi/a$ (points E, F, G, and H).

At intermediate angles, the Bragg condition for first-order diffraction off (1 0) atom rows (rows perpendicular to the x-axis) is simply $\lambda = 2\pi/k = 2a \sin\theta$, where θ is the angle between the direction of motion of the electron wave (the direction of the vector **k**) and the (1 0) row. But the component k_x normal to the (1 0) rows equals $k \sin\theta$. Thus, in our two-dimensional square-lattice crystal, we expect an energy gap at $k_x = \pm\pi/a$, regardless of k_y. Similarly, we will get diffraction off the (0 1) atom rows (rows perpendicular to the y-axis) when $k_y = \pm\pi/a$, regardless of k_x. The two conditions $k_x = \pm\pi/a$ and $k_y = \pm\pi/a$ define lines in k-space that form the square EFGH in Fig. 13.11, and we expect diffraction and an energy gap as the tip of the wave vector **k** approaches the sides of this square. This square represents the *first Brillouin zone* of our two-dimensional crystal, and it is the two-dimensional equivalent of the line segment shown in Fig. 13.4.

What about diffraction off other atom rows? The rows with the next largest d-spacings are the {1 1} rows, for which $d = a/\sqrt{2}$. A bit of geometry (smaller d corresponds to smaller λ, larger k) will convince you that the planes in k-space that define diffraction off {1 1} atom rows are *outside* the square defined by $k_x = \pm\pi/a$ and $k_y = \pm\pi/a$ (they're shown as dashed lines in Fig. 13.11). Within square EFGH, the components of k are too short (the wavelength components too long) for *any* diffraction—there are no energy gaps for **k** vectors *inside* the square. The various allowed states inside this square in k-space all correspond to the *same NFE energy band.*

How many allowed states are there within this energy band—within this square, the area of k-space corresponding to the first Brillouin zone? We can answer that in much the same way that we answered the analogous question for the one-dimensional NFE crystal a few pages ago. The area of this square in k-space is $(2\pi/a)^2$. Each allowed level corresponds to a tiny square of area $(2\pi/L)^2$ in k-space. Thus the total number of allowed levels in the band is $(2\pi/a)^2$ divided by $(2\pi/L)^2$, or $(L/a)^2$. But $(L/a)^2$ is just N, the number of atoms in our two-dimensional crystal. So our Brillouin zone in k-space, our square $2\pi/a$ on an edge that holds all the energy levels corresponding to one energy band, *can accommodate 2N electrons*. By now, that shouldn't be a surprise.

The four triangles surrounding square EFGH represent the *second* Brillouin zone, the energy band of higher energy, in a two-dimensional extended-zone scheme. It can be seen that the total area of these four triangles is the same as the area of square EFGH, so the second Brillouin zone contains the same number of electron states as the first zone. Thus the second energy band can also accommodate $2N$ electrons.

When counting electron states, it is convenient to remind ourselves that quantization of k results in a finite number of allowed states that can be represented schematically by an array of dots, as we did in Figs. 12.1, 12.2, 12.3, and 13.11. However, L is usually a macroscopic dimension of the order of centimeters, whereas the lattice parameter a is of the order of angstroms, a factor of 10^8 smaller. Thus the dimensions of the Brillouin zones in k-space (of the order of $1/a$) are 10^8 times greater than the spacing of allowed states in k-space (of the order of $1/L$), and we can consider k and E as continuous variables when we plot $E(k)$ and $Z(E)$, as we did in Figs. 13.9 and 13.10 and in many earlier figures.

It is not as easy to plot $E(k)$ for our two-dimensional crystal as it was for a one-dimensional crystal. One approach is the one we used in Fig. 13.9, plotting $E(k)$ in different directions, which we reproduce as Fig. 13.12a. Another popular approach is to draw, in two-dimensional k-space, *contours of constant energy*, an approach analogous to geological contour maps that show lines of constant elevation.

For free electrons in a two-dimensional crystal,

$$E(k) = \frac{\hbar^2(k_x^2 + k_y^2)}{2m} \qquad \textbf{(13.12)}$$

Thus for totally free electrons, or even if you modified equation (13.12) with an effective mass m^*, constant-energy contours would be *circles* in two-dimensional k-space, an approximation that should be pretty good near the bottom of an energy band. Suppose we have only a few electrons in the band and our Fermi level only reaches level 1 in Fig. 13.12a. Up to here, $E(k)$ is nearly the same in the $\langle 1\,0\rangle$ and $\langle 1\,1\rangle$ directions and any direction in between, and the corresponding energy contour is the circle labeled 1 in Fig. 13.12b.

Now suppose we add a few more electrons to this band until the Fermi level reaches level 2 in Fig. 13.12a. We are now approaching diffraction in the $\langle 1\,0\rangle$ directions, and $E(k)$ in these directions is flattening out—reaching to higher k values

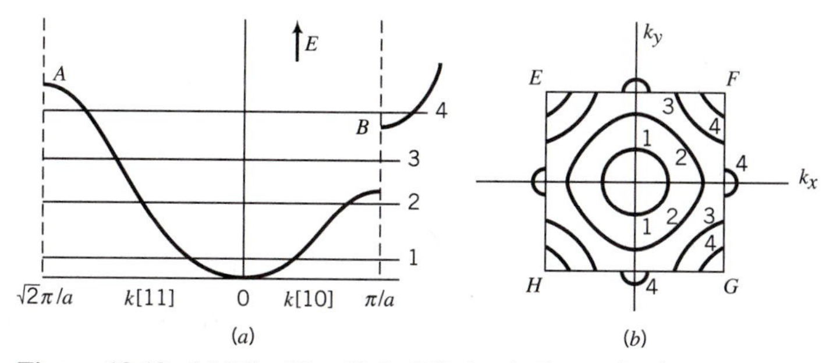

Figure 13.12. (a) Like Fig. 13.9, *E*(*k*) for ⟨1 0⟩ and ⟨1 1⟩ directions in the two-dimensional square-lattice crystal of Fig. 13.8. (b) Series of corresponding constant-energy contours within square in *k*-space (first Brillouin zone) of Fig. 13.11. Energy 4 is in the region of *band overlap.* The energy contour corresponding to the Fermi energy is called the *Fermi surface.*

in the ⟨1 0⟩ directions than in the ⟨1 1⟩ directions. Thus our energy contours in Fig. 13.12b are distorting from circles into a more square-like shape, labeled contour 2.

Add a few more electrons and you've reached the Brillouin zone boundaries at the midpoints of edges EF, FG, and so on, and you've produced an energy gap in the ⟨1 0⟩ directions. If you've added enough electrons to reach level 3 in Fig. 13.12a, your energy contour in Fig. 13.12b, labeled 3, has separated into four parts aiming towards the corners of the square. You're in the midst of an energy gap in the ⟨1 0⟩ directions but still have available states in the ⟨1 1⟩ and nearby directions.

Finally, add a few more electrons to reach level 4. You are now nearing the energy gap in the ⟨1 1⟩ directions, corresponding to energy contour 4, small arcs in the corners of the square. But you have exceeded the energy gap in the ⟨1 0⟩ directions and now have small bits of energy contour 4 out in the second Brillouin zone (the next energy band). If by now you have added a total of exactly $2N$ electrons (a divalent atom), the total area enclosed by contour 4 will exactly equal the area of the square EFGH. The bits in the corners of the square that are not yet filled therefore have the same area as the arcs in the second zone. Once again, your divalent 2-D material will indeed be a metal, with electron states near the ⟨1 0⟩ directions capable of carrying current and hole states near the ⟨1 1⟩ directions also capable of carrying current.

We have reached the same conclusion that we reached in the preceding section—it is the overlapping of energy bands in two or three dimensions that allows zinc and several other divalent elements to be metals, metals in which holes contribute to conduction (and, when dominant, lead to positive Hall coefficients). But if you've followed the arguments in two-dimensional *k*-space, you're a bit better prepared for more advanced texts that focus on three-dimensional *k*-space. And you've been introduced to the important concept of constant-energy contours in *k*-space. The constant-energy contour in *k*-space that corresponds to the Fermi energy is called the *Fermi surface*. We noted in Chapter 12 that it is the electrons near the Fermi energy that dominate many properties of metals, including the electronic specific heat and electrical and optical properties. Therefore many solid-state physics texts devote considerable attention to the Fermi surfaces of materials.

Sample Problem 13.3

Your LCAO treatment of a two-dimensional crystal with a square lattice led to $E(k) = \alpha + 2\beta\cos(k_x a) + 2\beta\cos(k_y a)$. (a) Select values of α and β so that $E(k)$ approaches (13.12) as k approaches zero. (b) At what value of E will the energy contour touch the boundaries of the Brillouin zone in the $\langle 1\,0\rangle$ directions? (c) At this value of E, what is the value of k in the $\langle 1\,1\rangle$ directions? (d) At what value of E will the energy contour reach the corners of the Brillouin zone (and therefore fill the zone)?

Solution

Near $k_x = 0$, $\cos(k_x a) \approx 1 - (k_x^2 a^2)/2$. Using this and the equivalent approximation for $\cos(k_y a)$, $E(k) \approx \alpha + 4\beta - \beta a^2(k_x^2 + k_y^2)$ near $k = 0$. To be equivalent to (13.12), $\alpha = -4\beta$ and $\beta = -\hbar^2/2ma^2$.
(b) At $k_x = \pi/a$, $k_y = 0$ (and equivalent points), $E = -4\beta = 2\hbar^2/ma^2$.
(c) In the $\langle 1\,1\rangle$ direction, $k_x = k_y$. $E = -4\beta = -4\beta + 4\beta\cos(k_x a)$, so $\cos(k_x a) = 0$, and $k_y = k_y = \pi/2a$. Hence $k = (k_x^2 + k_y^2)^{1/2} = \pi/(a\sqrt{2})$. (Since $k = \pi/a$ in the k_x and k_y directions for this energy, clearly the constant-energy contour is now far from a circle. In fact, at this value of E it's a square.)
(d) For $k_x = k_y = \pi/a$, $E = -4\beta - 4\beta = -8\beta = 4\hbar^2/ma^2$. (The energy contour first reached the zone boundaries at $E = 2\hbar^2/ma^2$, but the zone wasn't completely filled until $E = 4\hbar^2/ma^2$.)

We considered only the simple case of a two-dimensional crystal with a square lattice, for which the first Brillouin zone was a square, and the second was a set of four triangles surrounding the square. If we had instead a *three*-dimensional cubic lattice, the first Brillouin zone would have been a cube. Since the boundaries of Brillouin zones are determined by the Bragg diffraction condition (4.13), the shapes of Brillouin zones are of course different for different crystal structures. The boundaries of the first zone correspond to the crystal planes with the largest interplanar spacing d (largest d corresponds to largest λ, smallest k). In body-centered-cubic (bcc) crystals, the first planes to diffract are the (1 1 0) planes, and the resulting first Brillouin zone is a regular dodecahedron in k-space, with each of its 12 surfaces normal to $\langle 1\,1\,0\rangle$ directions in k-space. The first Brillouin zone in face-centered-cubic (fcc) crystals is instead a 14-sided figure bounded by 6 surfaces corresponding to (1 0 0) diffraction and 8 surfaces corresponding to (1 1 1) diffraction.

One more warning about three-dimensional band theory. For our two-dimensional crystal, we represented $E(k)$ in Figs. 13.9 and 13.12a by showing curves in two different directions in k-space. Full representation of $E(k)$ for a three-dimensional crystal requires showing more such curves, including [1 0 0], [1 1 0], and [1 1 1] directions, as well as intermediate k vectors. The results are very complicated diagrams with many wiggly lines, sometimes called "spaghetti diagrams." We spare you those here, but when you encounter them in other texts you should keep calm and focus on: (1) the *energy gaps*—regions of energy (if any) where there are no $E(k)$ curves, (2) the *Fermi level*—in particular, the location of the Fermi level with regard to energy bands and gaps, and (3) the *curvature of* $E(k)$, which determines the *effective mass* of electrons and holes. Since effective mass affects the mobility of charge carriers, it affects the speed of operation of various electronic devices.

13.7 PHASE STABILITY AND PROPERTIES

Our focus in Fig. 13.12 was on what happened when the electron concentration was increased until the Fermi surface approached the Brillouin zone boundary. If we started with a monovalent metal, the first zone and first energy band would be only half full. With $2N$ possible electron states in the first zone, N electrons would only fill half the states—half the area of the zone for a 2-D crystal or half the volume of the zone for a 3-D crystal. However, if we were to alloy this monovalent metal with increasing amounts of a divalent metal, electron concentration and Fermi energy would increase, and the Fermi surface would expand, growing closer to the zone boundaries. Since *the Brillouin zones of different crystal structures have different shapes*, the relation between the Fermi surface and the boundaries of the Brillouin zone would be different for different crystal structures, resulting in different total electron energies. Since the phase of lowest energy is the stable phase, such effects influence the relative stability of different crystal structures for a given alloy composition.

One of the very early uses of these concepts was in explaining the *Hume-Rothery rules*, empirical observations from binary phase diagrams that phase changes often correlate with special numbers of valence electrons per atom. For example, monovalent copper can accommodate a certain amount of divalent zinc in solid solution and maintain its equilibrium fcc structure. But the gradual increase of the number of valence electrons per atom as zinc (divalent) or aluminum (trivalent) or silicon (tetravalent) is added to copper leads to a gradual growth of the Fermi surface and increasing interaction of this growing Fermi surface with the Brillouin zone of the fcc lattice. This interaction affects the average electron energy and increases it above the average electron energy the alloy would have in the bcc structure, which has different Brillouin zones. Since a given volume in k-space enclosed by the Fermi surface corresponds to a given value of valence electrons per atom, fcc-bcc phase transitions in copper alloys occur at lower concentrations of silicon than aluminum and at lower concentrations of aluminum than zinc, each at roughly the same average number of valence electrons per atom.

In addition to influencing the relative stability of different crystal structures (and therefore equilibrium phase diagrams), the interaction of Fermi surfaces with Brillouin zone boundaries also affects material properties, including their variation with crystal direction (e.g., anisotropy of effective mass). We focused in the preceding sections on the effect of this interaction on band overlap and hole conduction in divalent metals. Fermi surface–Brillouin zone interactions also change the density of states function $Z(E)$ from the simple parabolic shape of Fig. 12.4 into complex curves with peaks and valleys. Numerous properties, including the electronic heat capacity, electrical and thermal conductivity, and paramagnetism of conduction electrons, are particularly sensitive to $Z(E_F)$, the density of states at the Fermi surface. Since the Brillouin zone boundaries strongly influence the shape of $Z(E)$, they can determine whether the Fermi energy E_F is located in a peak or a valley of $Z(E)$, and thereby influence these and other properties. In Chapter 14 we discuss several properties of metals that are sensitive to $Z(E_F)$.

Sample Problem 13.4

When the number of valence electrons per atom in copper-rich alloys exceeds 1.4, the equilibrium crystal structure changes from fcc to bcc. At what composition will that phase change occur in Cu-Zn, Cu-Al, and Cu-Si alloys?

Solution

Copper is monovalent. If an atom fraction x of an alloying element with a valency of V is added to copper, the average valence will be $xV + (1 - x)1 = 1 + x(V - 1)$. This will reach 1.4 when $x(V - 1) = 0.4$.
For Zn, $V = 2$, so $x = 0.4$. For Al, $V = 3$, so $x = 0.2$. For Si, $V = 4$, so $x = 0.13$.

SUMMARY

The nearly free electron (NFE) model of electrons in solids starts with the quantum-mechanical free-electron theory of Chapter 12 and then introduces a periodic potential energy function representing the ions. The periodic lattice potential has little effect on electron waves with wavelengths much longer than the interatomic spacing. However, energy gaps—ranges of energy forbidden to the electron waves—develop where wavelengths satisfy the Bragg diffraction condition, and traveling waves are replaced by standing waves. An NFE energy band for a one-dimensional crystal with N atoms has N closely spaced energy levels and can accommodate $2N$ electrons, a conclusion identical to that reached earlier for energy bands developed from the LCAO model and the principle of conservation of states.

The interaction of electron waves with the periodic potential of the lattice introduces a complex dependence of electron energy on wave vector k. The shape of the resulting $E(k)$ is similar to that derived from the LCAO-Hückel model. As the top of the energy band is approached, the group velocity approaches zero and the effective mass is negative. Near the bottom of the energy band, the effective mass is positive, but it may be quite different from the free-electron mass. Large AO overlap leads to wide bands (since the width of the band is proportional to the bond integral β) and low effective mass (light electrons), while small AO overlap leads to narrow energy bands and high effective mass (heavy electrons).

The density of states function $Z(E)$ for an energy band in a one-dimensional crystal peaks at the top and bottom of the band and goes through a minimum at midband, a prediction consistent with experimental x-ray photoelectron spectra (XPS) results on long-chain alkanes and polyethylene, molecules in which the energy band of the bonding electrons is completely full.

An energy band that is completely full cannot conduct electricity. However, a band that is *nearly* full, with unoccupied electron states near the top of the band, *can* conduct. These electron states have negative effective mass, and the conduction resulting from a missing electron with negative mass can be viewed as resulting from the motion of a positively charged particle of positive effective mass—an imaginary particle called a *hole*. When hole conduction is dominant, the Hall coefficient is positive.

From treatment limited to a one-dimensional crystal, we would expect a divalent element like zinc to have a completely full energy band. However, if we consider a two-dimensional or three-dimensional crystal, we find that the energy gaps occur at different wave numbers and different energies for electron waves in different directions. If the energy gaps are sufficiently small, the electron energies forbidden in one direction will be allowed in other directions and there will be no totally forbidden electron energies for the valence electrons. In other words, there will be *band overlap*. In divalent metals like zinc, the Fermi level is in the region of band overlap, and some of the conduction is by holes, which explains the positive Hall coefficients observed.

In our discussion of band overlap in a two-dimensional crystal with a square lattice, we focused on two concepts, the *Fermi surface* and the *Brillouin zone*. The Fermi surface is the constant-energy surface in k-space (reciprocal space) that corresponds to the Fermi energy. The Brillouin zones are the regions in k-space separated by boundaries at which the Bragg diffraction condition is satisfied and energy gaps occur. The interaction of Fermi surfaces and the boundaries of Brillouin zones affect band overlap and other properties of solids, including the relative stability of various crystalline phases.

PROBLEMS

13-1. A cyclotron resonance experiment (see Chapter 2) on high-purity germanium at low temperatures, using microwaves of 24 GHz, gave peaks at 0.19 and 0.3 tesla attributable to electrons, and peaks at 0.13 and 0.44 tesla attributable to holes. Calculate the effective mass corresponding to each peak, and express it in terms of a fraction of the normal electron mass.

13-2. The energy-momentum relationship for a one-dimensional crystal of interatomic spacing a, within the band corresponding to $0 < k < \pi/a$, is given by

$$E(k) = Ck^2\left(1 - \frac{2ak}{3\pi}\right)$$

(a) Derive expressions for the group velocity and effective mass of electrons in this band as a function of k. (b) If the width of this energy band is 5 eV, and the interatomic spacing is 0.3 nm, calculate the maximum value of group velocity and the effective masses at the top and bottom of the band.

13-3. If $E(k)$ in a two-dimensional square-lattice crystal equals $4C - 2C\cos(k_x a) - 2C\cos(k_y a)$, calculate values of k along the $\pm k_x$ axis ($k_y = 0$) corresponding to $E = C$, $2C$, $3C$, and $4C$ and along the $k_x = \pm k_y$ line corresponding to these four values plus $E = 5C$, $6C$, $7C$, and $8C$. Put the resulting data on an $E(k)$ plot and draw smooth lines indicating $E(k)$ for the [1 0] and [1 1] directions. (Reminder: along the $k_x = \pm k_y$ line, $k^2 = 2k_x^2 = 2k_y^2$.)

13-4. In Table 12.1, the experimental value of the Fermi energy of lithium from x-ray emission spectra is 3.9 eV. Setting this equal to the Fermi energy predicted from free-electron theory, what average effective electron mass does this correspond to? The atomic weight of Li is 6.94 and its density is 0.53 g/cc.

13-5. In the highest energy band of a one-dimensional crystal with interatomic spacing $a = 0.3$ nm, the group velocity of electrons in m/s is $10^6 \sin(ka)$. (a) Calculate the effective mass in kilograms at the top and bottom of this band. (b) Calculate the energy width of the band in electron volts. (c) This crystal has a total length of 300 μm. How many electron states are in this band? (d) If these atoms are divalent, will this crystal be a conductor or an insulator?

13-6. In the two-dimensional square-lattice crystal of Problem 13-3, what number of electrons per atom corresponds to the energy $E = 4C$ at which the Fermi surface first touches the boundaries of the first Brillouin zone? At that energy, what is the group velocity of electrons in the $\langle 1\ 0 \rangle$ and $\langle 1\ 1 \rangle$ directions?

13-7. (a) For a three-dimensional cubic-lattice crystal, write an $E(k)$ relationship analogous to that in Problem 13-3, with constants adjusted so that $E = 0$ at the origin of k-space. (b) At what value of energy will the Fermi surface first touch the boundaries of the first Brillouin zone? (c) How many electrons per atom does this correspond to?

13-8. For a one-dimensional crystal with the relation between electron energy and wave number given by equation (13.8), derive an expression for the density of states in this energy band.

13-9. From the $E(k)$ given in Problem 13-3 for a two-dimensional square-lattice crystal, calculate the group velocity of electrons in the [11] direction as a function of k.

13-10. (a) You have a two-dimensional square-lattice crystal, but the $E(k)$ relation is simply that of free electrons—equation (13.1). (Apparently, the lattice-electron interaction is very weak.) At what value of electrons per atom does the Fermi surface touch the boundary of the first Brillouin zone in this case? (b) Answer the same question for the case of free electrons in a *three-dimensional* cubic-lattice crystal.

Chapter 14

Metals and Insulators

14.1 METALS

We approached the complex problem of valence electrons in solids, those partly free, partly slave electrons that determine most of the properties of solids, by two methods: LCAO and NFE. In the LCAO approach (Chapters 10 and 11), we started with atomic orbitals (slave electrons) and gradually considered larger and larger molecules. We observed the changes in the electron wave functions and electron energies (Fig. 11.20) as the electrons became more free, that is, more delocalized. We also pictured these changes as we varied the interatomic spacing R (Fig. 11.21). In the NFE approach, we started instead with free electron waves (Chapter 12) and then observed the effects of introducing the atomic lattice (Chapter 13), which changed the electron waves to nearly free. Both approaches led us to the picture of *bands* of allowed energies for the valence electrons, separated by *gaps* of forbidden energies. For solids containing N atoms, both approaches gave us energy bands that contained N energy levels, with each band capable of accommodating $2N$ electrons.

We can now discuss why some materials are metals and some materials are insulators in terms of the *location of the Fermi level with respect to the energy bands and energy gaps*. The definition of the Fermi level appropriate here is not the maximum electron energy at $T = 0$, but instead the energy at which the probability of occupation of electron states at a finite temperature is 0.5. The two definitions are equivalent for a metal but not for an insulator, for which the Fermi level falls within an energy gap, that is, at a forbidden electron energy.

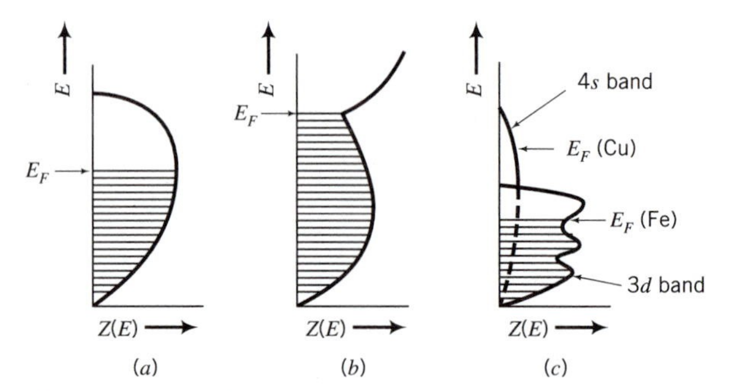

Figure 14.1. Schematic density-of-states curves $Z(E)$ for (a) monovalent metal like Li, (b) divalent metal like Be, (c) 3d transition metals.

Let's start with a simple monovalent metal like lithium ($Z = 3$), which we considered in Figs. 11.20 and 11.21. Since N monovalent atoms give us N valence electrons, the 2s energy band of lithium is *half full.* In Fig. 14.1a, we show, schematically, the density-of-states function $Z(E)$ of lithium. In three dimensions, $Z(E)$ will increase parabolically at the bottom of the band, and will decrease to zero, also parabolically, at the top of the band. We have shaded in the electron states that are occupied at $T = 0$, which is just half of the available states in the band. In lithium, the Fermi level is at *midband.* There are empty states immediately above full ones, allowing electrons to be accelerated to higher energies by electric fields and allowing absorption of even low-energy photons. Because the Fermi level of lithium is within the band of allowed electron energies, lithium is a conductor and is opaque. The other alkali metals and other monovalent metals also have Fermi levels at midband.

In contrast, N atoms of divalent beryllium ($Z = 4$) give us $2N$ valence electrons, which would fill the 2s energy band in a one-dimensional crystal. However, in three-dimensional Be, there is *band overlap*, which was represented, in different but equivalent ways, in Figs. 13.9, 13.10, and 13.12. As seen in Fig. 14.1b, the Fermi level is in the region of overlap of the 2s and 2p bands. There are empty states adjacent to full ones, and beryllium is an opaque metal, but since the Fermi level is near the top of the 2s band, much of the conduction in Be is by *holes* (empty electron states near the top of a band) and the Hall coefficient of Be is positive, like that of Zn and several other divalent metals.

Sample Problem 14.1

Do you expect aluminum, a trivalent metal, to have a positive Hall coefficient?

Solution

With $3N$ free electrons, and $2N$ electron states per band, we expect the Fermi level of aluminum to be midband, with no significant hole conduction. Like a monovalent metal, it should have a negative Hall coefficient because electron conduction is dominant.

The *transition metals* are more complicated. In the first transition-metal series, for example, the 3*d* and 4*s* energy bands overlap (Fig. 14.1c). Because the 4*s* AOs extend farther from the nuclei than the 3*d* AOs, their overlap and bond integrals are greater and the 4*s* band is much wider in energy than the 3*d* band. Because the 3*d* band is narrower in energy but contains more electrons, the density of states is much higher in the 3*d* band than in the 4*s* band. The $Z(E)$ curves for real materials are all a bit more complicated than shown schematically in these figures, but the *d* bands are especially complex, so I've shown a few extra wiggles in $Z(E)$.

In Fig. 14.1c, we estimate the location of the Fermi level of Fe and of Cu. In Cu, there is a large maximum in the density of occupied states a few eV below the Fermi level. The absorption of photons by 3*d* electrons near this peak ($3d \rightarrow 4s$ transitions) leads to a preferential absorption of light near the blue end of the visible spectrum, producing the familiar reddish color of copper (another example of color by subtraction). Alloying with divalent zinc raises the Fermi level and shifts the light absorption enough to produce the yellowish color of brass.

Figure 14.2 shows the variations of the cohesive energy, atomic volume, and bulk elastic modulus across the 3*d*, 4*d*, and 5*d* transition-metal series. Across each series, increasing atomic number corresponds to higher and higher Fermi levels. At first, the energy levels near the bottom of the band, corresponding to bonding COs, are gradually filled, leading to increasing cohesive energy, decreasing atomic volume, and increasing elastic modulus (as in our diatomic molecules of Chapter 10, bonds become shorter, stronger, and stiffer—the 3S rule—as bonding levels become more occupied). However, as the Fermi level begins to rise into the upper half of the *d*-band, antibonding levels are occupied, and the trends in properties are reversed, as they were for diatomic molecules when antibonding levels were filled (Fig. 10.9). (Trends are a bit more complicated in the 3*d* series because of the magnetic nature of Cr and Mn, which are antiferromagnetic, and of Fe, Co, and Ni, which are ferromagnetic.)

In contrast to simpler metals and to covalent and ionic solids, the bond strengths of transition metals in a given column of the periodic table generally increase as you descend to lower rows. Transition metals are held together by a complex mixture of *s* and *d* bonding, but *d* AOs apparently have more overlap and a greater contribution to bonding in the lower rows. The metal with the highest melting point of all (3683 K), and which serves as the filament in our incandescent bulbs, is tungsten, a 5*d* metal.

A variety of theoretical models, some close to the LCAO model, some close to the nearly free electron (NFE) model, some considerably more sophisticated, have been used to calculate the detailed band structures of common metals. The results of the calculations include $E(k)$ curves in various crystal directions, the Fermi surface, and $Z(E)$, the density of states. Although these results are beyond the scope of the present text, some can be found in the books listed in the Suggestions for Further Reading at the end of this book.

As noted in Chapter 13, many properties of metals are specifically dependent on $Z(E_F)$, the density of states at the Fermi surface. One important example is electrical conductivity. In band theory, the drift velocity of electrons is represented by the

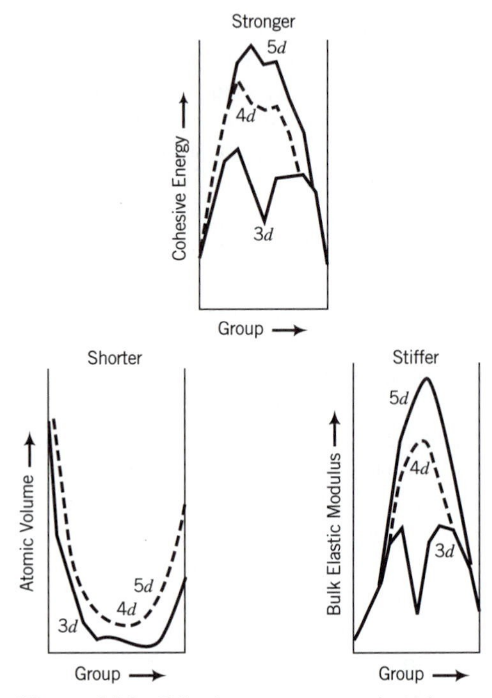

Figure 14.2. Cohesive energy, atomic volume, and bulk elastic modulus across 3*d*, 4*d*, and 5*d* transition-metal series. (Adapted from K. A. Gschneidner, *Solid State Physics*, Vol. 16, F. Seitz and D. Turnbull, eds., Academic Press, New York, 1965.) As bonding levels of the lower half of the *d*-band are filled, bonds get *stronger*, *shorter*, and *stiffer*, but trend is reversed as the antibonding levels at the top of the band are filled. (See Fig. 10.9.)

group velocity of electron waves, which in one dimension is given by equation (13.3): $v_g = (1/\hbar)(\partial E/\partial k)$. In three dimensions, it becomes a vector relation:

$$\mathbf{v}_g = (1/\hbar)\ \nabla_k E \qquad \textbf{(14.1)}$$

where $\nabla_k E$ is the gradient of electron energy in *k*-space (a vector directed normal to the Fermi surface). Analysis of the influence of an electric field on the distribution of electron velocities leads to the following expression for electrical conductivity (averaged over all directions):

$$\sigma = \frac{v_F^2 Z(E_F)}{3}\, e^2 \tau_F \qquad \textbf{(14.2)}$$

where v_F is the average Fermi velocity, $Z(E_F)$ is the density of electron states per unit volume at the Fermi energy, and τ_F is the collision time for electrons at the Fermi energy. In contrast, classical free-electron theory led us to equation (1.10), which we reproduce here:

$$\sigma = \frac{N_e}{m} e^2 \tau \tag{14.3}$$

Whereas classical theory considered conductivity as resulting from *all* of the free electrons (N_e is the total number of free electrons per unit volume), equation (14.2) shows that band theory focuses specifically on the electron states near the Fermi energy. Detailed comparison with experimental data shows that indeed equation (14.2) yields better agreement with experiment than (14.3). But, as we showed in Chapter 1, classical theory comes pretty close. At first glance, that's surprising, since equations (14.2) and (14.3) look very different. However, it can be shown that in quantum-mechanical free-electron theory (Chapter 12), the two equations are equivalent. Thus the only differences between (14.2) and (14.3) result from the departures of band theory from quantum-mechanical free-electron theory, which are modest for many metals.

Sample Problem 14.2

Show what the text says can be shown—that equations (14.2) and (14.3) are identical in quantum-mechanical free-electron theory.

Solution

From (12.16), $Z(E_F)$ (per unit volume) $= CL^{-3}E_F^{1/2}$
since $v_F^2 = 2E_F/\mathrm{m}$, $\{v_F^2 Z(E_F)\}/3 = (2CL^{-3}/3m)E_F^{3/2}$
For CL^{-3}, substitute the expression in (12.16) and for E_F, substitute (12.14)
Since $NL^{-3} = N_e$, $\{v_F^2 Z(E_F)\}/3 = (N_e/m)$ and thus (14.2) = (14.3).

Another property of metals that depends on $Z(E_F)$ is the critical temperature T_c of superconductors. According to the BCS theory of superconductivity, T_c depends exponentially on the density of electron states at the Fermi energy. Another example is the spin paramagnetism of the conduction electrons. An applied field H will lower the energy of electrons with spins parallel to H by $\mu_0\mu_B H$, and raise the energy of anti-parallel spins by the same amount. Electrons well below the Fermi energy are unable to respond, but those in the vicinity of the Fermi energy can flip their spins and align with the field. The result is a temperature-independent contribution to paramagnetic susceptibility given by:

$$\chi_m \approx \mu_0 \mu_B^2 Z(E_F) \tag{14.4}$$

Sample Problem 14.3

According to band theory, what path in k-space will an electron at the Fermi energy traverse in a cyclotron resonance experiment? (applied DC magnetic field, $\omega\tau \gg 1$)

Solution

As in classical physics, the Lorentz force is perpendicular to both magnetic field and electron velocity. In band theory, however, electron velocity is given by (14.1), directed normal to the Fermi surface. Thus the Lorentz force is *parallel* to the Fermi surface and lies within the plane normal to the magnetic field. The electron trajectory will follow the path created by the intersection of the Fermi surface with the plane perpendicular to the magnetic field. In quantum-mechanical free-electron theory, the Fermi surface is a sphere, and this path will be a circle (corresponding to a helical trajectory in real space, as in the classical theory). However, in general the Fermi surface will be nonspherical, and the trajectory may have a complex shape. The resulting resonance frequency is still usually interpreted in terms of an effective mass m^*.

14.2 INSULATORS

We turn next to a typical ionic solid like KCl. The bonding MO of the diatomic molecule was built primarily from the $3p_z$ AO of Cl, and the antibonding MO primarily from the $4s$ AO of K, as shown schematically in Fig. 10.11 for HF. Bringing many K and Cl atoms together to form a solid, the bonding MO level widens to a narrow but *full* energy band, while the antibonding MO level widens to a broad but *empty* (unoccupied) energy band, separated from the lower band by an energy gap of 7 eV (Fig. 14.3).

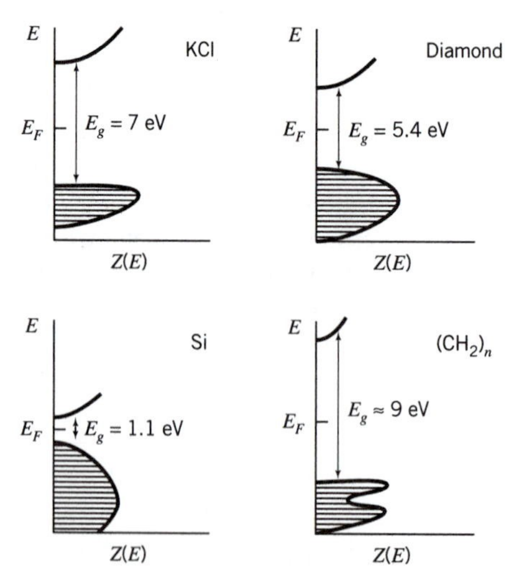

Figure 14.3. Schematic density of states $Z(E)$ for KCl, diamond, Si, and polyethylene.

Sample Problem 14.4

What AOs are dominant in the valence and conduction bands of RbBr? Will the energy gap be larger or smaller than that of KCl?

Solution

RbBr is also an alkali halide, with the 5*s* AO electron of Rb dominant in the upper (empty) band, and the $4p_z$ AO of Br dominant in the lower (full) band. Rb ions are larger than K ions (more electron shells), and Br ions are larger than Cl ions, so that the strength of the bond and the energy gap (energy required to pull an electron out of the bond) are *smaller* than in KCl.

Numerous other ionic solids (e.g., MgO) similarly have a fully occupied electron energy band associated primarily with the anion (in this case built primarily from the 2*p* electrons of oxygen) and an unoccupied energy band associated primarily with the cation (in this case, built from the 3*s* electrons of Mg), separated by a substantial energy gap (7.8 eV). LiF, the highly ionic alkali halide we specifically considered in Chapter 10, has an energy gap of 12 eV, which is the largest energy gap of any solid.

The Fermi levels of KCl, MgO, LiF, and similar solids are *within the gaps.* Neither the lower band, which is full, nor the upper band, which is empty, can conduct electricity. These ionic solids are *insulators.* Since visible photons have much less energy than the 7 eV energy gap of KCl, no photons can be absorbed, and pure crystals of KCl are transparent. Pure ionic solids with energy gaps above about 3.1 eV, including MgO and LiF, are *transparent insulators.*

The development of bands from bonds is more complex in diamond, a pure covalent solid, because of sp^3 hybridization. As many carbon atoms are brought together to form diamond, the *s* and *p* AO energy levels broaden in energy and overlap, and form into three energy regions: a lower band of bonding MO levels that can hold four electrons per atom, a gap of forbidden energies, and an upper band of antibonding MO levels (Fig. 14.3). At $T = 0$, the lower band is completely full, the upper band completely empty, and the Fermi level is located within an energy gap of 5.4 eV. Pure diamond, like pure KCl, is a transparent insulator.

Silicon also has sp^3 hybridization and the tetrahedral diamond structure, but the Si-Si covalent bonds are a bit weaker than the C-C bonds in diamond. (Silicon is below carbon in the periodic table, which, we noted earlier, leads to weaker bonds.) The energy gap between the occupied and unoccupied levels represents the energy required to remove an electron from the covalent bond, and is only 1.1 eV in Si (Fig. 14.3). The Fermi level of pure Si is within the energy gap, and at $T = 0$, Si is an insulator. (We'll discuss the effect of elevated temperature in Chapter 15.)

However, Si is *not transparent.* All visible photons have more energy than the 1.1 eV energy gap of Si, and therefore can elevate electrons from the lower occupied band across the gap and into the upper unoccupied band. Since visible photons are absorbed, pure Si is opaque and looks very much like a metal, but it certainly does not conduct like a metal. In Chapter 2, we learned how a plasma frequency in the infrared, as in indium tin oxide, can produce a transparent conductor. With Si and band theory, we now can see that an energy gap in the infrared can produce an opaque insulator. We've now explained another one of those mysteries that we listed in the Halftime Review at the end of Chapter 7. The metallic appearance of silicon can be explained in terms of photons and band theory but not by classical physics.

Figure 14.3 also shows the schematic energy band structure of *polyethylene*, a saturated carbon-based polymer with bonds also built from sp^3 hybrids. It also has a full band separated from an empty band, with an energy gap much larger than the energy of visible photons, and it is a transparent insulator. However, the one-dimensional nature of the polymer chain leads to a different shape of the $Z(E)$ curve, as discussed in Chapter 13.

If pure *polyacetylene* were ideally conjugated so that all C-C bonds were equivalent, as in benzene, the delocalized pi electrons would produce a half-filled band with a $Z(E)$ shaped like that of polyethylene. However, a subtle effect mentioned in Section 11.5 (a Peierls distortion) leads to a slight lowering of energy if the C-C bonds alternate somewhat in spacing and strength, leading to band splitting and a separation of about 1.8 eV between a full lower band and an empty upper band. This gap is small enough to make polyacetylene opaque, and polyacetylene is a semiconductor that can be made into a reasonable conductor with small additions of impurities (see Fig. 1.1).

14.3 INORGANIC COLOR

In Chapter 11, we discussed the optical properties and colors of various organic materials in terms of the magnitude of the HOMO-LUMO separation—the gap between the energy levels of the highest-occupied and lowest-unoccupied molecular orbitals. In this chapter we've been discussing optical properties of insulators in terms of the energy gap E_g. The top of the full band is the equivalent of the HOMO and the bottom of the empty band is equivalent to the LUMO. Optical excitation of electrons across the energy gap can occur for photons with $h\nu > E_g$, but not for photons with $h\nu < E_g$. Thus a plot of optical absorption vs. frequency will show an abrupt increase in absorption, called the *absorption edge*, at a frequency defined by

$$\nu = \frac{E_g}{h} \quad \textbf{(14.5)}$$

This transition from transparent to opaque with increasing optical frequency is opposite to the classical effect observed at the plasma frequency (Chapter 2) (where increasing frequency produces a transition from opaque to transparent). You may recall that indium tin oxide had a plasma frequency in the infrared, making it transparent in the visible and allowing it to be used for transparent electrodes. *A second requirement for a transparent electrode*, satisfied by indium tin oxide but not discussed in Chapter 2, *is that the energy gap be greater than about 3.1 eV* so that the absorption edge is in the ultraviolet. Indium tin oxide absorbs light in the infrared (below the plasma frequency) and in the ultraviolet (above the energy gap) but not in the intervening visible frequencies. Convenient!

If the energy gap of a pure insulator is greater than about 3.1 eV, no visible photons can be absorbed, and the material will be transparent. Examples we gave earlier included KCl, diamond, and polyethylene. On the other hand, if the energy gap is less than about 1.8 eV, *all* visible photons, from red to blue, will be absorbed, and the material will be opaque. Silicon, with $E_g \approx 1.1$ eV, is opaque and reflects like a metal. It looks like a metal, but its electrical and mechanical properties are very different from a metal.

How about insulators with energy gaps between 1.8 and 3.1 eV? They will absorb *some* light at the high-frequency end of the visible spectrum, and the result will be *color*. Several inorganic dyes and paint pigments are insulators with energy gaps in this range. With an energy gap above about 2.5 eV, the material will be yellow. (The transmitted or reflected light will include all the visible photon energies from 1.8 to 2.5 eV—but not from 2.5 to 3.1 eV—and our eyes perceive that mixture as yellow.) As energy gap is decreased further (e.g., by alloying with another semiconductor with a smaller gap), the color will change from yellow to orange to red—and eventually to black or opaque when all visible light is absorbed (see Table 14.1). The effect of decreasing energy gap in this energy range is to absorb more and more of the blue (high photon energy) end of the spectrum, so that the color sequence follows that of a sunset, when more and more of the blue end of the spectrum is scattered from the sunlight and the western sky goes from yellow to orange to red.

The interband absorption edge at $h\nu = E_g$ can produce red, orange, and yellow insulators by subtractive coloration, but not violet, blue, or green. However, impurities can give localized electron energy levels and associated absorption *peaks* rather than absorption edges. Thus impurities in insulators can produce a wide variety of colors by preferential absorption. For example, nitrogen impurities in diamond turn it yellow, but boron impurities preferentially absorb at the red end of the visible range and result in bluish diamonds. The famous Hope diamond, on display in the Smithsonian, is slightly blue because of boron impurities. It's one of the most valuable impure insulators in the world.

Table 14.1. Energy Gaps of Inorganic Compounds and Color Resulting from Interband Absorption

Material	Energy Gap (eV)	Color
ZnO	3.1	colorless
CdS	2.5	yellow
GaP	2.3	orange
HgS	2.0	red
CdSe	1.7	black

Sample Problem 14.5

Why can't the absorption edge in an insulator produce blue?

Solution

To produce blue by subtractive coloration, you must have a material that absorbs photons on the red (low photon energy) end of the visible spectrum but *does not absorb blue* (high photon energy). Any material with an energy gap small enough to absorb red photons will also absorb blue photons.

Coloring in other gems also results from absorption at localized impurity levels. The red of ruby comes from Cr impurities in the otherwise colorless and transparent Al_2O_3 (sapphire) crystal. (Electrons localized at Cr impurities in ruby produce red not only by absorbing blue but also by *emitting* red in a subsequent electron transition. Ruby crystals exposed to ultraviolet light emit a deep red light.) The blue of blue sapphires comes from Fe and Ti impurities. The purple of amethysts results from Fe impurities in otherwise colorless SiO_2 (quartz). The green of emeralds (preferential absorption at both the blue and red ends of the spectrum, like grass) comes from Cr impurity energy levels in otherwise colorless $Be_3Al_2Si_6O_{18}$ (beryl). Of course, localized electron states at impurities can also produce color in many other materials that are colorless when pure, like glass and polyethylene, which are considerably cheaper than rubies, sapphires, and emeralds.

Many insulators also show optical absorption peaks just below the absorption edge, that is, they absorb photons with energies *slightly less* than E_g. Whereas standard interband absorption ($h\nu > E_g$) creates an electron in the upper band and an independent hole in the lower band, photons with slightly less energy can produce bound electron-hole pairs called *excitons*. Electron transitions are not the only mechanisms of photon absorption in solids. For example, infrared photons can induce lattice vibrations in ionic crystals, that is, phonon transitions. For example, NaCl has a strong absorption and associated reflection for $\lambda \approx 60\ \mu m$. Such phenomena are called *reststrahlen* (German for *residual rays*) because they enable appropriate wavelengths to be separated from others by selective reflection. Photon absorptions can also produce simultaneous electron and phonon transitions. (The reverse process—generation of photons by coordinated electron and phonon transitions—is discussed briefly in Section 16.5.)

Our rough picture of energy bands and energy gaps associated with valence electrons in solids enables us to explain qualitatively why some materials are metals and some are insulators and why some materials are transparent, some opaque, some colored. In Chapters 15 and 16, we will apply the band picture in full force to consider the electrical and optical properties of semiconductors, the materials that led us into the remarkable world of microelectronics and that are now leading us into the fast-growing world of optoelectronics. But before that, a brief digression to warn you about some limitations of simple band theory.

14.4 BEYOND BANDS

Science can be frustrating. Just when you were beginning to get a feel for the complexities of Newtonian mechanics and Maxwellian electromagnetics, you were told that they're not quite right. To really understand the world, you have to learn quantum mechanics and relativity. And in the latter part of the 20th century, ever more complex theories have been developed in an attempt to explain the natural world.

The same complexity exists in the science of solids. In Chapters 1 and 2, we explained a few things about metals with Drude's classical free-electron theory. But we found that there were things we couldn't explain, and in Chapter 12, we explained a few more things about metals with quantum-mechanical free-electron theory. But there were still a few things we couldn't explain. In Chapter 13, we introduced nearly free electrons and developed some basic concepts of band theory, including energy gaps, effective mass, holes, Brillouin zones, Fermi surfaces, and so on. Simple band theory enabled us to explain several things about metals and insulators that neither classical free-electron theory nor quantum-mechanical free-electron theory could. Now, of course, it's time to tell you that band theory, in the simplified form in which we've presented it, also has its limits.

One limitation that is more apparent than real is that of interpreting the properties of *amorphous solids*. If your understanding of band theory is based entirely on the nearly-free electron theory and the arguments of Chapter 13, which stressed the importance of the periodic atomic lattice, Bragg diffraction, and Brillouin zones, it's not that easy to understand why the properties of crystalline SiO_2 (quartz) and those of amorphous SiO_2 (glass) are so similar. (They're both transparent insulators with band gaps in the ultraviolet.) But we remind you that the alternate approach to developing band theory, via LCAO as in Chapter 11, does not depend so critically on long-range crystalline order. The amorphous structure of glassy solids does indeed lead to varying interatomic distances and sometimes even a varying number of nearest neighbors, but these changes merely add a bit of fuzziness to the quantitative arguments. The general picture of energy gaps and energy bands for amorphous SiO_2 is not very different from that for crystalline SiO_2.

A more serious problem is that there are many compounds for which simple band theory predicts partly filled d or f bands, and therefore metallic behavior, but experiment shows these compounds to be semiconducting or insulating. For example, this is true of several transition-metal monoxides with the NaCl structure (MnO, FeO, CoO, and NiO), for which simple band theory predicts a partly-full band built primarily from the $3d$ electrons of the transition metals. Experiment instead indicates that these d bands are split into two, a full band and an empty band, with the Fermi level in the intervening gap. Magnetic interactions play a role.

Most such problems appear to stem from the fact that in simple band theory we have implicitly used what is called the *independent electron approximation*. We have not fully taken into account the mutual interactions between electrons. In the LCAO-MO approach, we considered electron-electron repulsion *in an averaged way* in the SCF approximation. In the NFE theory, electron-electron repulsion was considered implicitly in the periodic lattice potential produced by the ion cores, also only in an averaged way. More detailed treatments of the repulsive forces between individual electrons, which generally lead to greater localization of the electrons than assumed in simple band theory, have been able to remove most of these limitations of simple band theory. One such model is the Hubbard model, which focuses specifically on the repulsion between electrons on the same atom. This and related models, in which electron concentration, and therefore atomic volume, are important parameters, have also been used to explain transitions between metallic and insulating behavior that have been observed in several systems with sufficient changes in density (e.g., under high hydrostatic pressure).

Simple band theory can be a powerful tool to understand the properties of many solids, including many metals, insulators, and semiconductors. And with modern computers, advanced computational methods such as *density functional theory* can today successfully calculate, from first principles, many of the basic properties of solids. However, there is a huge and growing variety of materials of engineering interest, and each new material to be considered must at first be viewed cautiously to determine the validity of various theoretical approaches. It is useful to remember that all of these approaches, as impressive and powerful as they are, are still only approximations of the impossibly complicated problem of solving Schrödinger's equation for a system of 10^{23} interacting particles.

SUMMARY

In band theory, the properties of solids are interpreted in terms of the location of the Fermi level with respect to the energy bands and energy gaps. The conductivity and opacity of metals are explained by the presence of unoccupied electron states immediately above occupied states, that is, by the Fermi level being located within allowed energy bands. The Fermi level of monovalent metals is located in the middle of an energy band, while that of divalent metals is in the region of band overlap. In transition metals, the d and s bands overlap; the s bands are much wider than the d bands, and the density of states $Z(E)$ is much higher in the d bands. Absorption of bluish photons in $3d \rightarrow 4s$ electron transitions gives copper its reddish color. The variation of cohesive energy, atomic volume, and elastic modulus across the various transition-metal series can be explained by the sequential filling of bonding and antibonding COs.

Many properties of metals are determined primarily by electrons near the Fermi energy. For example, according to band theory, electrical conductivity is proportional to $v_F^2 Z(E_F)$. Although this looks rather different from the classical formula, the two equations are equivalent in quantum-mechanical free-electron theory, and many metals are close enough to free-electron behavior that the classical formula remains approximately valid. The critical temperature T_c of superconductors and the spin paramagnetism of conduction electrons are two other examples of properties sensitive to $Z(E_F)$.

Pure insulators like KCl, diamond, silicon, and polyethylene have full bands separated from empty bands by energy gaps, with their Fermi levels within the energy gaps. Neither full bands nor empty bands can contribute to conductivity. If the energy gap is larger than about 3.1 eV, as with KCl, diamond, and polyethylene, photons in the visible range cannot be absorbed and the materials are transparent. If the energy gap is less than 3.1 eV, as with silicon, visible photons are absorbed and the material is opaque.

Optical excitation across the energy gap yields an *absorption edge* (an abrupt increase in optical absorption with increasing frequency) at $h\nu = E_g$, and therefore the value of E_g compared to the visible range (1.8–3.1 eV) determines whether an insulator is transparent, opaque, or colored. Electron energy levels associated with impurities can also absorb photons and affect optical properties, including color. Many gems and other materials derive their colors from impurity absorptions.

The existence of energy bands and energy gaps in amorphous solids appears inconsistent with the NFE theory presented in Chapter 13, which emphasized the importance of the periodic lattice potential. The LCAO approach of Chapter 11, however, also leads to energy bands and energy gaps, and does not require long-range crystalline order. Although simple band theory can explain the properties of many solids, it fails in some cases, such as transition-metal oxides like NiO, because it does not fully take into account the mutual interaction between electrons.

PROBLEMS

14-1. Diamond is a hard transparent insulator, while graphite is a soft opaque conductor. Explain why these two crystalline forms of carbon have such different properties in terms of the AOs and MOs involved in their bonding and in terms of band theory.

14-2. Briefly explain why zinc and several other divalent metals have Hall coefficients opposite in sign to those of monovalent and trivalent metals.

14-3. Like diamond and silicon, germanium and grey tin have the diamond crystal structure and tetrahedral sp^3 bonds. How do you expect the energy gaps of germanium and grey tin to compare with those of diamond and silicon? How does the sequence of energy gaps in these four materials compare with the sequence of melting points?

14-4. In the spirit of Fig. 14.3, sketch $Z(E)$ for NaCl. Indicate from which AOs the wave functions of each band are built. How will the energy gap compare with that of KCl? Why?

14-5. Diamond has an energy gap of 5.5 eV and is transparent. Silicon has an energy gap of 1.1 eV and reflects all visible light, making it look much like a metal. What optical properties would you expect of an insulator with an energy gap of about 2.5 eV?

14-6. In Chapter 2, we considered the problem of developing a conductor that was transparent to visible light and concluded that a semiconductor with a plasma frequency in the near-infrared would work. Using simple band theory, can you now state another requirement for a transparent conductor?

14-7. Properties of metals that cannot be explained by classical free-electron theory include electronic specific heat, x-ray and photoelectron spectroscopy, effective mass, and positive Hall coefficients. Explain briefly which aspects of quantum-mechanical free-electron theory or simple band theory can be used to explain these properties.

14-8. List the following alkali halide crystals in order of increasing energy gap: NaCl, KCl, CsI, LiF, NaF, KBr, RbBr, RbI. Indicate briefly the reasoning behind your ordering.

14-9. A quantum-mechanical treatment of electrical conduction in metals shows that conductivity is proportional to $Z(E_F)$, the density of states at the Fermi energy. With this information, explain why divalent metals typically have lower conductivities than monovalent metals, a result which contradicts predictions of classical free-electron theory.

14-10. You are likely to become much more sunburned from sunlight coming through an open window than from sunlight coming through a closed window, even though the window glass transmits most of the visible light. Explain.

14-11. As a result of photon-induced electron transitions across the energy gap, cadmium sulfide is yellowish. Describe what color changes you would expect in the Cd(S,Se) series of ternary semiconductors as sulfur is gradually replaced by selenium.

14-12. A doped semiconductor is found that transmits the full spectrum of visible light but absorbs all infrared and ultraviolet light. The effective mass is the free electron mass, and the effective dielectric constant is 1. Calculate the energy gap in eV and the Hall coefficient.

Chapter 15

Semiconductors

15.1 CROSSING THE GAP

The Fermi level of a metal is within an energy *band*, a range of *allowed* electron energies. A metal has empty electron states adjacent to occupied states. The Fermi level of an insulator is within an energy *gap*, a range of *forbidden* electron energies. An insulator has a lower band of fully occupied states, separated by an energy gap from a higher band of empty allowed states. Neither the completely full lower band nor the completely empty higher band can conduct electric current. What about semiconductors like silicon or GaAs, which are so important to modern technology?

The Fermi level of semiconductors is nearly always within an energy gap. However, the energy gap is usually not very large. Either the lower energy band (called the *valence band*) is not completely full, allowing conduction of electric current by holes, or the higher energy band (called the *conduction band*) is not completely empty, allowing conduction by electrons, or both. (We should note at the outset that the traditional names of the two bands are misleading, since conduction can occur in *both* bands, and in some materials it occurs primarily in the valence band. However, it's too late to change the names now!)

We can get holes in the valence band, electrons in the conduction band, or both for three reasons:

1. *Thermal excitation.* We recall (Chapter 12) that at finite temperatures, the Fermi function tells us that some states below the Fermi level will be unoccupied and some states above the Fermi level will be occupied. At high enough temperatures, a significant number of electrons can be thermally excited across the energy gap, producing both holes in the valence band and electrons in the conduction band. This thermal excitation of electron-hole pairs is the major cause of the observed increase of the electrical conductivity of semiconductors with increasing temperature that we noted way back in Chapter 1.
2. *Optical excitation.* If photons with sufficient energy ($h\nu$ greater than the energy gap E_g) impinge on a semiconductor, they can excite electrons from the valence band across the gap and into the conduction band. This optical production of electron-hole pairs by "interband" electron transitions increases the conductivity of semiconductors, an effect called *photoconductivity*. Photoconductivity can be used in a wide variety of devices, including camera light meters and automatic door openers. Chapter 16 discusses other important devices based on optical excitation of charge carriers in semiconductors. Absorption of photons by interband electron transitions also determines whether the semiconductor will be transparent, opaque, or colored, as discussed in Chapter 14.
3. *Impurities.* Impurity atoms introduce allowed electron energy levels different from those that are present in the pure semiconductor. For example, if the electron energy levels associated with an impurity atom were just below the bottom of the conduction band, even a modest temperature could produce thermal excitation of electrons from these impurity levels into the conduction band. Similarly, if the impurity levels were just above the top of the valence band, even a modest temperature could produce thermal excitation of electrons from the valence band into the impurity levels, creating holes in the valence band. Such effects explain why even very small impurity concentrations can greatly increase the electrical conductivity of semiconductors, an experimental observation we noted in Chapter 1.

Semiconductors are simply insulators that with thermal or optical excitation and/or with impurity additions can achieve electrical conductivities intermediate between metals and insulators (roughly between 10^{-5} and 10^5 S/m; see Fig. 1.1) We will first discuss the electrical and optical properties of pure semiconductors (called *intrinsic* semiconductors) and later consider the properties of semiconductors containing impurities (called *extrinsic* semiconductors).

15.2 INTRINSIC SEMICONDUCTORS

We consider an intrinsic semiconductor with an arbitrary energy gap E_g, and want to calculate the number of electron-hole pairs thermally excited at a given temperature T. We assume that the electrons in the conduction band are near the bottom of the band, where they are nearly free, and the density of allowed states, from (12.16),

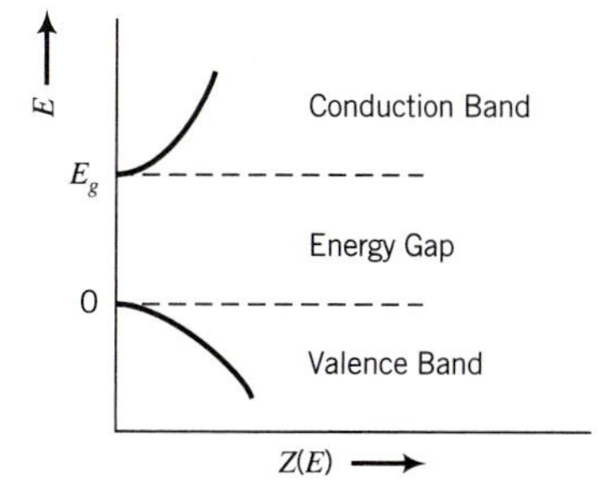

Figure 15.1. Schematic density of states $Z(E)$ for semiconductor in the vicinity of the gap. $Z(E)$ is parabolic at the bottom of the conduction band and the top of the valence band—equations (15.1) and (15.3).

is approximately proportional to the square root of the energy from the bottom of the band. Setting our zero for energy at the bottom of the gap (Fig. 15.1), the energy at the bottom of the conduction band is E_g, and the density of states (per unit volume) in the conduction band is

$$Z_e(E) = C_e(E - E_g)^{1/2} \tag{15.1}$$

where

$$C_e = \frac{4\pi(2m_e^*)^{3/2}}{h^3} \tag{15.2}$$

Here we have duplicated (12.16) except for substituting an effective electron mass m_e^* for the free-electron mass we had assumed in Chapter 12. (As a result of reading Chapter 13, we are now a bit more sophisticated about band theory than we were way back in Chapter 12.) What do we do about the valence band? Here, instead of focusing on all the occupied electron states, we'll focus on the holes. In the semiconducting game, we assume that we can treat holes in the valence band much as we treat electrons in the conduction band, except that everything is topsy-turvy. Near the top of the valence band, we expect the *density of hole states* to vary approximately as the square root of the energy from the *top* of the valence band, or

$$Z_h(E) = C_h(-E)^{1/2} \tag{15.3}$$

where

$$C_h = \frac{4\pi(2m_h^*)^{3/2}}{h^3} \tag{15.4}$$

where we recognize that m_h^*, the effective mass of holes in the valence band, may be very different from m_e^*, the effective mass of electrons in the conduction band. And we remind you that $Z(E)$ represents the density of *allowed* states, so it is zero in the energy gap. We also remind you that the actual density of *occupied* electron states is $Z_e(E)F(E)$, where the Fermi function $F(E)$, repeating (12.18), is

$$F(E) = \frac{1}{1 + e^{(E-E_F)/k_BT}} \tag{15.5}$$

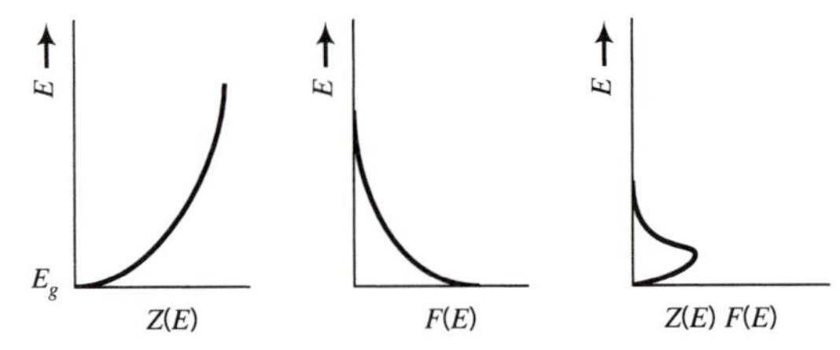

Figure 15.2. Density of available states $Z(E)$, Fermi function $F(E)$, and product $Z(E)F(E)$, the density of occupied states, near the bottom of the conduction band.

We're interested first in calculating, for a given temperature T, the total number of electrons N_e in the conduction band, which should be

$$N_e = \int_{E_g}^{?} Z_e(E)F(E)dE \qquad \textbf{(15.6)}$$

The upper limit should probably be the top of the conduction band, but we realize that, although $Z_e(E)$ increases with increasing energy, $F(E)$ *decreases exponentially* with increasing energy, so that their product will go through a maximum near the bottom of the band and then decrease rapidly (Fig. 15.2). So if we use infinity as the upper limit, it won't change things much.

Sample Problem 15.1

A semiconductor has a Fermi energy that is 0.5 eV below E_g, and an effective mass for electrons in the conduction band that is one-half the free-electron mass. At room temperature, what is the density of occupied electron states (a) at E_g? (b) At $E_g + k_BT$? (c) At $E_g + 3k_BT$?

Solution

The density of occupied electron states is $Z_e(E)F(E)$.

$m_e^* = 4.56 \times 10^{-31}$ kg, from (15.2), $C_e = 3.76 \times 10^{55}$ m^{-3}J$^{-3/2}$

(a) at E_g, $Z_e = 0$, so $Z_e(E)F(E) = 0$

(b) $k_BT = 4.14 \times 10^{-21}$ J, $Z_e(E_g + k_BT) = 2.42 \times 10^{45}$ m^{-3}J^{-1}

At $E = E_g + k_BT$, $E - E_F = 8.41 \times 10^{-20}$ J, $F(E) = 1.51 \times 10^{-9}$

$Z_e(E)F(E) = (2.42 \times 10^{45})(1.51 \times 10^{-9}) = 3.65 \times 10^{36}$ m^{-3}J^{-1}

(c) $3k_BT = 1.24 \times 10^{-20}$ J, $Z_e(E_g + 3k_BT) = 4.19 \times 10^{45}$ m^{-3}J^{-1}

At $E = E_g + 3k_BT$, $E - E_F = 9.24 \times 10^{-20}$ J, $F(E) = 2.03 \times 10^{-10}$

$Z_e(E)F(E) = (4.19 \times 10^{45})(2.03 \times 10^{-10}) = 8.51 \times 10^{35}$ m^{-3}J^{-1}

In going from $E = E_g + k_BT$ to $E = E_g + 3k_BT$, the decrease in $F(E)$ overwhelms the increase in $Z_e(E)$, so that the product $Z_e(E)F(E)$ decreases. By $E = E_g + 3k_BT$, we are apparently past the peak in Fig. 15.2.

The numbers are so huge because we are using meters and joules. In terms of centimeters and eV, the answer for (c) is 1.36×10^{11} (cm)$^{-3}$(eV)$^{-1}$.

One more approximation. We'll assume that the Fermi energy isn't too close to the bottom of the conduction band ($E = E_g$, the top of the gap). That will allow us to approximate $F(E)$ with a simple exponential, as in (12.21). With these approximations, and substituting (15.1) for $Z_e(E)$, (15.6) becomes

$$N_e = C_e \int_{E_g}^{\infty} (E - E_g)^{1/2} e^{-(E-E_F)k_BT} dE \qquad \textbf{(15.7)}$$

This can be evaluated if we substitute x for $(E - E_g)/k_BT$, which transforms (15.7) into

$$N_e = C_e(k_BT)^{3/2} e^{-(E_g-E_F)/k_BT} \int_0^{\infty} x^{1/2} e^{-x} dx \qquad \textbf{(15.8)}$$

The definite integral equals $\sqrt{\pi}/2$, so our final result for the total number of electrons per unit volume in the conduction band is

$$N_e = N_c e^{-(E_g-E_F)/k_BT} \qquad \textbf{(15.9)}$$

where

$$N_c = \frac{\sqrt{\pi}}{2} C_e(k_BT)^{3/2} = \frac{2(2\pi m_e^* k_BT)^{3/2}}{h^3} \qquad \textbf{(15.10)}$$

We can now mentally turn everything upside down and similarly calculate N_h, the total number of holes in the valence band. The probability of a hole state being occupied is $[1 - F(E)]$, the probability that the corresponding electron state is *un*-occupied. Assuming that the Fermi energy is not too close to the bottom of the gap ($E = 0$), we can use the simple exponential approximation (12.23) for $[1 - F(E)]$ and integrate $Z_h(E)[1 - F(E)]$ from $-\infty$ to 0, to get

$$N_h = N_v e^{-E_F/k_BT} \qquad \textbf{(15.11)}$$

where

$$N_v = \frac{\sqrt{\pi}}{2} C_h(k_BT)^{3/2} = \frac{2(2\pi m_h^* k_BT)^{3/2}}{h^3} \qquad \textbf{(15.12)}$$

With N_h proportional to $\exp(-E_F/k_BT)$ and N_e proportional to $\exp(+E_F/k_BT)$, the Fermi energy disappears from their product, which is simply

$$N_eN_h = N_cN_ve^{-E_g/k_BT} \tag{15.13}$$

In an intrinsic semiconductor, $N_e = N_h$, so that

$$N_e = N_h = (N_cN_v)^{1/2}e^{-E_g/2k_BT} \tag{15.14}$$

Sample Problem 15.2

If we assume that the effective masses of electrons and holes are both equal to the free-electron mass, what would be the electron and hole concentrations in a semiconductor with an energy gap of 0.6 eV (a) at 300 K? (b) At 600 K?

Solution

(a) At 300 K, $N_c = N_v = 2.52 \times 10^{25}$ m^{-3} and $e^{-E_g/2k_BT} = 9.22 \times 10^{-6}$
so $N_e = N_h = (2.52 \times 10^{25})(9.22 \times 10^{-6}) = 2.32 \times 10^{20}$ m^{-3} = 2.32×10^{14} cm^{-3}
(b) At 600 K, $N_c = N_v = 7.13 \times 10^{25}$ m^{-3} and $e^{-E_g/2k_BT} = 3.04 \times 10^{-3}$
so $N_e = N_h = (7.13 \times 10^{25})(3.04 \times 10^{-3}) = 2.16 \times 10^{23}$ m^{-3} = 2.16×10^{17} cm^{-3}
(Note that the change of the pre-exponential term with temperature is orders of magnitude less than the change of the exponential term.)

So we can see that the number of thermally activated electron-hole pairs in an intrinsic semiconductor will increase exponentially with increasing temperature and, at a given temperature, will decrease exponentially with increasing width of the energy gap. (For completeness, we should mention that the energy gap itself actually varies slightly with temperature, primarily as a result of thermal expansion, but that is usually a small effect that can be ignored.)

Where is the Fermi level E_F of an intrinsic semiconductor located? All we've assumed so far is that it is somewhere in the energy gap, and it is not too close to either the top or the bottom of the gap. By equating (15.9) and (15.11) and solving for E_F,

$$E_F = \frac{E_g}{2} + \frac{3k_BT}{4} \ln \frac{m_h^*}{m_e^*} \tag{15.15}$$

If the effective masses of the electron and hole are the same, the second term vanishes and the Fermi level is exactly at midgap. At room temperature, k_BT is only about 0.025 eV, far less than the energy gaps of materials of interest. (E_g = 1.1 eV for Si, for example.) So the Fermi level of intrinsic (pure) semiconductors at moderate temperatures is very close to midgap, halfway between the valence and conduction bands.

In Chapter 1, we introduced the concept of *mobility* and noted that conductivity depended on the product of charge carrier concentration and mobility. In intrinsic semiconductors, the densities of electrons and holes are equal, and both contribute to conductivity:

$$\sigma = e(N_e\mu_e + N_h\mu_h) = eN_e(\mu_e + \mu_h) \qquad \textbf{(15.16)}$$

Even though $N_e = N_h$, the electrons often carry the majority of the current because the mobility of electrons is usually greater than the mobility of holes. Typical values for Si at room temperature are $\mu_e = 0.15$ and $\mu_h = 0.05\ \mathrm{m^2V^{-1}s^{-1}}$.

Why do holes usually have lower mobilities than electrons? Mobility is inversely proportional to mass of the carrier, and the appropriate mass in a semiconductor is the *effective* mass. As we noted in Chapter 13, greater AO overlap leads to bands that are wider in energy, with greater curvature of $E(k)$, and therefore lower effective mass and higher mobility. In general, higher-energy bands correspond to greater overlap and greater absolute value of the bond integral β. Since electrons are associated with the higher-energy conduction band, and holes with the lower-energy valence band, it is not surprising that m_h^* is usually higher than m_e^*, a major source of the difference in mobilities. (We are oversimplifying a bit here, since $E(k)$ curves can have very different curvatures in different crystal directions, and semiconductors can sometimes have, for example, both heavy holes and light holes. But life is complicated enough.)

We have used in this text the symbols N_e and N_h for the concentrations of electrons (in the conduction band) and holes (in the valence band). In much of the semiconductor literature, these same quantities are instead simply represented by n and p, respectively. However, in this text n has already been used for quantum number and index of refraction, and p has been used for electric dipole moment and momentum—perhaps that is enough baggage for two lowercase letters to carry.

15.3 EXTRINSIC SEMICONDUCTORS

It is, of course, impossible to produce a perfectly pure semiconductor. The conductivity of silicon and other semiconductors can be sensitive to impurity concentrations far less than a part per million, and a major accomplishment of materials scientists in the early days of semiconductor technology was to develop techniques (notably zone refining) to purify silicon to a level where the conductivity approached intrinsic values. Having done that, they then intentionally added small amounts of known impurities (called *dopants*) to alter its electronic properties.

Suppose you add to intrinsic Si a very small amount of a dopant from *group V* of the periodic table, such as P ($Z = 13$, one more than Si). It enters substitutionally, replacing a Si atom on one of the sites in the tetrahedrally bonded diamond structure. The P atom has one more proton in its nucleus, and one more electron, than Si. Four of the valence electrons of P will form sigma bonds with its four Si nearest neighbors, leaving one valence electron that will be bound only loosely to the P atom. We can estimate this binding energy with a *hydrogenic* model, viewing the electron as attracted to the +1 excess charge of the nucleus shielded by the other 12 electrons. The binding energy of the 1*s* electron in hydrogen, from (9.10), was $(me^4)/(8\varepsilon_0^2h^2)$ or 13.6 eV. However, the outer electron of the P ion is in a dielectric medium, so the ε_0 in this formula should be multiplied by the dielectric constant of Si, which is about 12. The free-electron mass m should also be replaced by the effective mass m_e^* of an electron at the bottom of the Si conduction band, which is about $0.5m$. This oversimplified model gives about 0.05 eV as the binding energy of the outer electron of P in the Si lattice, which is very close to the experimental value.

This binding energy is much less than the energy gap in Si, so that it is much easier to thermally excite the outer electron of the P atom into the Si conduction band than to excite one of the electrons out of the covalent sigma bonds. In the band picture, this means that the energy level corresponding to the outer electrons of P dopant atoms is in the gap, only slightly below the bottom of the conduction band (Fig. 15.3). We label this donor level E_D.

At room temperature, nearly all of these outer electrons will be thermally excited from the donor level into the conduction band. With a high enough concentration N_D of these *donor* P atoms, N_e will be increased far above the intrinsic value given by (15.14). In the intrinsic semiconductor, we had a hole in the valence band for each electron in the conduction band, that is, $N_e = N_h$. In the phosphorus-doped Si, however, no holes have been created by ionizing the phosphorus. The negative charge of the thermally freed electron is counterbalanced by the positive charge of the ion it left behind, but the P ions are immobile and cannot contribute to conductivity. We thus have $N_e \gg N_h$. We can similarly add electrons to the conduction band by adding other group V elements like Sb or As, which have similar donor energy levels, about 0.05 eV below the conduction band.

What have these phosphorus impurity additions done to the Fermi level? Have they had any effect on the hole concentration N_h? We derived expressions (15.9) and (15.11) for N_e and N_h, respectively, without knowing exactly where the Fermi level was within the gap. It was only when we assumed that $N_e = N_h$ that we learned in equation (15.15) that the Fermi level of an intrinsic semiconductor is near midgap. As long as the Fermi level isn't too close to the conduction band, equation (15.9) is valid. And since we know that N_e is much increased above intrinsic values in our

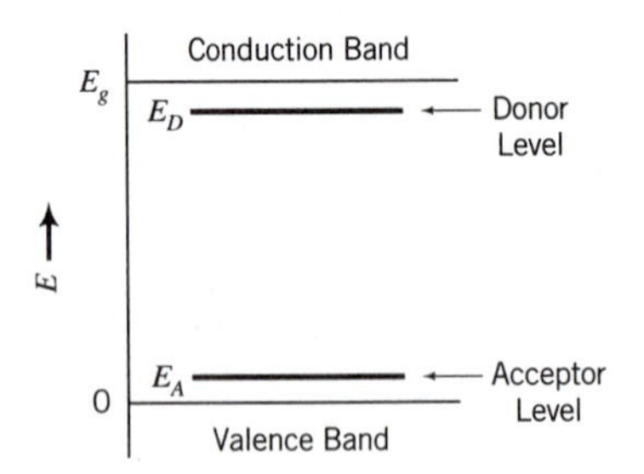

Figure 15.3. The donor energy level E_D lies slightly below the conduction band edge and the acceptor energy level E_A lies slightly above the valence band edge.

phosphorus-doped Si, it is clear from (15.9) that, in an n-type semiconductor, E_F *is above midgap.*

For Si at room temperature, it is a good approximation that essentially *all* of the P atoms have ionized and contributed an electron to the conduction band. Thus, to a good approximation, $N_e \approx N_D$, the donor concentration. What about N_h? Equation (15.13) shows that the product of N_e and N_h depends only on temperature and energy gap and should not be changed by doping. From (15.13) and (15.14),

$$N_e N_h = N_i^2 \tag{15.17}$$

where N_i is the *intrinsic charge concentration* given by equation (15.14). A good approximation for donor-doped Si at room temperature is therefore

$$N_e \approx N_D \qquad \text{and} \qquad N_h \approx \frac{N_i^2}{N_D} \tag{15.18}$$

(We have assumed that the donor concentration is much larger than N_i.) Thus by increasing the concentration of our donor impurity, we increase the electron concentration and *decrease* the hole concentration, their product remaining constant. Semiconductors like phosphorus-doped Si, in which $N_e \gg N_h$ and therefore the dominant charge carriers are *n*egatively charged, are called *n-type semiconductors.* In an n-type semiconductor, electrons are the *majority carriers* and holes are the *minority carriers.*

Typically, the concentrations of majority and minority carriers in extrinsic semiconductors differ by *many* orders of magnitude. For Si at room temperature, N_i is about 10^{10} per cm^3. Suppose we add 10^{16} P atoms per cm^3, which amounts to only about one P atom for every 10^7 Si atoms. Not very many. Yet, from (15.18), at room temperature N_e is about 10^{16} per cm^3, while N_h is only about $10^{20}/10^{16} = 10^4$ per cm^3. For each minority carrier (each hole in the valence band), there are a *trillion* (10^{12}) majority carriers (electrons in the conduction band). That's a substantial majority!

The mere addition of one P atom for every 10 million Si atoms (this is like adding one tourist to the population of New York City) has increased the electron concentration in the conduction band from 10^{10} to 10^{16} per cm^3, and thereby increased the room-temperature conductivity of Si by a factor of nearly a million. The properties of semiconductors are *very* sensitive to dopants!

Suppose we had instead added a *group III* atom like Al ($Z = 11$, one *less* than Si) as our dopant. The Al atoms have one less proton in their nuclei, and one less electron, than the Si atoms. The three valence electrons of Al can form bonds with three of their four Si neighbors but come up one electron short. We now just invert all of our previous thinking for group V dopants. There is a low-energy state associated with the Al atoms that can, with only a small energy cost, accept an electron from a nearby Si-Si bond to form a fourth Al-Si bond. This creates a hole—a missing electron—in the valence band of Si. The local electron energy level associated with the Al atoms is therefore in the gap, just slightly above the top of the valence band (Fig. 15.3). We label this *acceptor* level E_A.

At $T = 0$, each P donor atom would hold onto its extra electron, and each Al acceptor atom would hold onto its *lack* of enough electrons to fully bond with its Si neighbors. However, by room temperature, just as most P atoms will have ionized and contributed their extra electron to the conduction band, most Al atoms will have ionized and *grabbed an electron from the valence band, thereby contributing a hole to the valence band.* Measurements indicate that the ionization energies of group III acceptor impurities in Si, like Al, Ga, and B, are not much different from the ionization energies of group V donor impurities, approximately 0.05 eV.

In semiconductors in which acceptor dopants are dominant, $N_h \gg N_e$, and the dominant charge carriers are *positively* charged holes. Such materials are called *p-type semiconductors*. At room temperature in Si, most acceptor atoms are ionized, so that

$$N_h \approx N_A \qquad \text{and} \qquad N_e \approx \frac{N_i^2}{N_A} \tag{15.19}$$

where N_A is the concentration of acceptor dopants (assumed to be much larger than N_i). Of course, *p*-type semiconductors have *positive* Hall coefficients.

(For our donor and acceptor atoms, we chose P and Al as our examples, atoms adjacent to Si in the periodic table. We could equally well have chosen to use As or Sb as examples of donors, or Ga or In as examples of acceptors. However, we then would not have had the seeming paradox, which should keep you alert, of P-doping, that is, phosphorus-doping, producing an *n*-type semiconductor. In this game, *P*s and *N*s are often different from *p*s and *n*s. As it turns out, however, N-doping, that is, nitrogen-doping, also yields *n*-type behavior, since N and P are both in group V.)

Like formula (15.9), formula (15.11) is valid both for intrinsic and extrinsic semiconductors. Thus it is clear that *in a p-type semiconductor the Fermi level is below midgap*. Adding donor atoms to an intrinsic semiconductor raises E_F above midgap (increasing N_e and decreasing N_h), while adding acceptor atoms lowers it below midgap (increasing N_h and decreasing N_e). The changes in E_F are approximately proportional to the logarithm of the dopant concentrations (Fig. 15.4).

To be concrete, we focused in this section on silicon, the most commonly used semiconductor. Most of our equations and qualitative conclusions, however, also apply to other semiconductors, including a variety of ceramics and polymers with

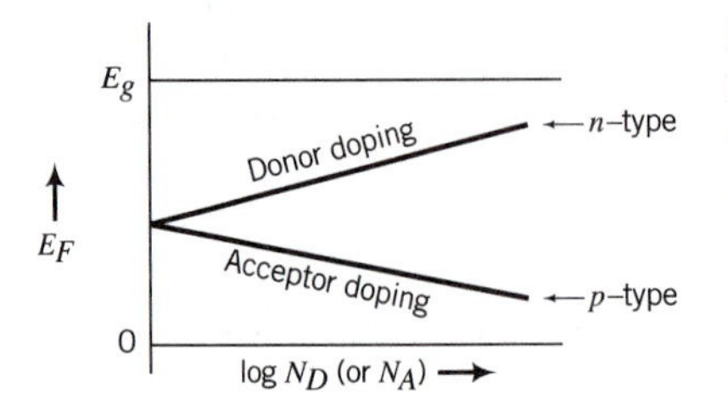

Figure 15.4. Variation of room-temperature Fermi energy of *n*-type (donor doping) and *p*-type (acceptor doping) semiconductors with log of dopant concentration.

moderate energy gaps. For simplicity, we considered semiconductors that contained only donor atoms or only acceptor atoms. If a semiconductor contains *both* donors and acceptors, they compensate each other, electrons from the donor levels falling into the acceptor levels, so that the net effect on charge carrier concentrations will depend only on the *excess* of one dopant over the other.

Sample Problem 15.3

(a) If you add 10^{17} donor atoms/cm^3 to the semiconductor described in Sample Problem 15.2, what will be the charge carrier concentrations in the valence and conduction bands at $T = 300$ K? Assume the donors are essentially fully ionized. (b) Where is the Fermi level before and after doping?

Solution

(a) At 300K, $N_i = 2.32 \times 10^{14}$ cm^{-3}, so $N_i^2 = 5.38 \times 10^{28} = N_e N_h$
$N_e \approx 10^{17}$ and $N_h \approx 5.38 \times 10^{28}/10^{17} = 5.38 \times 10^{11}$ cm^{-3}
(b) In this material, the effective masses of holes and electrons are equal, so, according to (15.15), the Fermi energy before doping must be exactly at midgap, that is, 0.3 eV below E_g.

After doping, from equation (15.9),
$e^{-(E_g - E_F)/k_BT} = N_e/N_c = (10^{17})/(2.52 \times 10^{19}) = 3.97 \times 10^{-3}$
$(E_g - E_F) = 5.53\ k_BT = 0.143$ eV, so the Fermi energy is 0.143 eV below E_g, or 0.157 eV above midgap.

15.4 RECOMBINATION AND LIFETIME

As discussed earlier, absorption of photons with $h\nu > E_g$ excites electrons from the valence band across the energy gap into the conduction band, creating electron-hole pairs. Because an equal number of electrons and holes are created, interband absorption in *extrinsic* semiconductors has a much greater effect on the concentration of minority carriers than on the concentration of majority carriers. Consider the example we cited earlier, an *n*-type P-doped Si that had $N_e = 10^{16}$ and $N_h = 10^4$ (per cm^3). If we irradiate this Si with enough light to create 10^{10} electron-hole pairs per cm^3, the majority electron concentration will be increased by only one part in a million, but the minority hole concentration will be increased by a *factor* of a million. In Chapter 16, we will see how the substantial effect of optical absorption on minority carrier concentration can be harnessed to produce useful devices, such as solar cells and sensitive light detectors.

In the absence of radiation, the equilibrium carrier concentrations of semiconductors are established by a dynamic balance between thermal generation of electron-hole pairs and annihilation of electron-hole pairs by *recombination*. (There may be several mechanisms of recombination, but a direct electron-hole annihilation would be expected to emit a photon with $h\nu = E_g$.) The recombination rate is proportional to the product of the electron and hole concentrations, much as a chemical reaction rate is proportional to the product of the concentrations of the reactants (law of mass action). Thus the *equilibrium* product $N_e N_h$ remains constant at a given temperature, as stated in equation (15.17).

Exposure to light with $h\nu > E_g$ creates *additional* electron-hole pairs, and the concentrations of both majority and minority carriers will be greater than their equilibrium values. If the light is then turned off, the electron-hole recombination rate will be greater than the rate of thermal generation, and concentrations will relax back to their equilibrium values. As noted earlier, since light creates an equal number of electrons and holes, the major change is in the minority carrier concentration. After the light is turned off, the excess minority carrier concentration (the amount above equilibrium) decreases as $\exp(-t/\tau)$, where τ is the *minority carrier lifetime*. This lifetime can depend not only on the dopant concentration but also on the presence of other defects that may affect the recombination process. Minority carrier lifetime is an important parameter in the operation of some of the semiconductor devices discussed in Chapter 16.

15.5 III-V AND II-VI

In addition to elemental semiconductors like silicon, there are numerous compound semiconductors. In particular, a number of *III-V* semiconductors are formed as equiatomic compounds of group III and group V elements, and *II-VI* semiconductors are formed from group II and group VI elements. Examples of III-V semiconductors are GaAs and InSb, and examples of II-VI semiconductors are ZnS and CdSe. Table 15.1 lists the energy gaps between the valence and conduction bands in a number of group IV, III-V, and II-VI semiconductors. Si and other elemental semiconductors are held together with pure covalent bonds, but the bonds are of course polar in III-V compounds, and even more polar in the II-VI compounds.

Table 15.1. Energy Gaps of Various Semiconductors (in eV)

IV	III-V	II-VI
C-5.4	AlN-6.2	ZnS-3.6
SiC-2.9	GaN-3.4	ZnSe-2.6
Si-1.11	GaP-2.3	CdS-2.5
Ge-0.66	AlAs-2.1	CdSe-1.7
Sn-0.1	InN-1.9	CdTe-1.4
(gray Sn)	GaAs-1.4	PbTe-0.29
	InP-1.3	
	GaSb-0.7	
	InAs-0.35	
	InSb-0.17	

Sample Problem 15.4
From interband absorption and subtractive coloration, as discussed in Chapter 14, what colors do you expect across the Zn(S,Se) series of semiconductors?

Solution
ZnS, with an energy gap of 3.6 eV, will be colorless because no visible photons have enough energy to excite electrons across the gap. However, in ZnSe, which has a gap of 2.6 eV, blue photons will be absorbed and the compound will appear yellow. Color will begin to appear when enough Se is added to reduce the energy gap below 3.1 eV.

The upper half of the *d*-band in transition metals represents antibonding crystal orbitals (COs), which is why the bond strength decreases with increasing atomic number as that part of the band is filled (Fig. 14.2). In most ionic and covalent solids, however, the bonding and antibonding orbitals form into separate bands, with an energy gap between the valence band (the bonding COs) and the conduction band (the antibonding COs). *The energy gap itself can then be pictured as a band-theory measure of the strength of the interatomic bonds.*

Since larger E_g corresponds to stronger bonds, the relative values of the energy gaps listed in Table 15.1 are consistent with our earlier discussion (Chapter 10) of the dependence of the strength of interatomic bonds on the atom positions in the periodic table. As in diatomic molecules, the bond strengths of analogous elemental or compound semiconductors decrease as atomic number increases. The energy gap of diamond (C) is higher than that of Si, which in turn is higher than that of Ge. Similarly, the energy gap of GaAs is higher than that of GaSb, InAs, and InSb (but lower than that of GaP and AlAs), and that of ZnS is higher than that of ZnSe, CdS, and CdSe.

The energy gap also increases with increasing ionicity or polarity of the bond, which we can see if we compare elemental, III-V, and II-VI semiconductors of the same row of the periodic table. The energy gap of Ge is less than that of GaAs, which in turn is less than that of ZnSe. The energy gap of gray Sn is less than that of InSb, which is less than that of CdTe. (Below about 13°C, the equilibrium form of pure tin is not the familiar metallic phase but semiconducting gray tin, which has the diamond structure. However, the metallic phase is usually stabilized at low temperatures by impurities.)

For various optoelectronic applications, some of which will be mentioned in Chapter 16, it may be necessary to produce a semiconductor with a specific energy gap, and *alloying* of various elemental or compound semiconductors can be used to produce energy gaps intermediate between those of the end members. Examples are (Ge,Si), (Ga,Al)As, Cd(S,Se), and (Ga,In)(As,P). Selecting semiconductor alloy compositions to develop specific energy gaps (often in conjunction with other requirements, such as specific lattice parameters required to avoid lattice strains at interfaces between different phases) has been termed *band-gap engineering*.

SUMMARY

Charge carriers can be produced in semiconductors in three ways: thermal excitation, optical excitation, and impurity doping. The equilibrium carrier concentrations can be calculated by integration over the conduction band of the density of occupied electron states, that is, the product $Z_e(E)F(E)$, and integration over the valence band of the density of occupied hole states $Z_h(E)[1 - F(E)]$. In an intrinsic semiconductor, the electron concentration and the hole concentration are equal, and vary as $e^{-E_g/2k_BT}$. The Fermi level of an intrinsic semiconductor lies approximately at midgap, about halfway between the top of the valence band and the bottom of the conduction band.

Addition of a low concentration of a group V dopant to a group IV semiconductor like Si produces electron energy levels in the gap, slightly below the bottom of the conduction band. These *donor* levels are easily ionized by thermal excitation, and greatly increase the equilibrium density of electrons in the conduction band. The ionization energy of the donor atoms can be estimated by a *hydrogenic* model, modified by the dielectric constant of the matrix and the effective mass of the electron. In equilibrium, the product of electron and hole concentrations is a constant, so addition of group V dopants also greatly decreases the equilibrium concentration of holes in the valence band. The Fermi level of these *n*-type semiconductors is above midgap, and increases with increasing donor concentration.

Addition of group III dopants also produces electron energy levels in the gap, but these are slightly above the top of the valence band. These *acceptor* levels are also easily ionized, producing holes in the valence band. Increasing acceptor concentration increases the equilibrium hole concentration and decreases the equilibrium electron concentration. The Fermi level in these *p*-type semiconductors is below midgap, and it decreases with increasing acceptor concentration.

In the absence of radiation, the equilibrium concentrations of minority and majority carriers in semiconductors are determined by a balance between thermal excitation and recombination of electron-hole pairs. In extrinsic semiconductors, optical production of electron-hole pairs produces a much greater change in the minority carrier concentration than in the majority carrier concentration. After the light is turned off, the excess minority carrier concentration decreases at a rate characterized by the minority carrier lifetime, a parameter of importance in the operation of several semiconductor devices.

The relative energy gaps of elemental, III-V, and II-VI semiconductors reflect the relative strengths of their interatomic bonds. Elemental and compound semiconductors can be alloyed to produce energy gaps and lattice parameters intermediate between those of the end members, allowing *band-gap engineering*.

PROBLEMS

15-1. Calculate the intrinsic carrier concentration at $T = 300$ K for (a) diamond, (b) zinc sulfide, (c) cadmium sulfide, (d) gallium arsenide, and (e) indium antimonide. Assume that the effective masses of electrons and holes are equal to the free electron mass.

15-2. What fraction of the current is carried by holes in intrinsic silicon?

15-3. Exposure meters on many cameras formerly used cadmium sulfide as the photoconductive material. Explain why photographs of sunsets taken with such cameras often came out overexposed.

15-4. As shown in Fig. 15.2, the density of occupied states in the conduction band goes through a maximum slightly above the bottom of the band. Calculate the energy separation (in eV) between the position of this maximum and the bottom of the band at $T = 300$ K. You may assume that the density of states is of the form shown in equation (15.1).

15-5. A sample of extrinsic semiconductor is 5 cm long, 5 mm wide, and 1 mm thick. When the sample is placed in a magnetic field of 0.6 T normal to the wide surface, a Hall voltage of 8 mV develops at a current of 10 mA. (a) What is the carrier concentration? (b) If the longitudinal voltage driving the 10 mA current is 1 mV, what is the mobility of the dominant carrier?

15-6. A group IV semiconductor is doped with 10^{19} donor atoms per cm^3. The intrinsic carrier concentration at $T = 300$ K is 2.3×10^{13} per cm^3, and $N_c = N_v = 7 \times 10^{16}$ per cm^3. (a) At 300 K, the donor atoms are essentially fully ionized. What is the hole concentration at 300 K? (b) This semiconductor has a dielectric constant of 3. What is its plasma frequency at 300 K? (c) Calculate the position of the absorption edge corresponding to interband transitions. (d) Sketch the light absorption of this semiconductor as a function of photon energy, identifying quantitatively the energies of any opaque-transparent and/or transparent-opaque transitions.

15-7. A semiconductor with $N_i = 10^{16}$ m^{-3} (at 300 K) is doped with acceptor impurities to a concentration of 10^{23} m^{-3}. (a) What are N_e and N_h? (b) Assuming that the effective masses of electrons and holes are equal to the free electron mass, calculate E_g and E_F. (c) Donor impurities are then added to a concentration of 5×10^{22} m^{-3}. What are the new values of the four quantities calculated above?

15-8. You decide to double the donor concentration of an n-type semiconductor that you are selecting for a particular application. How much will that increase or decrease the following quantities: hole concentration, electron concentration, intrinsic carrier concentration, energy gap, Fermi energy, conductivity, Hall coefficient, plasma frequency, and skin depth?

15-9. In metals and intrinsic semiconductors, the Fermi energy is nearly independent of temperature. In extrinsic semiconductors, however, the Fermi energy varies with temperature in the low-temperature range in which the donor or acceptor atoms become ionized. In an n-type semiconductor, what is the position of E_F with respect to E_D at the following temperatures: (a) temperature T_1, where only 10% of the donor atoms are ionized, (b) temperature T_2, where 50% are ionized, and (c) temperature T_3, where 90% are ionized?

15-10. Using the hydrogenic model, how much energy is required to ionize a donor atom in a semiconductor with a dielectric constant of 10 and an electron effective mass that is only 30% of the free electron mass?

15-11. A group IV semiconductor with $N_i = 10^{11}$ per cm^3 is doped with 10^{17} atoms per cm^3 of a group III element. (a) What are the equilibrium electron and hole concentrations? (b) The material is irradiated with light that produces 10^9 electron-hole pairs per cm^3. By what factor does this irradiation change the electron and hole concentrations? (c) The minority carrier lifetime is 3×10^{-4} seconds. What will the electron and hole concentrations be one millisecond after the light is turned off?

15-12. (a) You wish to alter the energy gap of GaAs by substituting a group III element for some of the Ga. Identify one group III element that would decrease the gap and one that would increase it. (b) You decide instead to change the energy gap of GaAs by substituting a group V element for some of the As. Identify one element that would decrease the gap and one that would increase it. Explain your choices.

15-13. Some optoelectronic devices are based on the ternary semiconductors (In,Ga)N. If you start with InN and gradually replace the In with Ga, what sequence of colors would you expect your series of semiconductors to exhibit from interband absorption?

Chapter 16

LEDs, Photodetectors, Solar Cells, and Transistors

16.1 WHERE *p* MEETS *n*

Light-emitting diodes (LEDs). Sensitive light detectors. Solar cells. Rectifiers. Semiconductor lasers. Bipolar and field-effect transistors (FETs). Tunnel diodes. Most important semiconducting devices depend on the details of what happens at the junction between a *p*-type semiconductor and an *n*-type semiconductor: the *p-n junction*.

In Chapter 12, we discussed the *contact potential* produced at the junction between two different metals. When the two metals were brought into contact, electrons flowed from one metal to the other until the Fermi energies of the two metals reached the same level. Suppose instead that we bring into contact two pieces of Si, one doped with P (or another donor element from group V) to make it *n*-type, the other doped with Al (or another acceptor element from group III) to make it *p*-type. As we learned in Chapter 15, the Fermi level in the *n*-type semiconductor is *above* midgap, while the Fermi level in the *p*-type semiconductor is *below* midgap. When the two semiconductors are brought together, charge will flow to equilibrate the two Fermi levels, and a contact potential will develop, as it did with the two metals. And what a tremendous impact on modern technology the contact potential at a *p-n* junction has produced! It's worth analyzing in some detail.

Figure 16.1 shows how the edges of the energy gap—the bottom of the conduc-

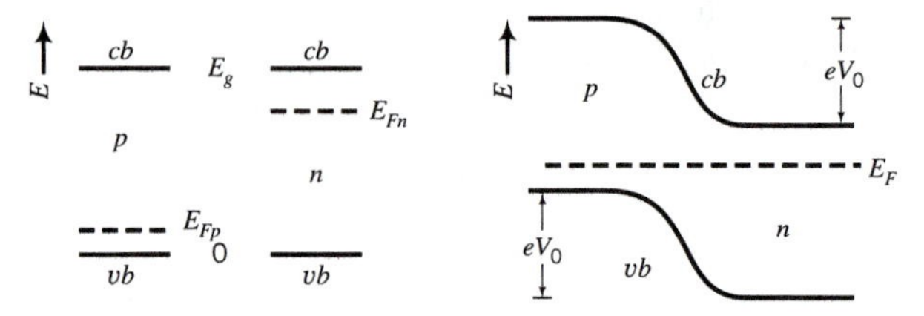

Figure 16.1. Electron energy levels of p-type and n-type semiconductors (a) before and (b) after being brought into contact to form a p-n junction. Equilibration of the Fermi levels produces a contact potential V_0.

tion band and the top of the valence band—will look after the Fermi levels have equilibrated. As in all of our other band diagrams, what is plotted vertically is electron energy. The energy step at the junction corresponds to $V_0 e$, the equilibrium contact potential V_0 times the electronic charge. The slope of the band edges in the junction region corresponds to a built-in electric field ($\mathbf{E} = -\nabla V$) in this region.

Suppose that N_D is the donor concentration on the n side and N_A is the acceptor concentration on the p side. Assuming that approximations (15.18) and (15.19) hold away from the junction, the equilibrium electron and hole concentrations *on the n side*, away from the junction, are

$$N_{en} \approx N_D \qquad \text{and} \qquad N_{hn} \approx \frac{N_i^2}{N_D} \tag{16.1}$$

and the equilibrium electron and hole concentrations *on the p side*, away from the junction, are

$$N_{ep} \approx \frac{N_i^2}{N_A} \qquad \text{and} \qquad N_{hp} \approx N_A \tag{16.2}$$

Remembering that the majority and minority carrier concentrations will differ by many orders of magnitude, we plot in Fig. 16.2 the logarithm of the electron and hole concentrations across the junction. From the carrier concentrations in (16.1) and (16.2), we can calculate the equilibrium built-in voltage V_0, since we know that

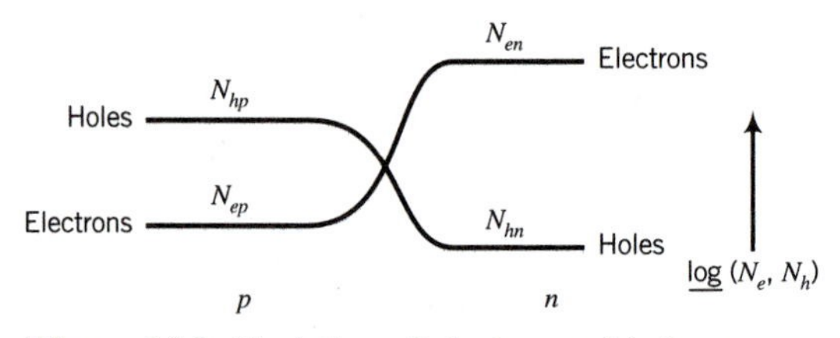

Figure 16.2. Variation of electron and hole concentrations across an equilibrium p-n junction.

the energy step equals the difference in the two Fermi levels before equilibration, that is,

$$eV_0 = E_{Fn} - E_{Fp} \tag{16.3}$$

For each side of the junction, we can use equation (15.11) to relate the Fermi level to the equilibrium hole concentration:

$$E_{Fn} = -k_BT \ln(N_{hn}/N_v) \quad \text{and} \quad E_{Fp} = -k_BT \ln(N_{hp}/N_v) \tag{16.4}$$

Plugging (16.4) into (16.3), we get

$$V_0 = \frac{k_BT}{e} \ln(N_{hp}/N_{hn}) = \frac{k_BT}{e} \ln(N_AN_D/N_i^2) \tag{16.5}$$

(The second expression is valid if approximations (16.1) and (16.2) are valid.) For Si at room temperature, $N_i \approx 10^{10}$ per cm^3. If our dopant concentrations are, for example, $N_A = 10^{16}$ cm^{-3} and $N_D = 10^{18}$ cm^{-3}, equation (16.5) yields a built-in voltage of 0.83 volts.

Alternatively, we could have used equation (15.9) relating the electron concentration to the Fermi level and gotten two equations

$$E_{Fn} - E_g = -k_BT \ln(N_{en}/N_c) \quad \text{and} \quad E_{Fp} - E_g = -k_BT \ln(N_{ep}/N_c) \tag{16.6}$$

Plugging (16.6) into (16.3), we get

$$V_0 = \frac{k_BT}{e} \ln(N_{en}/N_{ep}) = \frac{k_BT}{e} \ln(N_AN_D/N_i^2) \tag{16.7}$$

or, exponentiating (16.5) and (16.7),

$$\frac{N_{en}}{N_{ep}} = \frac{N_{hp}}{N_{hn}} = e^{eV_0/k_BT} \tag{16.8}$$

Equations (16.5), (16.7), and (16.8) directly relate the equilibrium built-in voltage of a *p-n* junction to the ratio of the majority carrier concentration on one side of the junction to the minority carrier concentration on the opposite side of the junction.

Sample Problem 16.1

If *both* the *n* and *p* sides of a silicon *p-n* junction are doped to the level of 10^{18} per cm^3, what will be the built-in voltage and the hole and electron concentrations on each side of the junction?

Solution

$N_eN_h = N_i^2 \approx 10^{20}$, $N_{en} = N_{hp} = 10^{18}$ cm^{-3}, so $N_{ep} = N_{hn} = 10^2$ cm^{-3}
From (16.5), $V_0 = (k_BT/e)\ln(10^{18}/10^2) = 0.95$ volts.

We mentioned in Chapter 15 that in much of the semiconductor literature the electron and hole concentrations are represented simply by n and p, respectively, rather than by N_e and N_h, as we have done. For example, using this notation equation (16.8) becomes instead

$$\frac{n_n}{n_p} = \frac{p_p}{p_n} = e^{eV_0/k_BT} \tag{16.8a}$$

16.2 JUNCTION WIDTH

In Fig. 16.1, we have drawn the transition between the p and n sides as having a finite width, not an abrupt step. Even if we assume that the change in chemical composition from acceptor doping to donor doping is abrupt (which it usually is not in practice, but we'll assume that it is), basic electrostatics dictates that the p-n junction has a finite width. It will be instructive to calculate that width by considering in more detail the local effects of the charge flow that resulted in an equilibration of the Fermi levels. (If nothing else, it will remind you of the connections between net electric charge, electric field, and voltage.)

Before coming into contact, the n-type semiconductor had a much higher concentration of electrons than the p-type semiconductor. Thus we would expect that the direction of electron flow after contact was from n to p. (That is consistent with the direction of the resulting energy step in Fig. 16.1, since the electrons flowing into the p side would shift the entire band structure to higher electron energies.) Similarly, before contact the p-type semiconductor had a much higher hole concentration than the n-type semiconductor. On contact, holes would therefore flow from p to n (resulting in a lowering of electron energies on the n side).

The electrons and holes flowing into the junction from opposite sides will meet and annihilate. But the electrons that departed the n side left behind a net positive charge in the form of *immobile* donor ions, and the holes that departed the p side left behind a net negative charge in the form of *immobile* acceptor ions. We have, as we had in the metal-metal contact discussed in Chapter 12, a double layer of electrical charge (Fig. 16.3a). To simplify our mathematics, we'll assume that the

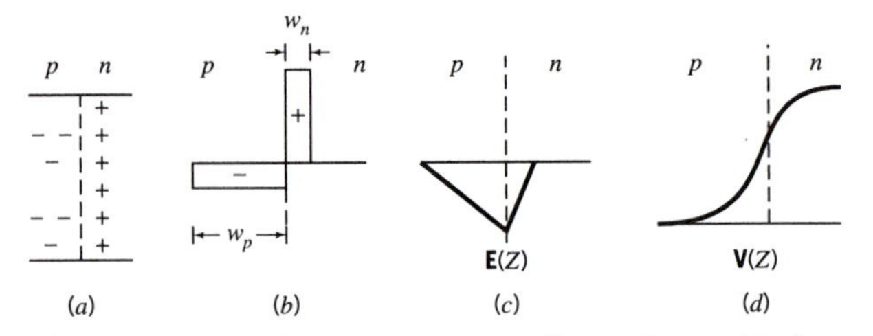

Figure 16.3. (a) Dipole charge layer, (b) net charge, (c) electric field, and (d) potential across equilibrium p-n junction. Shown for $N_D \gg N_A$, so that $w_p \gg w_n$.

junction region is *completely depleted of mobile charge* (electrons and holes). As shown schematically in Fig. 16.2, this is an exaggeration. But recalling that Fig. 16.2 plots the *logarithm* of the electron and hole concentrations, we can see that these concentrations in the junction region are indeed much less than the majority carrier concentrations away from the junction.

Assuming that all donor and acceptor atoms are completely ionized, and that the junction region is completely depleted of electrons and holes, the net charge concentration on the n side is $+eN_D$, and the net charge concentration on the p side is $-eN_A$ (Fig. 16.3b). If w_n and w_p are the widths of the charge layers on the n and p sides, respectively, by charge conservation we know that

$$w_n N_D = w_p N_A \tag{16.9}$$

In the specific case considered earlier ($N_A = 10^{16}$ cm^{-3} and $N_D = 10^{18}$ cm^{-3}), $N_D \gg N_A$, so that $w_p \gg w_n$. The depletion region is largely on the p side of the junction. Through Gauss's law ($\varepsilon \nabla \cdot \mathbf{E} = \rho$), the constant charge densities lead to linear variations of electric field across the junction, and the field reaches a maximum value at the boundary between the layers of positive and negative charge (Fig. 16.3c) given by

$$E_{\text{max}} = -\frac{e w_p N_A}{\varepsilon} \tag{16.10}$$

where ε is the dielectric constant of the material, in this case silicon. By convention, the field is directed from positive toward negative charge, hence from the n side towards the p side.

We got $E(z)$ by integrating $\rho(z)$, and, since $\mathbf{E} = -\nabla V$, we can get $V(z)$, the variation of potential across the junction, by integrating $E(z)$. With our simplifying assumption of a completely depleted junction region with constant charge densities, $E(z)$ is linear and $V(z)$ is therefore parabolic, with the curvature changing sign at the boundary (Fig. 16.3d). (Note by comparing Figs. 16.1 and 16.3 that $V(z)$ curves in the opposite direction to electron energy because of the negative charge of electrons.) The potential drop across the barrier V_0 is equal to the width of the depletion region times the *average* electric field in the region, which is just $E_{\text{max}}/2$. Thus

$$V_0 = w_p \frac{e w_p N_A}{2\varepsilon} = \frac{e w_p^2 N_A}{2\varepsilon} \tag{16.11}$$

For the specific case we treated earlier ($N_A = 10^{16}$ cm^{-3} and $N_D = 10^{18}$ cm^{-3}), this formula indicates that the width $w \approx w_p$ of the depletion region is 0.33 μm. Thus even if the chemical composition changes abruptly, the p-n junction has a finite width. Since the mobile charge carriers are depleted in this region, this width is often called the *depletion width*.

16.3 DIFFUSION MEETS DRIFT (FICK VS. OHM)

We are really more interested in the properties of a *p-n* junction when it is *not* in equilibrium, that is, when a voltage is applied and a net current is flowing. However, it is instructive to first consider the various components of charge flow in a junction at equilibrium—with no applied voltage.

Just because a physical system is in equilibrium doesn't mean that nothing is happening. When a liquid is in equilibrium with its vapor, molecules are evaporating from the liquid into the vapor, and molecules are condensing from the vapor into the liquid, but the rates of evaporation and condensation are exactly equal, and no *net* phase change occurs. Similarly, in an equilibrium *p-n* junction, electrons and holes are flowing *in both directions* across the junction, but the electron flow from *n* to *p* is exactly equal to the electron flow from *p* to *n*, and the hole flow from *n* to *p* is exactly equal to the hole flow from *p* to *n*. There is no *net* flow of either electrons or holes. (You will notice that we have now abandoned the approximation that we just used to estimate the junction width—that there were absolutely no mobile carriers in the junction. We have returned to the reality of Fig. 16.2.)

As we noted earlier, the built-in electric field in an equilibrium *p-n* junction is directed from *n* toward *p*. Thus any holes that find their way into the junction region will be pushed in that direction ($n \rightarrow p$). Charge flow in response to an electric field leads to what we called in Chapter 1 a *drift velocity*, and we now call the resulting current a *drift current*. The drift velocity of the holes equals the electric field times the hole mobility μ_h, so the hole drift current at each position in the junction is, by Ohm's law,

$$J_{\text{drift}} = eN_h(z)\mu_h E(z) \tag{16.12}$$

But in equilibrium, there must also be an equal current of holes in the opposite direction, from *p* toward *n*. Those holes are flowing *against* the force from the electric field. What is driving them?

Using the dopant densities we quoted earlier, the equilibrium hole concentration on the *p* side of the junction is 10^{16} per cm^3. On the *n* side of the junction, a few tenths of a micron away, the equilibrium hole concentration, given by (16.1), is only $10^{20}/10^{18} = 10^2$ per cm^3. We have in the junction region a *huge gradient* in the hole concentration (dN_h/dz). Concentration gradients lead to *diffusion*, and there is a *diffusion current* of holes from *p* toward *n* across the junction given, from Fick's first law of diffusion, by

$$J_{\text{diffusion}} = -eD_h \frac{dN_h(z)}{dz} \tag{16.13}$$

The diffusion coefficient for holes, D_h, is related to the hole mobility through the *Einstein relation:*

$$\frac{D_h}{\mu_h} = \frac{k_B T}{e} \tag{16.14}$$

(Einstein published a number of important papers in 1905. In addition to his papers on relativity and the photoelectric effect, he analyzed the statistics of random, or Brownian, motion and derived the above relation.)

Sample Problem 16.2

If the hole mobility in a semiconductor is 0.05 m²/V-s, what is the hole diffusion coefficient at $T = 300$ K?

Solution

$D_h = \mu_h(k_BT/e) = (0.05\ \text{m}^2/\text{V-s})(0.0259\ \text{V}) = 1.30 \times 10^{-3}\ \text{m}^2/\text{s}$

In equilibrium, the drift and diffusion currents of holes across the junction, given by (16.12) and (16.13), are equal and opposite. There is no net hole current. Similarly, there is a diffusion current of electrons from n toward p, proportional to the gradient (dN_e/dz) in the electron concentration, and a drift current of electrons from p toward n, driven by the electric field. In equilibrium, these two currents, like the two hole currents, are equal and opposite. Figure 16.4 shows the direction of the particle flow and current for holes and electrons (remembering that electron flow is opposite to the current direction). For both electrons and holes, *diffusion current is* $p \rightarrow n$ *and drift current is* $n \rightarrow p$.

These various components of current in an equilibrium p-n junction not only cancel each other out, but are each individually rather small. For example, the electron drift current is small because it results from electrons that reach the junction from the p-side, and there are very few electrons on that side. The electron diffusion current is small because, despite the huge concentration gradient across the junction, only a few electrons have enough energy to surmount the energy barrier eV_0, that is, overcome the electric field. We're now ready to see what happens to drift and diffusion currents when we apply a voltage across the p-n junction.

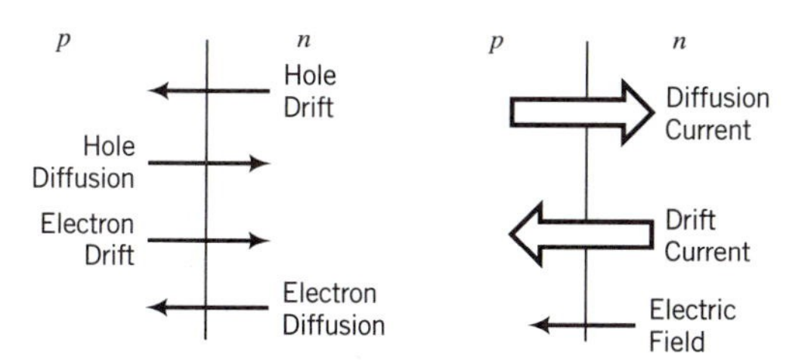

Figure 16.4. Directions of particle flow and currents across an equilibrium p-n junction, for both drift and diffusion of both electrons and holes.

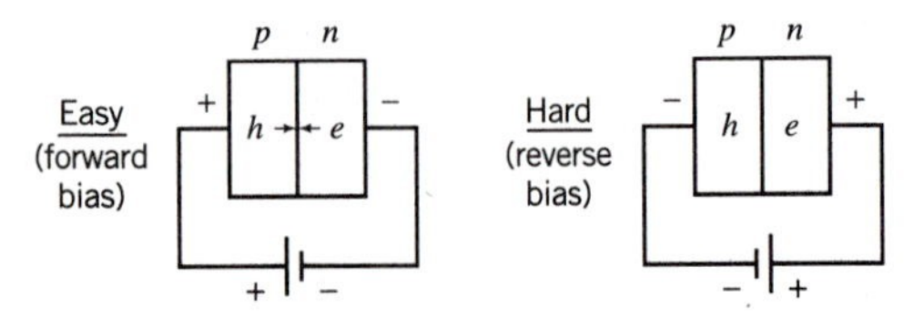

Figure 16.5. Current flow is easy when a *p-n* junction is forward biased (attracting majority carriers to junction) and hard when the junction is reverse biased (attracting minority carriers to junction).

16.4 BIAS AND RECTIFICATION

Before getting into mathematical details, we can see schematically from Fig. 16.5 that the ability of a *p-n* junction to carry current will depend on the sign of the applied voltage. If we apply the positive voltage to the *p* side ("forward bias"), *majority carriers* will be attracted toward the junction—electrons from the *n* side and holes from the *p* side. If we instead apply the negative voltage to the *p* side ("reverse bias"), *minority carriers* will be attracted toward the junction—holes from the *n* side and electrons from the *p* side. But there are very few minority carriers, so we would expect it to be comparatively difficult to conduct current in this direction. Even this very qualitative approach shows that a *p-n* junction will conduct current better in one direction than the other. It will be a *rectifying* junction.

Figure 16.6 shows the effect of forward and reverse bias on the energy barrier at the junction. *Forward bias* (plus on *p*) lowers the electron energy on the *p* side and raises it on the *n* side, thereby *reducing the energy barrier* (and the electric field in the junction). The energy barrier was eV_0 in the equilibrium junction, and is reduced to $e(V_0 - V)$ with a forward-bias voltage of V. In the equilibrium junction, diffusion

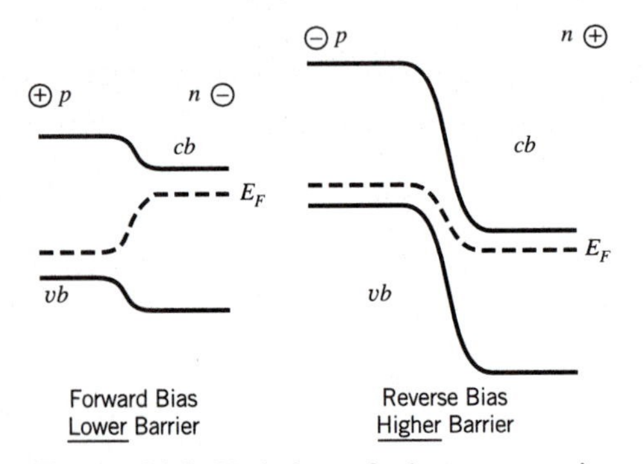

Figure 16.6. Variation of electron energies across a *p-n* junction under (a) forward bias (plus on *p*) and (b) reverse bias (plus on *n*). Most of the voltage drop occurs directly across the junction.

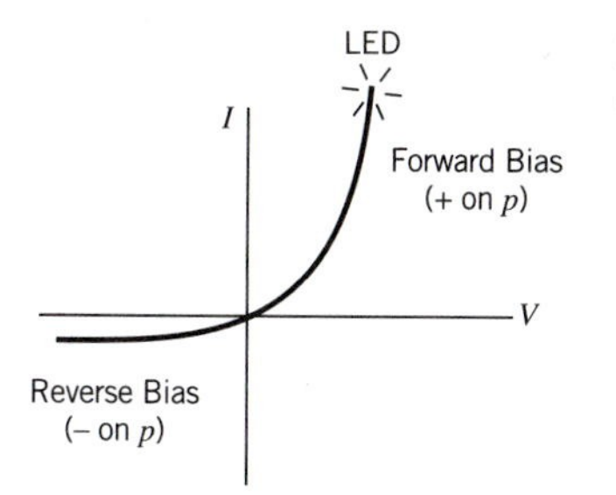

Figure 16.7. Current-voltage curve of a *p-n* junction diode, showing rectification.

current was very limited because few electrons or holes had enough energy to overcome the energy barrier. By lowering the energy barrier, more electrons and holes can diffuse across the junction. Since we are dealing with the Boltzmann tail of the energy distribution, an applied voltage V should *increase the diffusion current* approximately as $\exp(eV/k_BT)$. Diffusion current will now exceed the small reverse drift current, and there will be rapidly increasing net current across the junction.

On the other hand, *reverse bias* (plus on n) lowers electron energy on the n side and raises it on the p side, thereby *increasing the energy barrier* (and the electric field in the junction). This will *decrease the diffusion current.* The drift current, which equalled the diffusion current in equilibrium, will now exceed the diffusion current, and there will be net current across the junction. It will remain very limited, however, because of the very limited supply of minority carriers on each side of the junction. Figure 16.7 shows the current-voltage behavior of a single *p-n* junction, with the current increasing exponentially with forward-bias voltage, but limited to a very small drift current with reverse-bias voltage. The *diode equation* (a diode is a two-terminal electrical device) describing this rectifying behavior is

$$J(V) = J_0\{e^{eV/k_BT} - 1\} \qquad \textbf{(16.15)}$$

Sample Problem 16.3

In the limit of high reverse bias, the limiting current for a *p-n* junction diode at $T = 300$ K is -10^{-5} amps. How much current will this junction carry at this temperature and (a) a reverse bias of 25 millivolts, (b) a forward bias of 25 millivolts, (c) a reverse bias of 100 millivolts, (d) a forward bias of 100 millivolts?

Solution

From (16.15), the limiting current density for high reverse bias is $-J_0$. So apparently $J_0A = 10^{-5}$ amps

(a) $I = 10^{-5}(0.381 - 1) = -6.19 \times 10^{-6}$ amps
(b) $I = 10^{-5}(2.63 - 1) = 1.63 \times 10^{-5}$ amps
(c) $I = 10^{-5}(0.021 - 1) = -9.79 \times 10^{-6}$ amps
(d) $I = 10^{-5}(47.7 - 1) = 4.67 \times 10^{-4}$ amps

The constant J_0 in (16.15) is the limiting drift current density in reverse bias. It is limited by the equilibrium minority carrier densities on each side of the junction, N_{ep} and N_{hn}, plus other parameters related to the minority carriers:

$$J_0 = e\left(\frac{D_e}{L_e} N_{ep} + \frac{D_h}{L_h} N_{hn}\right) \tag{16.16}$$

where L_e and L_h are the *diffusion lengths* for the electrons and holes, respectively. The diffusion length of a minority carrier is related to the diffusion coefficient and the *minority carrier lifetime* τ, the average time excess minority carriers can exist before recombining (which we discussed in Chapter 15) via

$$L_e = (D_e \tau_e)^{1/2} \qquad \text{and} \qquad L_h = (D_h \tau_h)^{1/2} \tag{16.17}$$

(The rectifying property shown in Fig. 16.7 and described by equation (16.15) results from the effect of bias voltages on the energy barrier at the *p-n* junction, and on the resulting changes in diffusion currents. Similar behavior can be obtained in some junctions between metals and semiconductors, called *Schottky diodes*. If the semiconductor is *n*-type, forward bias involves positive voltage on the metal, negative on the semiconductor. As in the *p-n* junction, there is a dipole charge layer (although the "layer" in the metal is infinitely thin) and an energy barrier. Schottky diodes behave much as *p-n* junction diodes, except that under forward bias, minority carriers are not injected into the semiconductor. Although Schottky diodes have numerous applications, in many cases an ohmic, nonrectifying contact between a metal and a semiconductor is desired instead. With appropriate choice of the metal and the semiconductor doping type and level, metal-semiconductor contacts can be made either rectifying or ohmic.)

With the picture developed in this section of the effects of forward and reverse bias on currents across a *p-n* junction, we can now gain a qualitative understanding of some of the *optoelectronic* devices based on *p-n* junctions, which we consider next.

16.5 CREATING LIGHT

We've noted that in forward bias, holes travel across the junction from the *p* side, where they are the majority carrier, to the *n* side, where they are the minority carrier. Similarly, electrons move from the *n* side to the *p* side. Forward bias thus "injects" minority carriers into the regions on both sides of the junction. These are *excess* minority carriers—minority carriers in excess of the equilibrium values approximated by (16.1) and (16.2). They won't get far—only about a diffusion length from the junction—before they will be annihilated by recombination with majority carriers moving toward the junction.

If the semiconductor is a *direct gap* semiconductor, one in which the minimum of the conduction band and the maximum of the valence band occur at the same value of k (Fig. 16.8a), the electron and hole can directly combine with no loss or gain of momentum and give off a photon with $h\nu = E_g$. Our forward-biased *p-n* junction is therefore a *light-emitting diode* (LED). By pushing forward current through a *p-n* junction, we inject excess electrons into a *p*-type semiconductor and excess holes into an *n*-type semiconductor, and the resulting annihilation of these excess minority carriers creates light. We have converted electrical energy into light energy.

In "indirect gap" semiconductors (Fig. 16.8b), such as Si, the maximum of the valence band and the minimum of the conduction band occur at different values of k. In this case, electron-hole recombination or generation also requires energy transfer with the lattice, that is, with phonons, to conserve momentum. This multibody process is much less efficient than the direct recombination possible with direct-gap semiconductors, so direct-gap semiconductors like GaAs are favored for LEDs.

When you change TV channels with your remote control, you are forward-biasing a *p-n* junction and turning on an LED with a semiconductor that has an energy gap in the infrared. The resulting infrared beam carries coded signals that are picked up by detectors in the TV. Infrared LEDs (and lasers) are also used in optoelectronic communication systems, where the photon energy can be chosen to minimize attenuation and/or dispersion in optical fibers, as discussed in Chapter 4. Of course, LEDs used in displays instead produce visible light. GaAs and InP and related alloys have been the basis for red, orange, and yellow LED technology, and (Ga,In)N and related alloys have recently been developed for efficient blue and green LEDs, allowing full-color displays. The shorter wavelengths of blue, violet, and ultraviolet light-emitting devices also allow optical patterning at smaller dimensions.

LEDs of various colors have also been produced with conjugated polymers, including PPP (polyparaphenylene) and PPV (polyphenylene vinylene). As discussed in Chapters 11 and 14, the delocalized pi MOs of conjugated polymers can produce conducting or semiconducting behavior, and polymer LEDs have the advantage of easy fabricability of large-area thin-film devices at low cost. However, these and other electronic devices based on conjugated polymers are still in a very early stage of development.

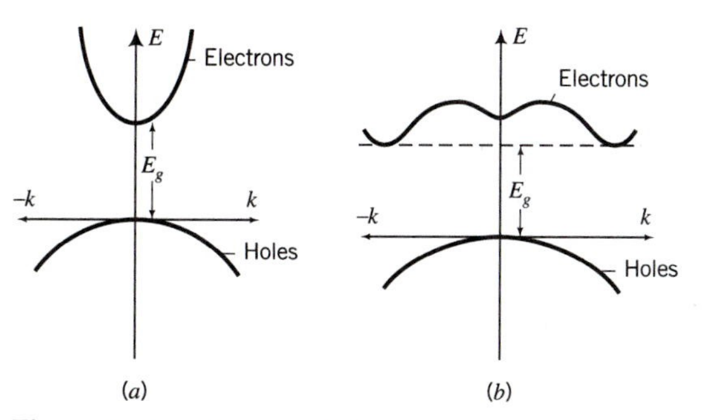

Figure 16.8. Schematic $E(k)$ curves for (a) direct-gap semiconductor and (b) indirect-gap semiconductor.

Sample Problem 16.4

Explain how a semiconductor that appears orange-yellow in normal light can emit green as an LED.

Solution

The color of a material in reflected or transmitted light is produced by subtractive coloration. A semiconductor with an energy gap near 2.4 eV will, by interband absorption, remove the blue end of the visible spectrum from white light and appear orange-yellow. (A *mixture* of light from 1.8 eV to 2.4 eV appears orange-yellow.) In an LED, however, the color is produced not by absorption and subtractive coloration, but directly by emission of 2.4 eV (green) photons resulting from electron-hole recombination across the energy gap.

Forward-biased *p-n* junctions are also the basis of semiconductor *lasers*. We noted in Chapter 8 that photons can not only stimulate upward electron transitions (absorption), they can also stimulate downward electron transitions (stimulated emission). The stimulated photon is in phase with the stimulating photon, the two in-phase photons can stimulate more in-phase photons, and light amplification via this optical chain reaction is possible. The resulting laser light is coherent (in phase) and, with appropriate design, can be made very intense.

For laser action to occur, it is necessary to have more electrons in the higher-energy state than in the lower-energy state (population inversion), because otherwise absorption overcomes stimulated emission. With semiconductor lasers, this is achieved through the injection of minority carriers. It is also necessary to have partial confinement of the photons (photons in a box), because if they escape too rapidly they can't continue to stimulate the emission of other photons. This confinement is often accomplished with a *heterojunction*, a thin layer of semiconductor with a different composition, different energy gap, and different index of refraction, allowing it to confine both the photons and the injected electrons and holes. A thin layer of GaAs (about 100 nm) between layers of *p*-type and *n*-type AlGaAs is commonly used. GaAs has a higher index of refraction than AlGaAs, allowing the photons to be confined via total internal reflection, a phenomenon we discussed in Chapter 4. AlGaAs has a larger energy gap than GaAs, thereby confining the electrons in an energy well (electrons in a box) of the type discussed in Chapter 8. Some semiconductor lasers are based on single or multiple quantum wells (each only a few nanometers wide) and contain a dozen or more layers of different compositions and dopings. A direct-gap material is necessary for semiconductor lasers, as it is for LEDs.

16.6 DETECTING LIGHT

Suppose instead that we put the *p-n* junction into *reverse* bias. With no light applied to the junction, the "dark current" density is limited to J_0, which with proper design

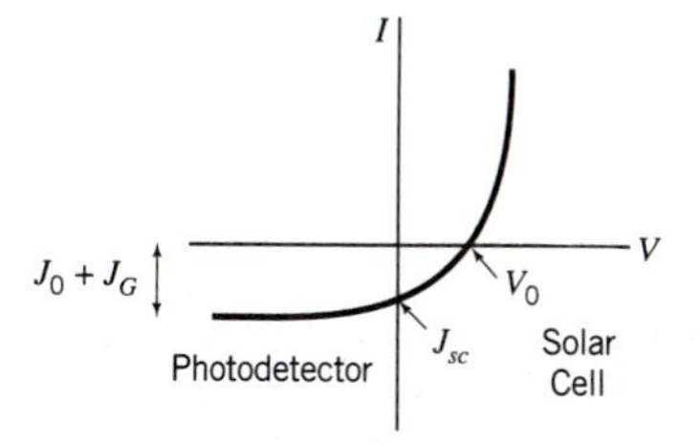

Figure 16.9. Current-voltage curve for a *p-n* junction diode under illumination that creates excess minority carriers near the junction.

can be kept very small. The small drift current under reverse bias is limited because of the low concentration of minority carriers on each side of the junction.

However, suppose we now expose the junction area to light with $h\nu > E_g$. Photons can be absorbed, creating electron-hole pairs. As we noted in the previous chapter, the electron-hole pairs created by light absorption have a much greater effect on the minority carrier concentration than on the majority carrier concentration. So even if the resulting increase in majority carriers is modest, the increase in minority carrier concentration near the junction, holes on the *n* side and electrons on the *p* side, can be substantial. These excess minority carriers can be "collected" by the reverse-biased junction, and the drift current will greatly increase. If the optically generated current density is called J_G, the $J(V)$ behavior of the junction will be changed to approximately

$$J(V) = J_0\{e^{eV/k_BT} - 1\} - J_G \tag{16.18}$$

The $J(V)$ curve for an illuminated junction is shown schematically in Fig. 16.9. A *p-n* junction in reverse bias is a very sensitive *photodetector*. An optical communication system can be constructed with an LED or semiconductor laser to generate light, an optical fiber to transmit the light, and a *p-n* junction photodetector to detect it. Thus *p-n* junctions commonly serve at both ends of a modern communication link. The minimum attenuation of the optical fiber determines the optimum choice of photon energy, which in turn determines the desired energy gaps of the semiconductors used to create and detect the light.

16.7 ENERGY FROM THE SUN

A *p-n* junction photodetector is operated in the southwest (lower left) quadrant of the $J(V)$ field in Fig. 16.9. But we now call your attention to the interesting south*east* portion of the $J(V)$ field in that same figure. In this region, the current is in the direction associated with reverse bias, but the voltage is in the direction associated with forward bias. In this region, the current is flowing across the junction from negative to positive voltage, contrary to the direction observed in resistors and observed in the northeast and southwest portions of the $J(V)$ field, but *consistent with the direction of current inside a battery* (Fig. 16.10). Apparently in this region the *p-n* junction is not a sink for electrical energy, but, like a battery, a *source of electrical energy*.

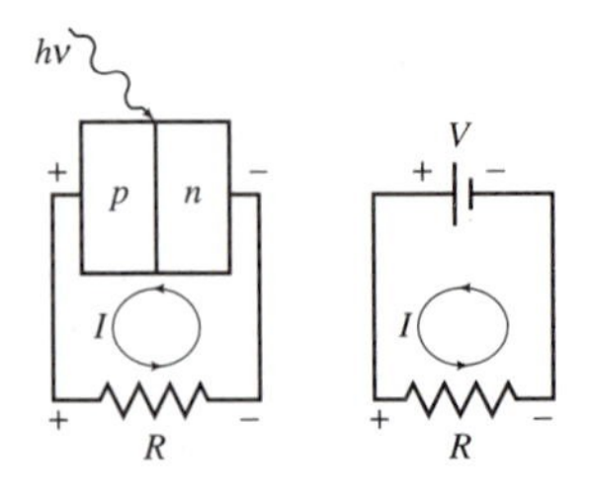

Figure 16.10. The direction of current flow (minus to plus) through a *p-n* junction in the southeast region of Fig. 16.9 is the same as the direction of current flow through a battery, indicating that the junction is acting as a power source.

Rather than receiving electrical energy from the external circuit, as in LEDs and photodetectors, the *p-n* junction in this case is delivering electrical energy. Such a device is called a photovoltaic, or, if our source of light is the sun, a *solar cell.* Solar cells today produce electrical power on satellites and in numerous specialized situations on earth. Materials research continues to increase their efficiency and to decrease their cost, increasing the hope that some day solar cells will make a major contribution to the generation of electrical power.

The intersection of the $J(V)$ curve with the current axis in Fig. 16.9 is the *short-circuit current* density J_{sc}, and its intersection with the voltage axis is the "open-circuit voltage" V_∞. Since the wattage (per unit area) delivered by the solar cell is given by J times V, it is zero at these two points, but will reach a maximum somewhere between them. The ratio of this maximum JV product to the product $J_{sc}V_\infty$ is called the *fill factor*, and is a figure of merit for solar cell design.

16.8 TRANSISTORS—AMPLIFICATION AND SWITCHING

The equilibrium concentration of minority carriers in most doped semiconductors is very small compared to the majority carrier concentration. But as we've seen in the preceding sections, in the vicinity of *p-n* junctions an *excess* concentration of minority carriers can have very important effects. In the LED, excess minority carriers are *injected* across a forward-biased junction, for example, electrons into the *p* side, and these excess minority carriers recombine with the majority carriers to produce light. In the photodetector, light impinging on the semiconductor near the *p-n* junction creates excess minority carriers, which are *collected* by the reverse-biased junction. In the solar cell, the excess minority carriers produced by the light are *collected* across the junction, allowing the built-in junction voltage to deliver electric power.

So the operation of each of these optoelectronic devices depends on the injection or collection of excess minority carriers across a *p-n* junction. They are *minority-carrier devices.* The same can be said for one common class of transistors, *bipolar junction transistors* (BJTs).

The core of a BJT is a sandwich structure. For example, an *n-p-n* BJT has a thin *p* region sandwiched between two *n* regions, with electrical leads attached to each of the three regions. (The transistor is thus a three-terminal device, whereas rectifiers, LEDs, detectors, and solar cells are *diodes*, i.e., two-terminal devices.) The central region is commonly called the *base*, and it is sandwiched between an *emitter* and a *collector*. Like the optoelectronic devices, the BJT is a minority-carrier device, but it depends on *both* the injection and collection of excess minority carriers.

The *p-n* junction between the emitter and the base is normally forward-biased, allowing injection of excess minority carriers into the base (electrons in the *n-p-n* case). The *p-n* junction between the base and the collector is normally reverse-biased, allowing collection of most of those excess minority carriers if the two junctions are not far apart (separated by less than a diffusion length). Even small changes of the bias across the emitter-base junction can produce large changes in current into the base, and therefore in the collector current. Depending on the details of the external circuit, the BJT can be used either as a simple switch, with small changes in base voltage turning large emitter-collector currents on or off, or as an amplifier, with small changes in currents in the emitter-base circuit producing large changes in the currents in the emitter-collector circuit.

In BJTs, the current to be controlled flows *across p-n* junctions. In the other major category of transistor, *field-effect transistors* (FETs), the current to be controlled instead flows *parallel* to *p-n* junctions. Like the BJT, the FET is a three-terminal device, but now the three terminals have different labels: *source*, *gate*, and *drain*. In one type of FET, a thin channel of *n*-type semiconductor near the surface of a *p*-type semiconductor carries current between the source and the drain (by electrons in the *n*-channel). However, via an intermediate third terminal, the gate, a voltage can be applied that changes the width of the *n*-channel. This allows small changes of the gate voltage to create large changes in the source-to-drain current, again allowing the transistor to be operated either as a switch (as in computer logic circuits) or as an amplifier. In contrast to the BJT, the current being controlled, that is, the current flowing in the channel, is due to the flow of the local *majority* carrier (in this case, the electrons within the *n*-channel).

In the metal-oxide-semiconductor field-effect transistor (MOSFET), an oxide layer (SiO_2) separates the metal gate from the semiconducting channel. (One of the great advantages of Si for semiconducting circuits is the ease of producing thin oxide layers with appropriate electrical properties.) Some MOSFETs are depletion-type "normally on" devices, with a negative gate voltage decreasing the source-drain current by constricting the *n*-channel. Others are enhancement-type "normally off" devices, with a positive gate voltage creating an *n*-channel (an *inversion layer*) in a formerly *p*-type substrate. The latter type, shown schematically at the left in Fig. 16.11, is dominant in today's integrated circuit technology. The regions labeled n^+ under the source and drain contacts are heavily doped *n*-regions (see the next section).

In some circuits using *p*-type substrates, deep *n*-type wells are also produced, upon which *p*-channel enhancement MOSFETs are constructed, as seen at the right in Fig. 16.11. Here, a negative gate voltage will create a *p*-channel inversion layer along the surface. The *n*-type well is produced by introduction of donors into the *p*-type substrate, and the p^+ regions at the source and drain are produced by subsequent introduction of acceptors into the *n*-type wells. Circuits using both types of MOSFETs are called *complementary metal-oxide semiconductor* (CMOS) structures.

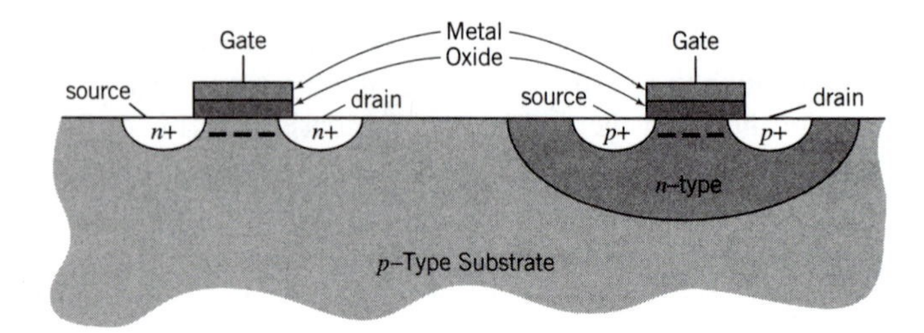

Figure 16.11. (left) Schematic structure of an enhancement MOSFET (metal-oxide-semiconductor field-effect transistor), in which positive gate voltage creates an n-channel (bounded by dashed line) along the surface, an *inversion layer*. (right) A p-channel enhancement MOSFET produced on a p-type substrate by formation of an n-type well. A structure using both n-channel and p-channel MOSFETs on the same substrate is called a *complementary MOS* (CMOS) structure.

16.9 TUNNEL DIODES

If the doping levels of semiconductors are extremely high (above about 10^{25} m^{-3}, corresponding to a dopant concentration of about 0.01%), the Fermi level of an n-type semiconductor moves out of the gap and up into the conduction band—and the Fermi level of a p-type semiconductor moves down into the valence band. Semiconductors doped this heavily are called *degenerate* semiconductors and are often designated as n^+ or p^+ semiconductors.

Fig. 16.12a shows an equilibrium p^+-n^+ junction. Here the equilibrium built-in voltage V_0 is greater than the energy gap E_g. Since doping is heavy, the junction width (see Section 16.2) is very thin. Only that thin barrier separates the electrons at the Fermi level on the n-side from available empty electron states at the top of the valence band on the p-side. So in addition to normal drift and diffusion currents,

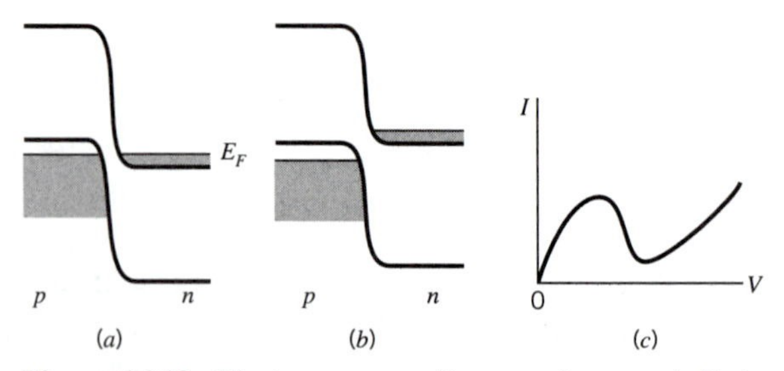

Figure 16.12. Electron energy diagrams for tunnel diode (a) in equilibrium, (b) under forward bias that brings Fermi levels in each side against the gaps in the other side, leading to decreased tunneling current. (c) Resulting current-voltage curve, showing region of negative differential resistance.

a small forward bias will allow electrons to *tunnel* from right to left across the barrier. At first, increasing forward bias leads to increasing tunneling current, but eventually the forward bias raises the *n*-side Fermi level above the top of the *p*-side valence band, and the electrons find themselves opposite the forbidden energy gap. Tunneling currents then fall, reaching zero when the bottom of the *n*-side conduction band coincides with the top of the *p*-side valence band (Fig. 16.12b). The resulting *I-V* curve for this "tunnel diode" has a region in which the differential resistance dV/dI is negative (Fig. 16.12c). At higher forward voltages, the normal rectifying diffusion current returns the diode to normal positive dV/dI.

In that region of "negative resistance" the tunnel diode can be used in a resonant circuit (see Problem 5-3) to cancel the effect of real positive resistances and allow electrical oscillation to continue, the negative resistance of the tunnel diode replenishing the energy lost as heat in the real resistance. If the negative resistance is greater than the positive resistance, the electrical signal will not only persist, it will grow. Thus tunnel diodes can be used in oscillators and amplifiers. Because the junction width is so thin in tunnel diodes, electron transit times across the junction are shorter than in normal transistors, and operation at very high frequencies (up to 100 GHz, or 10^{11} Hz) is possible. Unlike the solar cell, the tunnel diode is not a source of net electrical energy; it draws electrical energy from the battery that keeps it biased into the region of negative resistance.

Another interesting device based on tunneling is the *resonant tunnel diode*. Figure 16.13 shows the conduction band structure of an AlGaAs-GaAs quantum well structure. The GaAs layer sandwiched between two layers of higher-bandgap AlGaAs (each layer about 5 nm thick) is a quantum well, like the electron-in-a-box problem we considered in Chapter 8, and electrons in the well have only a limited number of discrete allowed energy levels. Suppose that the ground state of electrons in the well, E_1, is slightly above the initial Fermi levels of the outer layers of GaAs, which are more heavily doped than the well. The two layers of AlGaAs are thin enough to tunnel through, but significant tunneling does not occur until enough voltage is applied to raise the Fermi level at the left into alignment with E_1, when the allowed electron level becomes available to tunnel into. This is called *resonant tunneling*. Electrons can then also tunnel out of the well through the second AlGaAs barrier, and the diode will carry current. However, if the voltage is further increased, the Fermi level at the left will then become higher than E_1, no allowed energy level is

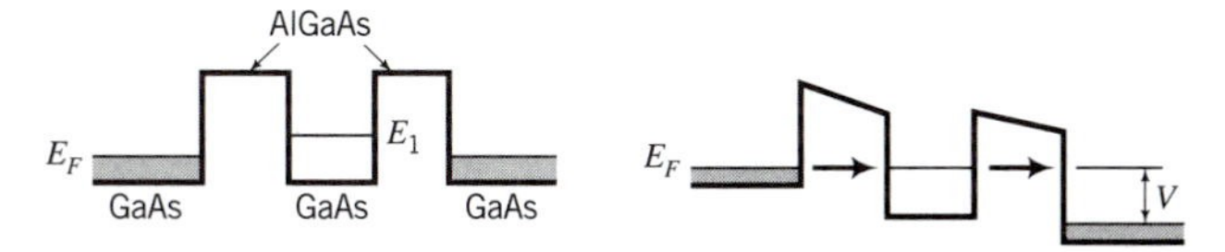

Figure 16.13. (a) Resonant tunnel diode containing a thin *quantum well* of GaAs between two thin layers of AlGaAs, which has a higher energy gap. Tunneling current peaks when the bias voltage V is sufficient to align the Fermi level of the GaAs at the left with the ground state energy level E_1 within the well (b), thus producing a "negative resistance" region like that in Fig. 16.12c.

available in the well to tunnel into, and the tunneling current will decrease. The diode will thus have a negative-resistance region like that of Fig. 16.12c.

We have briefly described in this chapter only a few of the many electronic and optoelectronic devices that can be made with *p-n* junctions, quantum wells, and other arrangements of semiconductors of different dopings and energy gaps. With complex multistep sequences of deposition, oxidation, irradiation, etching, ion implantation, diffusion, and so on, and the use of masking to localize the various compositional changes, extremely complex three-dimensional structures of semiconductors, insulators, and metals have been produced. As processing progresses steadily to smaller and smaller structures, microelectronics is moving toward *nanoelectronics*, with effects of quantized electron energy levels becoming more and more prominent, as in the resonant tunnel diode. For further details about the structure and operation of various electronic and optoelectronic devices, see the suggestions for further reading following this chapter.

SUMMARY

At an equilibrium junction between an *n*-type semiconductor and a *p*-type semiconductor, there is a built-in voltage resulting from an equilibration of Fermi levels. Within the junction, there is an electrical field (directed from *n* toward *p*) and an associated double layer of immobile charge from which mobile charge carriers (electrons and holes) have been largely depleted. The width and built-in voltage of an equilibrium *p-n* junction depend on the doping of the two sides of the junction. In equilibrium, the diffusion currents ($p \rightarrow n$) and drift currents ($n \rightarrow p$) across the junction are small, equal, and opposite, so that there is no net current.

Applying a forward bias to the junction (plus on *p*) lowers the energy barrier at the junction and increases the diffusion current exponentially, so that it becomes much greater than the drift current. Applying a negative bias (plus on *n*) raises the energy barrier and decreases diffusion current exponentially, so that current in reverse bias is dominated by drift. The maximum drift current is limited by minority carrier concentrations (and the diffusion coefficients and diffusion lengths of the minority carriers). The current-voltage curve of a *p-n* junction is highly directional, allowing the junction to serve as a diode rectifier.

Applying forward bias to a *p-n* junction injects excess minority carriers into each side of the junction, and the annihilation of these excess carriers with the majority carriers can generate photons with an energy equal to the energy gap. Thus a forward-biased junction can serve as a light-emitting diode (LED) or laser. Direct-gap semiconductors like GaAs produce light more efficiently than indirect-gap semiconductors, which require the involvement of phonons to balance the momentum change of the electrons.

Applying reverse bias to a *p-n* junction yields little "dark current," but photons with energy greater than the energy gap can generate electron-hole pairs in the vicinity of the junction, greatly increasing the minority carrier concentration and the reverse-bias current. This allows a reverse-biased *p-n* junction to act as a sensitive photodetector. The collection of light-generated excess minority carriers in a *p-n* junction also allows the junction to be used to generate electrical power in a photovoltaic solar cell. The operation of transistors as switches or amplifiers in electronic circuits also depends on the properties of *p-n* junctions. Tunnel diodes and resonant tunnel diodes yield regions of negative differential resistance, and they can be used in oscillators and amplifiers.

PROBLEMS

16-1. A semiconductor doped with 10^{21} donor atoms per m^3 is joined to a sample of the same material that is doped instead with 10^{23} acceptor atoms per m^3. If the intrinsic carrier concentration is 10^{16} per m^3, calculate the electron and hole concentrations on each side of the *p-n* junction and the equilibrium voltage across the junction.

16-2. For the junction described in the previous problem, calculate the maximum electric field in the junction and the width of the junction. Assume that the dielectric constant is equal to that of silicon. In which direction does the electric field point?

16-3. In the example given in the text, the *n*-side of the junction is doped much more heavily than the *p*-side, so that the junction width is dominated by w_p. Derive a general formula for the junction width in terms of N_A and N_D that is applicable even when the doping levels are similar on both sides of the junction.

16-4. In a sample of silicon, the electron and hole mobilities are 0.15 and 0.05 $m^2V^{-1}s^{-1}$, respectively. Calculate the room-temperature diffusion coefficients for electrons and holes, and calculate the diffusion lengths if the lifetime for both types of carriers is 10^{-4} seconds.

16-5. We form a *p-n* junction in GaP ($E_g = 2.25$ eV) using $N_A = 10^{18}$ cm^{-3} and $N_D = 10^{16}$ cm^{-3}. The dielectric constant is 9, and the effective masses of electrons and holes are 0.35 and 0.5 times the free electron mass, respectively. Calculate the equilibrium junction voltage.

16-6. For the material described in the previous problem, calculate the wavelength at which a transition from opacity to transparency will occur, and the wavelength at which the reverse transition will occur. Sketch the reflectivity vs. wavelength, with these two wavelengths indicated.

16-7. If you produce LEDs with Ga(P,N) ternary semiconductors, how do you expect the color emitted to change as you gradually replace P with N?

16-8. In an equilibrium *p-n* junction, the sum of electron drift and diffusion currents is zero. From this equality and the Einstein relation, derive equation (16.7). (Hint: From the equality mentioned, find an expression for dN_e/N_e, and integrate across the junction.)

16-9. There are four different components of current across a *p-n* junction—electron drift and diffusion and hole drift and diffusion. Explain briefly the effect on the various components of current produced by (a) forward bias, (b) reverse bias, and (c) irradiation of the junction with light having $h\nu > E_g$.

16-10. Would a *p-n* junction built from ZnS be suitable for a solar cell? How about GaP?

16-11. If samples of *p*-type and *n*-type silicon of the doping levels given in section 16.1 ($N_A = 10^{16}$ cm^{-3}, $N_D = 10^{18}$ cm^{-3}) are brought into contact, roughly how many electrons and holes flow across the junction to reach equilibrium?

16-12. In producing a rod of *n*-type germanium, you have inadvertently introduced a gradient of dopant composition which has produced a built-in voltage between the ends of the rod of 0.125 volts. At the high-doping end, the local resistivity is 10 ohm-cm. What is the local resistivity at the other end of the bar? Assume all dopants are ionized.

Final Review

The Story So Far

Although there is still much about the properties of engineering solids that we have not been able to discuss in this text, the foregoing chapters should provide a good base for further studies. Throughout, our major focus was on the valence electrons, which bind atoms together in solids and determine most of their engineering properties.

In Part I, we looked at some properties of solids that can be treated at least partly without specific consideration of quantum mechanics. We discussed the effects of electric fields and electromagnetic waves on the free electrons of metals (Chapters 1 and 2) and on the bound electrons of insulators (Chapters 3 and 4). Although the basic origin of both ferromagnetism and superconductivity is deeply imbedded in quantum mechanics, we described in Chapters 5 and 6 some of the classical phenomenology of magnetic and superconducting materials and the relation between microstructure and some important technological properties. Finally, in Chapter 7, we very briefly considered elastic waves in solids and summarized many of the properties of solids that we were unable to explain without quantum mechanics.

In Part II, Chapter 8 reviewed some of the experiments of the early 20th century that led to the development of quantum mechanics and Schrödinger's wave equation and applied that equation to several model problems—free electrons, electrons and barriers, an electron trapped in an energy well (electron in a box), and a harmonic oscillator. We then applied the ideas of quantum mechanics to isolated atoms (Chapter 9), diatomic molecules (Chapter 10), and larger and larger molecules (Chapter 11), and converged on solids, which we viewed as very large molecules. The wave functions that described the probability distributions of electrons in atoms (AOs) were used to construct approximate wave functions for molecules (MOs) and crystals (COs). With this LCAO approach, the *s-p-d-f* electron energy levels of atoms broadened into continuous *bands* of energy levels in solids, and electrons were at least partly delocalized and shared by the many atoms of the solid. In this approach, we started with the bound electrons of atoms and let them become more delocalized and freer as molecular size increased.

In Chapters 12 and 13, we approached the quantum mechanics of valence electrons in solids from the opposite direction. We started with totally free electron waves in a very large box (Chapter 12) and a resulting continuous band of energy levels. Within this band, the distribution of electron energies was described with a density of allowed states $Z(E)$ and a Fermi function $F(E)$ that represents the probability of occupation of these states at finite temperatures. In Chapter 13, we finally put atoms into the box and developed the *nearly free electron* (NFE) approach to valence electrons, with Bragg diffraction introducing gaps of forbidden electron energies. From $E(k)$, the relation between the energy and momentum of the valence electrons, the concepts of effective mass and holes were developed, and in Chapter 14 we presented a summary of the band-theory view of why some materials are metals and some are insulators, and why some are opaque, some transparent, and some colored.

Finally, in Chapters 15 and 16 we applied the results of band theory—energy bands and energy gaps, electrons and holes, and so on—to semiconductors. We considered the electrical and optical properties of semiconductors and ended with some of the remarkable properties of *p-n* junctions (rectification, light emission, light detection, photovoltaic energy generation, amplification and switching, negative resistance). All of that abstract stuff about light particles (photons) and electron waves, AOs, MOs, and COs, densities of electron states and Fermi functions, energy bands and energy gaps, effective masses, and holes finally paid off! Quantum mechanics is not only one of the most important intellectual developments of the 20th century, it's also very useful!

Suggestions for Further Reading

Numerous books cover some or all of the material discussed in this book. A few are listed below, and students and other readers will find it helpful to consult them or others to learn more about specific topics or to see alternate presentations of similar material.

GENERAL REFERENCES ON ELECTRONIC PROPERTIES

Nicholas Braithwaite and Graham Weaver, *Electronic Materials*, Butterworths, London, 1990. Organization reverse of that of most texts. Starts with specific engineering applications, then concisely tackles the materials science necessary for each. Covers circuits, magnets, transducers, memories, displays.

Richard H. Bube, *Electrons in Solids: An Introductory Survey*, Third Edition, Academic Press, New York, 1992. With wave properties as an integrating theme, introduces the properties of lattice waves (phonons), light waves (photons), and matter waves (electrons) before tackling energy bands and the properties of solids.

P. A. Cox, *The Electronic Structure and Chemistry of Solids*, Oxford University Press, Oxford, 1987. Builds band theory with LCAO approach. Good coverage of spectroscopic methods and numerous advanced topics (excitons, polarons, lattice distortions, electron repulsion, surfaces, etc.)

Rolf E. Hummel, *Electronic Properties of Solids*, Second Edition, Springer-Verlag, Berlin, 1993. After introducing band theory and $E(k)$ diagrams, presents a concise but broad treatment of electrical, optical, magnetic, and thermal properties of solids, with frequent reference to devices. More advanced than the present text.

David Jiles, *Introduction to the Electronic Properties of Materials*, Chapman & Hall, London, 1994. Broad coverage of many electronic properties of materials. Final five chapters focus on applications, including more treatment of magnetic recording materials than most texts.

M. Omar, *Elementary Solid State Physics: Principles and Applications*, Addison-Wesley, Reading, MA, 1978. Extensive in-depth discussion of many topics (e.g., diffraction, phonons, dielectrics, defects) dealt with more concisely in most electronic materials texts.

Daniel D. Pollock, *Physics of Engineering Materials*, Prentice-Hall, Englewood Cliffs, NJ, 1990. Includes extensive treatment of statistics (Maxwell-Boltzmann, Bose-Einstein, and Fermi-Dirac) and of thermal, thermoelectric, and galvanomagnetic effects.

L. Solymar and D. Walsh, *Lectures on the Electrical Properties of Materials*, Fifth Edition, Oxford University Press, Oxford, 1993. Amusing and philosophical asides help make this a popular and readable text. Strong in semiconductor devices, lasers, and optoelectronics.

Adrian P. Sutton, *Electronic Structure of Materials*, Clarendon Press, Oxford, 1993. An excellent follow-up to this book. A balanced chemical-physical approach to the fundamentals, extended to a variety of specialized topics, including *s-p* bonding in silicon, transition metals, structural stability of compounds, and modern quantitative theory.

SPECIAL TOPICS

Siegmund Brandt and Hans Dieter Dahmen, *The Picture Book of Quantum Mechanics*, John Wiley & Sons, New York, 1985. Contains many illustrations aimed at improving the student's feeling for electron waves and their time evolution. Also available on CD-ROM.

Robert M. Eisberg, *Quantum Physics of Atoms, Molecules, Solids, Nuclei, and Particles*, Second Edition, John Wiley & Sons, New York, 1985. A good introduction to the principles and broad applications of quantum mechanics.

Anthony P. French, *Vibrations and Waves*, Norton, New York, 1971. A good place to become more familiar with the mathematics and physics of vibrations and waves. A sophomore-level text, part of the MIT Introductory Physics Series.

Larry Gonick and Art Huffman, *The Cartoon Guide to Physics*, Harper Collins, New York, 1990. Our students have found this a useful and entertaining means of reviewing their introductory physics, particularly electricity and magnetism.

David Jiles, *Introduction to Magnetism and Magnetic Materials*, Chapman & Hall, London, 1991. Broad coverage of a topic thinly covered in most electronic materials texts.

William L. Jorgensen and Lionel Salem, *The Organic Chemist's Book of Orbitals*, Academic Press, New York, 1973. Includes a collection of illustrations showing the shapes and energies of bonding and antibonding molecular orbitals of over 100 organic and inorganic molecules.

James W. Mayer and S. S. Lau, *Electronic Materials Science: For Integrated Circuits in Si and GaAs*, Macmillan, New York, 1990. Principles of the operation and fabrication of microelectronic devices from a materials viewpoint.

Roy McWeeny, *Coulson's Valence*, Oxford University Press, Oxford, 1979. Authoritative treatment of molecular orbitals and chemical bonding.

J. N. Murrell, S. F. A. Kettle, and J. M. Tedder, *The Chemical Bond*, Second Edition, John Wiley & Sons, New York, 1985. Quantum-mechanical treatment of the chemical bond, including LCAO/MO and the Hückel model.

Kurt Nassau, *The Physics and Chemistry of Color: The Fifteen Causes of Color*, John Wiley & Sons, New York, 1983. A fascinating treatment of the physics and chemistry of color, including effects of molecular orbitals and band theory.

West, Anthony R., *Solid State Chemistry and Its Applications*, John Wiley & Sons, New York, 1984. Broad treatment of solids from the chemical viewpoint.

OTHER BOOKS IN MIT SERIES IN MATERIALS SCIENCE AND ENGINEERING

Samuel M. Allen and Edwin L. Thomas, *The Structure of Materials*, John Wiley & Sons, 1999. Structure of crystalline and noncrystalline solids and liquid crystals. Text for a core course usually taken in the sophomore year.

Yet-Ming Chiang, Dunbar Birnie III, and W. David Kingery, *Physical Ceramics: Principles for Ceramic Science and Engineering*, John Wiley & Sons, New York, 1997. Covers crystal structure, defects, transport, phase diagrams, and microstructure of ceramics. Specialized undergraduate elective course.

David V. Ragone, *Thermodynamics of Materials, Vols. I and II*, John Wiley & Sons, New York, 1995. Thermodynamics and kinetics from the materials viewpoint. Texts for two department core courses, usually taken in the sophomore year.

David Roylance, *Mechanics of Materials*, John Wiley & Sons, New York, 1996. Text for a core course in our department, although not formally part of the MIT Series. Covers elastic and plastic behavior of materials, creep, fracture, and fatigue.

Index

absorption edge, 262–263
absorption peaks, 263
acceleration, 9, 32, 240
acceptor, 277–278, 286
acetone, 59, 60
acetonitrile, 193
acetylene, 191, 192, 193
Alice, 122
alkali metals, 33, 215
alkanes (saturated chains, C_nH_{2n+2}), 191, 207, 242
alkenes (conjugated chains, C_nH_{n+2}), 197, 207, 242
alloying elements
  and phase stability, 251
  converting type I to type II superconductor, 98
  decreasing conductivity in metals, 7, 14
alternating current (AC), 16–17, 32, 69
  and complex conductivity, 32, 34–36
  in capacitors, 43–45, 71, 72
  in inductors, 71, 72
aluminum, 13, 224, 226, 256
  as dopant in silicon, 277
  in copper, 251
aluminum gallium arsenide, 296, 301
aluminum oxide, 121, 264
amethyst, 264
ammonia, 190
amorphous magnetic alloys, 85, 86
amorphous solids, 265
Ampere's law, 25
angular frequency, 16, 27, 56, 129
anisotropy, 5–6, 7, 80–86
anisotropy field, 81, 82
anti-bonding orbitals, 171–173
  in conjugated carbon compounds, 195–198
  in covalent molecules, 171–173, 176–180
  in diamond and silicon, 261
  in polar molecules, 181–182, 260
  in transition metals, 258
antiferromagnetism, 76

argon, 163
atomic orbitals (AOs), 156–160, 164, 175
  hybrid, 188–190
  in LCAO approximation, 169, 177, 180, 187
attenuation, 30, 62–64

band-gap engineering, 282
barium iron oxide, 85
barium titanate, 43
base, 298
BCS theory of superconductivity, 102–103
beat, 115
benzene, 194–196, 199, 205
beryllium, 176, 224, 256
beryllium dihydride, 188
bias, 292–294
binding energy, 180, 226
bipolar junction transistor (BJT), 298–299
blackbody radiation, 125–127
Bloch functions, 234
Bohr magneton, 74, 121
Bohr radius, 155
bond energy (bond strength), 171
  and cohesive energy, 175
  and energy gap, 281
  carbon-carbon bonds, 193
  carbon-hydrogen bonds, 193
  covalent molecules, 176, 178, 179
  polar molecules, 182–183
  3S rule, 174, 178, 193, 258
bonding, 167 *ff*
bonding orbitals, 171–172
  conjugated carbon compounds, 195–198
  covalent molecules, 175–180
  hydrogen, 171–172
  polar molecules, 181–183
  transition metals, 258
bond integral, 170
  and band width, 197, 204, 240–242, 275
  and effective mass, 241, 275
  in Hückel model, 194
  of hybrid atomic orbitals, 188
bond length, 171
  and dipole moment, 182
  and lattice parameter, 175
  carbon-carbon bonds, 193
  carbon-hydrogen bonds, 193
  diatomic molecules, 171, 174, 178–179, 182
  hydrogen, 171, 174
  transition metals, 257, 258
  3S rule, 174, 178, 193, 258
bond order, 174, 179
Born-Oppenheimer approximation, 168
boron, 121, 177–178, 179, 263
boron trihydride, 189
boundary conditions
  on electromagnetic waves, 56, 58, 125
  on electron wave functions, 136
    in energy well, 138, 141, 152
    in hydrogen atom, 154
    in metal, 210, 213, 217
box, electron in, 138–142, 151–152, 210, 296
Bragg diffraction, 64–65
  and Brillouin zone boundaries, 115, 238, 247
  and energy gaps, 235–236, 244
  of electromagnetic waves, 64–65
  of electron waves, 129, 235, 238
  of sound waves, 113
brass (copper-zinc), 257
Brillouin zone, 115, 238, 246–252
bronze (copper-tin), 14
buckeyballs, 206
butadiene, 196, 199

calcium, 161
capacitor, 40–41, 43–44, 70–72
carbon, 178, 189, 190 *ff*
  buckeyballs, 206
  diamond, 59, 108, 189, 205, 260, 261, 263
  graphite, 205
carbon compounds
  conjugated, 193–201, 242, 262
  saturated, 189, 190–191, 199, 242
carbon monoxide (CO), 183
carotene, 199, 200
cavity radiation, 125–127
cesium, 225
charge carrier density (concentration)
  and plasma frequency, 32–33
  in metals, 9–10, 12–13, 259
  in semiconductors, 13, 15, 272–278
chlorine, 179

chlorophyll, 199–201
chromium, 76, 258
Clausius-Mosotti relation, 51
CMOS, 299, 300
cobalt, 76, 85, 86, 258
cobalt samarium, 86
coercivity, 80, 84–87, 101, 109
cohesive energy, 175, 257, 258
collector, 299
collision time, 9–10, 11, 14, 15, 32–34, 46
color, 121, 199, 262
  conjugated organic compounds, 199–201
  copper and brass, 257
  gems, 65, 263–264
  inorganic compounds, 262–264
  light-emitting diodes (LEDs), 295–296
  neon lights, 164, 199
  opal, 65, 199
  rainbow, 62, 199
  sky (Rayleigh scattering), 64, 199, 263
compensation (of dopants), 279
complex conductivity, 32, 34–36, 44, 72
complex conjugate, 132, 145
complex dielectric constant, 36, 45–50, 59, 63
complex numbers, 29, 30, 44, 132
composite, 5, 7, 92–93
Compton scattering, 129
conduction band
  in metal, 203, 256
  in semiconductor, 269, 271–273, 276
conductivity (electrical) 5
  complex, 32, 34–36, 44, 72
  from band theory and free-electron theory, 258–259
  insulators, 39
  metals, 5–15, 16–17, 203, 256, 258–259
  semiconductors, 15, 120, 270, 275, 277
conductivity (thermal), 17, 18, 24
conjugated carbon molecules, 193–201, 262
conservation of states, 173, 177, 194, 202, 237
contact potential, 228–229, 285–287
Cooper pairs, 15, 102, 103, 107
copper, 14, 31, 92–93, 108, 111, 222, 224, 257
copper alloys, 251
copper-niobium, 14
Coulomb integral, 170, 181, 194
covalent bonds, 172, 175–180, 280
critical angle, 57–58
critical current, 87, 94–95, 100–101, 102, 109
critical field, 94–98, 102
critical temperature, 93, 101–102
crocin, 199, 200
crystal orbitals (COs), 160, 203, 209, 233, 236, 237
Curie temperature, 76, 83, 84
curl, 25, 26
cyclotron resonance, 34, 120

Debye model (of orientational polarization), 45–46
degeneracy, 151–153
  and symmetry, 151–153, 159, 177
  diatomic molecules, 177
  hydrogen atom, 159
  MO levels in benzene, 195
  removed by differential screening, 161
  spin, 152, 159
degenerate semiconductor, 300
del, 24
delocalization, 193 *ff*, 209, 233
  pi orbitals, 193–201, 233, 237
  sigma orbitals, 202–203, 233, 237
delta orbitals, 176
demagnetizing field, 90
density functional theory, 266
density of occupied states, 220, 227, 271–272
density of states, 216–218
  at Fermi surface, 252, 258–259
  divalent metals, 246, 256
  in one dimension, 217, 242
  insulators, 260
  metals, 256
  semiconductors, 271, 272
depletion width, 288–289
diamagnetism, 74, 95–97, 178, 183
diamond, 6, 59, 108, 121, 189, 205, 260, 261, 263
dielectric constant, 40–42, 45–46, 55, 56, 276
  complex, 36, 45–50, 59, 63
  variation with frequency, 46, 48, 50–51, 59–60
dielectrics, 39 *ff*

dielectric strength, 51
diffraction, 64–65
  and Brillouin zone boundaries, 115, 238, 247
  and energy gaps, 235–236, 244
  of electromagnetic waves, 64–65
  of electron waves, 129, 235, 238
  of sound waves, 113
diffusion coefficient, 290, 294
diffusion current, 290–291, 293
digonal (*sp*) hybrids, 188, 191, 193
dipole layer, 229, 288
dipole moment (electric), 39, 42, 47, 73, 75, 182
dipole moment (magnetic), 73–76
direct gap, 294, 295
dislocations, 14, 79, 101, 109
dispersion, 61
  of electromagnetic waves, 61–62
  of electron waves, 138
  of sound waves, 113–114
displacement, 47, 49, 109–113
divalent metals, 13, 120, 244–246, 249, 256
divergence, 24
domain walls, 77–80, 82, 83–87
donor, 276–277, 278, 285, 288
dopants, 275
double bond, 178, 191, 193
drain, 299, 300
drift current, 9, 10, 290–291, 293, 297
drift velocity, 9, 10, 11, 290
Drude theory, 8–10, 12–13, 17–18, 210, 215, 259, 265

easy axis, 80–82
effective mass, 34, 120, 239–242, 243, 251, 271, 274, 275
Einstein relation, 290
elastic modulus, 108–109, 175, 205, 257, 258
elastic waves, 109–116, 146
electrode, transparent, 23, 33, 262–263
electromagnet, 72
electromagnetic waves, 26–28, 111
  in a cavity, 125–127
  in conductors, 29–33
  in insulators, 55–64
electron diffraction, 129
electronegativity, 183
electron-hole pairs, 264, 270, 279, 295, 297
electronic heat capacity, 17–18, 120, 220–222
electronic polarization, 41–42, 47–51, 60
electron-phonon interactions, 116, 295
emerald, 264
emitter, 298
energy band, 197–198, 202–204, 211
  conjugated chain, 197–198, 203
  energy-momentum relation in, 236, 238, 239–242
  insulators, 260
  metals, 202–204, 211, 223, 256
  semiconductors, 269, 271
energy barrier
  electron incident on, 134–137
  for escape from metal (work function), 225, 227–228
  in p-n junction, 291, 292–293
  tunneling through, 137, 228, 301
energy gap, 236–238, 255–261, 281
  and optical properties of insulators, 260–263
  as measure of bond strength, 261, 281
  direct and indirect, 294–295
  excitation across, 269–270
  in different directions, 245
  of semiconductors, 280–281
energy well, 138–142, 151–152, 155
  bulk metal as, 210, 225, 234
  in resonant tunnel diode, 301
ethane, 191
ethylene, 191, 193
Euler's formula, 29, 35, 44
exchange, 75, 76, 83–84, 162
excited state, 139
exciton, 264
expectation value, 145, 158, 169
extended-zone scheme, 238
extinction coefficient, 59
extrinsic semiconductors, 275–279

Faraday's law, 25, 56, 72, 95, 96
Fermi-Dirac statistics, 218
Fermi energy (Fermi level), 212, 214, 255, 269
  and contact potential, 228–229, 285–286
  and Fermi surface, 249
  divalent metals, 246, 256

insulators, 260–262, 269
measuring with spectroscopy, 223–224, 225–226
metals, 255–257, 269
semiconductors, 269, 274, 278, 286
Fermi function, 219–220, 221, 227, 271–272
Fermi surface, 249, 251, 252, 258–259
Fermi velocity, 215
Fermi wave number (Fermi wave vector), 212, 214–215
ferrimagnetism, 76
ferrites, 76, 86
ferroelectrics, 43, 50, 51
ferromagnetism, 76
Fick's law of diffusion, 24, 290
field effect transistor, 299–300
field emission, 228
fluorine, 178–179
flux lattice, 98–99
fluxoid, 99–101
flux quantum, 99, 121
Fourier series, 137, 138
fractional ionic character, 182
fracture, 109
free-electron theory,
classical (Drude), 8–10, 12–13, 17–18, 210, 215, 259, 265
quantum-mechanical, 209–218, 259, 265
frequency, 16

gallium arsenide, 280, 281, 295, 296, 301
gallium indium nitride, 295
gate, 299, 300
Gauss' laws, 25, 289
germanium, 280, 281
glass, 23, 50, 57, 58, 61, 63, 264, 265
gold, 224
gradient, 24, 258
grain boundaries, 43, 84–85, 101, 109
graphite, 7, 205
grass, 199–201
ground state, 139
group velocity, 114–115, 138, 239–241, 258

Hall angle, 15
Hall effect, 11–13
applications of, 15–16
in divalent metals, 13, 120, 244, 246, 256
in semiconductors, 13, 120, 278
Hamiltonian, 169, 170, 181
hard magnetic materials, 80, 84–87
hardness, 109
harmonic oscillator, 47, 116, 142–146, 174
Hartree-Fock treatment, 162
heads, 87
heat capacity
electronic, 17–18, 120, 220–222
lattice, 116–117, 120, 146, 222
helium, 75, 91, 162, 164, 175–176
helium-neon lasers, 164, 199
heme, 199–201
hemoglobin, 200
heterojunction, 296
high-temperature superconductors, 101–103
holes, 13, 243–244
across p-n junction, 286–288, 290–291, 292, 294, 297
divalent metals, 246, 249, 256
extrinsic semiconductors, 277–279
intrinsic semiconductors, 270, 271–275
HOMO, 195, 196, 199, 203, 262
Hooke's law, 108, 110
Hubbard model, 266
Hückel model, 194–198
Hume-Rothery rules, 251
Hund's rule, 178
hybridization, 187–193, 205, 261, 262
hydrogen atom, 153–159
hydrogen fluoride, 181–182
hydrogenic model (for dopants), 276
hydrogen liquid, 59
hydrogen molecular ion (H_2^+), 167–173
hydrogen molecule (H_2), 75, 173–175
hysteresis loop, 80, 100, 101

impedance, 72
impurities
and color, 263–264
and conductivity in metals, 7, 14
dopants, 270, 275–279
independent electron approximation, 265
index of refraction, 56–59, 136, 296
indigo, 199, 201
indirect gap, 295
indium antimonide, 15, 280, 281
indium phosphide, 295

indium tin oxide, 33, 262–263
inductor, 69–72
insulators, 3, 7, 39, 255, 260–262
intensity (of electromagnetic wave), 30, 58, 128, 131
intrinsic semiconductor, 270–275
ionic bond, 180–183, 260, 280
ionic conductivity, 39
ionic polarization, 42, 49, 50, 60
ionization energy, 225
ionosphere, 33
iron, 76, 85, 86, 256, 257, 258
iron neodymium boride, 85, 86
iron-nickel (permalloy), 85, 86
iron-silicon, 85, 86

kinetic energy, 8, 18, 116, 133–135, 139, 143, 145, 158
k-space, 212–216, 246–250

lanthanum barium copper oxide, 102
Laplacian, 26, 131, 153–154
laser, 51, 140–141, 164, 199, 296
lattice heat capacity, 116–117, 120, 146, 222
lattice spacing (lattice parameter), 112
  and bond length, 175
  diffraction, 64,
  in nearly-free electron theory, 234–235, 244, 248
  transition metals, 257
lattice vibrations, 14, 49, 107, 112–114, 116, 146
lead, 97, 98, 108
lead-indium, 98
lead zirconium titanate, 43
lifetime (of minority carriers), 280, 294
light-emitting diode (LED), 295–296
linear combination of atomic orbitals (LCAO), 169
  amorphous solids, 265
  conjugated carbon molecules, 194–198
  covalent molecules, 175–180
  Hückel model, 194–198
  hybrid AOs, 187–193
  hydrogen, 169–175
  polar molecules, 180–183
  solids, 203, 233, 236–237, 240–242, 250, 255, 265
lithium, 12–13, 176, 202–203, 215, 224, 256
lithium fluoride, 182–183, 261
lithium hydride, 182
longitudinal wave, 111
Lorentz force, 11, 34, 100
loss tangent, 50, 51, 63
lucite, 23
LUMO, 195, 196, 199, 203, 262

magnesium, 224
magnesium oxide, 261
magnetic anisotropy, 80–86
magnetic dipole, 73
magnetic domains, 77–79, 83, 84–86
magnetic pressure, 78
magnetic recording media, 80, 86–87
magnetization, 73, 76, 96
magnetization curve, 79, 80, 82, 97, 98, 100
magnetoresistance, 87
majority carrier, 277, 299
  concentration, 277, 278, 279, 286
manganese, 76, 258
Maxwell's equations, 25–26, 29, 110
mean free path, 10, 14, 15
mechanical properties, 107, 109
mercury, 92, 199
metal-insulator transitions, 266
methane, 189, 190, 193, 199
microwaves, 27, 31, 46, 60
minority carrier, 277
  collection across *p-n* junction, 297, 298, 299
  concentration, 277, 278, 286, 297
  devices, 298
  excess, 279–280
  injection across *p-n* junction, 294–296, 299
  lifetime, 280, 294
mixed state, 97–101
mobility, 9, 10, 12, 15, 251, 275, 290
  holes vs. electrons, 275
modulus, 108, 113, 175, 257, 258
molecular orbitals (MOs), 169–170
  carbon compounds, 190–198
  covalent molecules, 175–180
  hydrogen, 170–173
  lithium chain, 202
  pi, 176–178, 191–193, 194*ff*, 205

polar molecules, 180–183
polyatomic molecules, 187–190
sigma, 172, 176–178, 191, 202
molecular vibrations, 145, 174, 180
momentum, 129, 133, 134, 138, 139, 145, 240, 295
angular, 154, 155, 157
momentum space, 216
MOSFET, 299–300

nanoelectronics, 302
nearly free electron (NFE) model, 233–238, 255, 265
negative resistance, 301, 302
neon, 121, 205, 164, 199
nickel, 76, 258
niobium-copper, 14
niobium tin, 98, 101, 102
niobium-titanium, 92–93, 98, 101, 102
nitrogen, 121, 178, 179, 180, 205, 263
nodes
in atomic orbitals, 158, 159
in crystal orbitals, 237
in electromagnetic waves, 28
in molecular orbitals, 172, 196, 197, 198, 202
in pi orbitals, 176
in wave functions, 144
nonbonding orbitals, 181, 190, 197
normalization, 142, 156, 169, 188
n-type semiconductor, 277, 278, 285

Ohm's law, 3–11, 16, 24, 29–30, 32, 47
opal, 65, 199
optical fibers, 55, 58, 61–64, 297
optical properties (in quantum mechanics), 140
optoelectronics, 281, 297, 302
orbital magnetism, 74
orientational polarization, 43, 45–46, 50, 60
overlap integral, 170, 192–193, 194
overlapping energy bands, 245–246, 249, 256
oxygen, 178, 179

paramagnetism, 74, 75, 76
of molecules, 178, 183
of metals, 259

Pauli exclusion principle, 141, 161, 211, 212
Peierls distortion, 203, 262
periodic boundary conditions, 210–211, 213, 217
permanent magnets, 77, 80, 84, 86
permeability, 25, 29–30, 70, 71, 73–74
permittivity, 25, 29–30, 40–41, 70
phase angle, 35–36, 44–45, 72
phase stability, 251
phase velocity, 27, 114, 138
phonons, 116–117, 120, 146, 174, 264, 295
phosphorus, 179
as dopant in silicon, 275–278
photoconductivity, 270
photodetector, 297
photoelectric effect, 128, 224
photoelectron spectra, 180, 224–226, 242
photon, 128–129
absorption and emission spectra
and phonon transitions, 264
of atoms, 164, 180
of copper and brass, 257
of electron in a box, 140
of harmonic oscillator, 145
of hydrogen atom, 155–156
of insulators, 261–264
of metals, 256
and electron transitions, 140
emission, in LEDs and semiconducting lasers, 295–296
interband absorption in semiconductors, 270, 279, 281, 297–298
momentum of, 129
wavelength of, 130, 295
photovoltaic, 298
pinning
of dislocations, 101, 109
of domain walls, 85, 101
of fluxoids, 100–101
pi orbitals, 176–178, 191–193, 194 *ff*, 205
Planck's constant, 74, 99, 121, 127–129
plasma frequency, 32–33, 47, 119, 262
plastic deformation, 14, 101, 109
platinum, 225
p-n junction, 229, 285 *ff*
polar bond, 180–183, 260, 280
polarizability, 42, 47, 51, 58, 205
polarization, 39–45

polyacetylene, 6, 7, 191–192, 203, 262
polyethylene, 59, 60, 191–192, 199, 242, 260, 262, 264
polymerization, 191
polymers, 42, 51, 108, 191–193, 205, 262, 295
polymethylmethacrylate (PMMA), 23
polyparaphenylene (PPP), 295
polyparaphenylene vinylene (PPV), 295
population inversion, 296
potassium, 161, 176, 224
potassium chloride, 260, 261
probability distribution, 132, 145, 146
p-type semiconductor, 278, 285

quantization, 127, 139, 155
quantum dot, 142, 152
quantum number, 139, 143, 154–155, 158, 159
quantum well, 141, 209, 296, 301
quantum-well laser, 296
quantum wire, 142, 151
quartz, 59, 264, 265

radial probability function, 156, 157, 163
radon, 75
rainbow, 62, 199
rare-earth compounds, 84
Rayleigh scattering, 64, 199
reciprocal space, 216
recombination, 279–280, 295
rectifier, 4, 292–294
reduced-zone scheme, 238
reflection, 31, 33, 57–59, 135, 199–201, 264
  of electron waves, 135–136, 235
  of sound waves, 115
refraction, 56, 57
relativity, 121, 130, 149, 215, 265
relaxation time, 9, 45
remanent magnetization, 80
resistance, 3, 4, 301, 302
resistivity, 4, 6, 14
resistor, 69, 72
resonance, 34, 47–48, 143
resonant tunnel diode, 301–302
reststrahlen, 264
rotation (of magnetization), 81–82

rubbers, 108, 109
rubidium bromide, 261
ruby, 121, 264

sapphire, 264
saturated carbon molecules, 190, 191, 199
saturation magnetization, 76, 79, 80, 82
scattering, 64, 199
Schottky diode, 294
Schrödinger's wave equation, 131–133, 142, 153, 169, 266
screening, 160–161
secondary bonds, 204–206
self-consistent field (SCF), 162–164, 167, 173, 180, 265
semiconductors, 3, 7, 13, 15, 269 *ff*
  compound, 280–282
sigma orbitals, 172, 176–178, 191, 199, 202
silica (SiO_2), 58, 59, 61, 63, 65, 264, 265
silicon, 6, 7, 189, 260, 261–262, 295
  energy gap, 260, 274, 280
  in copper, 251
  intrinsic carrier concentration in, 287
  mobilities in, 275
  optical properties, 261–262
silver, 225
skin depth, 30–31, 119, 137
Snell's law, 57
sodium, 160–161, 176, 204, 222, 223–224
sodium chloride, 49, 59, 60
soft magnetic materials, 80, 84–87
solar cell, 297–298
sonic waves, 110–111, 113–114
sound velocity, 111, 113, 114
source, 299, 300
space, interstellar, 15
specific heat
  electronic, 17–18, 120, 220–222
  lattice, 116–117, 120, 146, 222
spectra
  absorption (conjugated compounds), 200, 201
  atomic, 155–156, 164, 180
  molecular, 180
  of electron in a box, 140
  of harmonic oscillator, 145
  photoelectron, 180, 224–226, 242
  x-ray, 223–224

spin, 74–76, 141, 152, 161, 178, 183, 216, 259
  degeneracy, 152, 161
spring constant, 47, 112, 142, 145, 174–175, 179, 180
standing waves, 28
  electromagnetic, 28, 125–126
  electron, 139, 210–211, 217, 235
  sound, 115
state, 216
stationary states, 133
steel, 6, 7, 31, 86
stiffness, 109, 179, 257
  and 3S rule, 174, 178, 193, 258
stored energy, 40, 70, 108
strain, 108–110
strength, 14, 101, 109
stress, 108, 110
structure-sensitive properties, 87, 101, 109
sulfur, 179
sulfur hexafluoride, 190
superconductors, 7, 15, 91–103
  type I, 97
  type II, 97–101
superparamagnetism, 77
susceptibility (dielectric), 41
susceptibility (magnetic), 73, 81
symmetry, 139–140, 144, 152, 159, 172, 177

Teflon, 6, 59, 60
tensor, 5, 7
terminal velocity, 11
tesla, 16
tetrahedral (sp^3) hybrids, 189–191, 193, 205, 261
texture, 5
thermal conductivity, 17, 18, 24
thermal excitation, 141, 218–221, 270, 274, 279
thermal velocity, 8–10, 215
thermionic emission, 227–228
thermoelectric effect, 18
thermopower, 18
tight binding, 167
tin, 95, 96, 280, 281
  in copper, 14
total internal reflection, 57–58, 296
toughness, 109
transistors, 298–300
transition-metal oxides, 265
transition metals, 222, 256, 257–258
transverse waves, 27, 111–112
traveling waves
  electromagnetic, 27–29
  electron, 134, 139, 211, 235
  sound, 111, 112, 115
trigonal (sp^2) hybrids, 189, 193, 194, 196, 205
triple bond, 178, 183, 191–193
tungsten, 125, 225, 258
tunnel diodes, 300–302
tunneling, 137, 143, 228, 301

ultraviolet catastrophe, 127
uncertainty principle, 134, 138, 139, 143, 158, 168

vacancies, 243
valence band, 243, 269, 271, 273, 277, 281
Van der Waals bonds, 205
variation method, 169
vibrational levels (of molecules), 145, 174, 180
viscosity, 11, 32, 47

water, 43, 45, 46, 59, 60, 190
wave equation
  damped, 30
  elastic, 110, 112
  electromagnetic, 26, 30
  Schrödinger, 131–133
wave function, 131–133
wavelength
  change in dielectric, 56
  deBroglie, 129–130
  electron vs. photon, 130, 131
wave number, 27, 111, 129, 135, 213, 216
  complex, 30, 137
  imaginary, 137
  real, 32–33
wave packet, 137–138
wave-particle duality, 128–129
wave vector, 213, 246, 247
White Queen, 122, 135

Wiedemann-Franz Law, 17
work function, 225, 227–229

x-rays, 27, 64–65, 223–224

Young's modulus, 108, 109, 113
yttrium barium copper oxide, 7, 102

zero-point energy, 143, 145
zinc, 13, 224, 244, 246, 256
  in copper (brass), 251, 257
zinc oxide, 6, 7
zinc selenide, 280, 281
zinc sulfide, 280, 281